GIS, Human Geography, and Disasters

by Andrew Curtis and Jacqueline W. Mills

University of Southern California
California State University, Long Beach

University Readers™
San Diego, CA

First published in the United States of America in 2009 by University Reader Company, Inc.

13 12 11 10 09 1 2 3 4 5

Printed in the United States of America

ISBN: 978-1-934269-86-2

University Readers™
800.200.3908 | www.universityreaders.com

Contents

Introduction 1

An Introduction to Thinking Spatially About Disasters 9

An Introduction to Geographic Information Systems 39

Disaster Connections—The Johnstown Flood: Part One 65

The Johnstown Flood—The Days and Weeks After: Part Two 87

Wildfires in the Wildland–Urban Interface:
The Case of San Diego County, California 123

The Geography of Vulnerability 151

Health Vulnerability 173

Art Therapy and Hurricane Katrina 195

Life Interrupted:
The Stress Process in the Face of Natural Disasters 207

The Post-Disaster Landscape 233

The Geography of Recovery 261

Using the Spatial Video for Neighborhood Recovery 283

Conclusions 309

Introduction

"Their cardboard city was destroyed by a storm on the afternoon it was completed. What they had built was left in a soggy, formless mess like the city they left behind."—From Chapter Eight: *Art Therapy and Hurricane Katrina*

THE FRONT COVER is a powerful image, and at first glance a casual observer may assume this book is only about racial issues in disasters or surrounding Katrina in particular. This book is about race, but it is also about health, environment, community, vulnerability, policy, planning, mitigation, response, and recovery. In essence, this book is about the human consequences of disasters. The hurt, pain, grief, and fight for normalcy that a disaster causes are part of this story as is the longevity of this suffering. The photography on the front cover was taken one year after Katrina, and yet this same degree of anger and desperation is found today, in the summer of 2009. Disasters have a long-term impact on people and places. As geographers, we are interested in both. We are interested in the spatial aspects of disasters. Often, studies of disasters occur at the scale of a region, or a city, or even at a planning district level. However, you will find that we focus on much finer scales of the human geography of disasters, such as the neighborhood, the street, a block, and even a single house. Returning to the front cover, although the man in the photograph is making a plea to "stop ethnic cleansing," he is focused not on the State of Louisiana, or the city of New Orleans, but on his neighborhood, the Lower Ninth Ward. This micro-scale geography is a theme that we will return to throughout the book.

This book was written primarily for undergraduates taking a class about disasters, but where the emphasis is placed not so much on the physical processes but on how people shape these events and how these events affect people. Although many other texts focus on the sociology of disasters, there are relatively few books on the human geography of these events. This book has been written to partly fill this gap. It has primarily been written for our students at the University of Southern California, drawing on our extensive field experiences of working with disasters, from the Emergency Operations Center (EOC) to neighborhood recovery. As geographers, we emphasize the spatial, showing how every aspect of a disaster can be

Figure I. *An abandoned building in the Lower 9th Ward—one door is open, vegetation has almost reached the roof on one side, and several search and rescue markings are daubed on the front.* Source: Curtis, June 2009.

Figure II. *A burned out car is abandoned in the road, while at the other end a water- and vegetation-filled pothole make this road in the Lower 9th Ward completely impassible.* Source: Curtis, June 2009.

mapped to more fully understand the processes at work. But again, in keeping with the theme of this book, these maps focus more on the distribution of social, rather than physical processes. In so doing, we also present a different side to many other texts as we stress the local, or using geographic terminology, the fine scale.

This is not to say it isn't important to understand the physical processes at work—it is. We also teach these in our classes, but there are many books that cover these processes, so there is no need for further replication. Although the subject matter draws on all disaster types, both from the past to the present, much of our focus is on Katrina. This is for two reasons. Firstly, this single event has generated more disaster-related discussion—especially from a social perspective—than any other event in recent times. Secondly, we have extensive research experience from the time of landfall to our ongoing collaboration with different New Orleans neighborhoods as they struggle to rebuild their shattered communities.

From this perspective, we are writing about what we know, and not just synthesizing the experiences of others (though obviously this occurs as well). Having said that, we have written this book in such a way that instructors need not have a personal experience with these events; nor is it vital that an instructor has experience with different geospatial technologies. The exercises included in this book can be applied by Geographic Information Systems (GIS) users, but anyone with access to Google Earth can also benefit. In our classes we use these as a spring board for more sophisticated GIS exercises, but these are not included here—we are stressing the human and the spatial, not just data and techniques. We also focus our attention entirely on the United States—and we are the first to admit that a similar book is needed for the many disasters that occur globally every year. But we have tried to concentrate on what we know, and in so doing help improve the students understanding of the complexities involved rather than covering too much material at a shallow level. From the student's perspective, this is not a text full of dates or numbers to memorize for exams. We want you to understand the social processes at work—linked by their geography.

From our perspective, we continue to visit our neighborhood collaborators, taking spatial video, turning these videos into maps for their ongoing fight to return to normalcy. Our last trip in June 2009 covered four of these communities. Unfortunately, for each the return rate is still less than 50% (far lower for some). Some streets are in terrible condition; in some places they are completely impassible. Badly damaged homes, some almost covered with vegetation, fading but still visible search and rescue markings on the walls are evidence their decline, while next door a family bravely tends to a well manicured flower garden. Crime is rampant in many areas. We are often told of the health problems that follow the returnees who are now forced to travel further afield for clinic visits, pharmacies, and some of the basic necessities we all take for granted. The physical anchors of the social networks so important to these communities, churches, community centers, corner stores, still too often are abandoned. And yet through it all, through the incredible hardship and daily suffering we are constantly amazed by their resolve—to never give up. That is why we continue to go back and do all we can. It is also the reason why we take students with us, because not only is labor needed, but so is education. More people need to learn about this ongoing disaster.

And that is our hope for this book—to not only provide a useful classroom tool to provoke discussion but also educate about the many social aspects of Katrina and disasters of its kind ... and maybe stimulate a few to carry on with work in this area. After every semester we have a few who want to be involved, to spend a spring break rebuilding the homes of Louisiana, to become part of our research team, or even to learn more about art therapy. In this spirit, any royalties from these books will go to a research fund to continue our collaborations with the communities of New Orleans.

Figure III. *Our spatial video system captures an abandoned church, a wooden cross wired to the door. A contact telephone number has been sprayed on the wall.* Source: Curtis, June 2009.

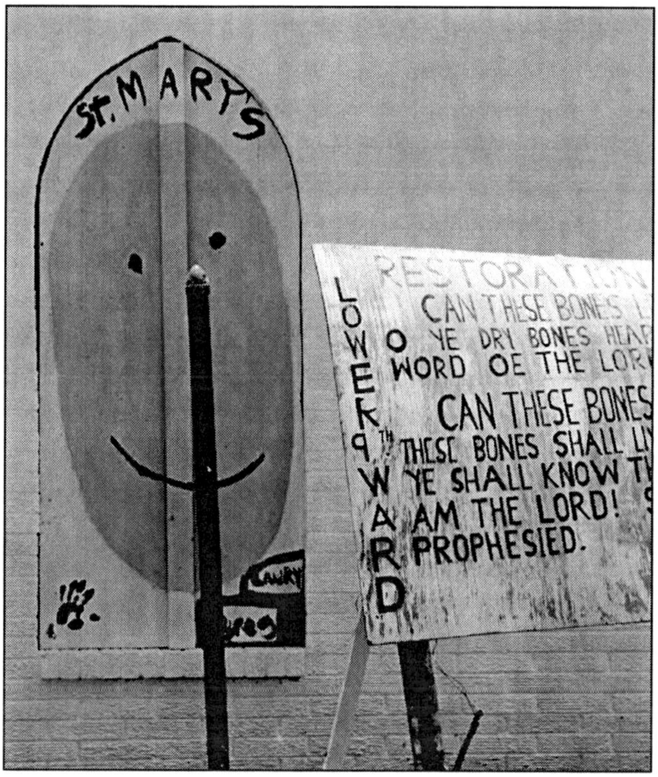

Figure IV: A message of support outside a church is now weather worn. *A smiley face has been painted on the window shutter behind. Readers will find this same scene elsewhere in the book (Figure 9:6), though taken two years earlier—the message of hope a little brighter, and search and rescue markings where the smiley face is.*
Source Curtis, June 2009.

The book is split into five general themes, though these meld between the chapters
1) Understanding why geography is important
2) Comparing the past to the present
3) Issues of vulnerability
4) The psychological impact
5) The road to recovery

In Chapter One, we introduce why geography matters—why various aspects of a disaster are spatial in nature and therefore should be investigated for their underlying spatial processes, and of course be mapped. Some basic spatial concepts are introduced, and an example of application of these concepts is provided through a case study on how to spatially predict the surge potential for a hospital after an earthquake in Los Angeles. The impact Google Earth and Street View have had on information dissemination are mentioned, as is our spatial video system which allows for the collection of post-disaster landscape data. By the end of this chapter, readers unfamiliar with geography, or spatial thinking in general, should have a sense of why a book on *GIS, Human Geography and Disasters* has relevance.

In Chapter Two, we continue this initial exploration of a spatial approach to the human side of disasters by concentrating on the technologies we have at our disposal, especially Geographic information Systems (GIS). The basic components of a GIS are described, with focus being placed on methods of spatial data input, analysis, and mapping. This chapter spends some time providing examples of different disaster cartographies, including mapping recovery. Both recovery mapping and the spatial video will be returned to in chapter twelve—in effect providing bookends for the reader. By this point the concepts will be familiar, and enriched by the discussion of human impact in intervening chapters.

Chapter Three begins a two part sequence that compares the events of the Johnstown Flood of 1889. The events of this terrible disaster, which have echoes of Katrina in terms of a natural event being compounded by human mistakes, are primarily told using the exact language of a Board of Public Health report. The approach of this chapter gives the reader exposure to another key disaster and allows for comparative discussion points. In order to facilitate this, photographs from Katrina with accompanying descriptions are scattered throughout the chapter at suitable comparative "moments."

Chapter Four continues our comparison of the Johnstown flood with Katrina using the same style as the previous chapter. However, while chapter three concentrated on the description of the actual disaster, chapter four focuses on the post-event response and clean-up stages. The chapter starts with a personal narrative of our first flight over New Orleans and the bayous beyond; it is included to add that human touch, a geographer's description of a devastated landscape.

Chapter Five changes disaster type to California Wildfires, though it maintains a similar presentation approach as Chapters Three and Four in that primary descriptive sources (in this case newspaper stories) are used to convey the human consequences. This chapter also segues between the opening chapters (spatial concepts, use of GIS) to the next section on disproportionate impacts.

The next sequence of chapters describes the social environment that underlies the disproportionate impact of many disasters.

In Chapter Six we further elaborate on the concept of social vulnerability, particularly which groups are known to suffer the brunt of a disaster, why this occurs, and the implications of this spatial pattern. Concepts such as "hazardscape" are introduced through examples that range from a nuclear detonation to Katrina. Cartographic overlays are used as an approach to understand how different geographies can be combined. These overlays compare pre-Katrina vulnerabilities with surfaces depicting various aspects of the disaster; including 911 call data, flood depth and building damage. The result from a more sophisticated analysis of vulnerability measures and flood height is also presented. This chapter provides details on the traditional measures of social vulnerability, even though three of particular relevance to New Orleans are emphasized: race, poverty, and females as head of household.

Chapter Seven continues the discussion of vulnerability by introducing a less well described pre-disaster concern, that of health vulnerability. Although tied closely to social vulnerability, the urban poor—and especially African Americans—carry a high debilitating health burden that not only increases their likelihood of disaster exposure but may also hinder the return to normalcy. Indeed, these chronic conditions may actually worsen and lead to a further widening of the health gap between minorities and whites. Emphasis in the chapter is placed on how single female headed families are of particular concern—using examples of how Katrina may cause future poor birth outcomes in Louisiana, and in so doing leave a health legacy to be borne by future generations. The chapter ends in considerable detail regarding the types of health and social services that were mobilized for evacuees in Baton Rouge after Katrina—these providing the basis for students to understand the post-event complexity of vulnerability.

Continuing with this progression, two guest chapters are included. Chapter Eight describes the work of Karla Leopold, who was part of a team that used Art Therapy to help evacuee children in Renaissance Village, a FEMA trailer park located just outside of Baton Rouge, Louisiana. Every semester Karla gives a guest lecture for me—and she never fails to move the students. From a geographer's perspective, I am always amazed at the spatial nature of the pictures the children produce, and how spatial order is important. I find the stories of the children that accompany these pictures to be extremely moving and important for understanding the mental health consequences of disasters.

Chapter Nine is also primarily written by a guest, Louisa Holmes, who is a senior doctoral student in the Department of Geography at USC. She has considerable experience working in the field of stress and the built environment. She provides a detailed overview of different mental health problems, their causes, manifestations, and coping strategies employed. This is not an easy chapter to read; it is dense and purposefully so. For the sake of brevity we often have to synthesize and therefore simplify information. However, this chapter will give the reader a glimpse of the complexity involved and how the information presented here is actually a thread that can be followed and expanded like a spider's web.

Chapter Ten also contains a section on disaster related psychopathology, using one of the models (conservation of resources) to help explain why so many people spray painted messages across their homes. This chapter looks at the post-disaster landscape from this perspective, not in terms of the damage wrought, but by the human markings left behind. We show how a GIS can be used to collate all these messages, especially the search and rescue crosses. Very little has been written about this marking of a landscape which is so visually arresting.

Chapter Eleven provides an overview of our current understanding of recovery, the multiple spatial scales where this process occurs, and the importance of planning for recovery. In addition, new forms of geospatial data collection are described as tools that allow researchers or communities to collect recovery data that can be linked to a GPS, mapped, and used as a management and advocacy tool.

Chapter Twelve brings together several of the themes described through the book in a description of our collaborative work with Churches Supporting Churches in New Orleans. We again return to the spatial video, describing in more detail how it works, from a technical perspective to the way in which we encode information from the video into a GIS. Different classification schemes are presented, as are examples of our neighborhood maps. This chapter ends with as series of exercises utilizing Google Street View which mirror our spatial video-to-map methodology. This allows students to experience how fine-scale geographic research can be conducted.

ONE

An Introduction to Thinking Spatially About Disasters

DISASTERS ARE INHERENTLY spatial—both in terms of the physical processes as well as the human implications. Hurricane tracks, the location of fault lines, how tornadoes are generated—these are patterns or processes that have or leave spatial footprints. Where people live in relation to potential hazards or the societal impact left after a disaster can again be described in terms of spatial patterns. Within these patterns are human places, cultures, and interactions. An earthquake-devastated city is not just a landscape of damage, morbidity, and mortality but it also comprises lost (and recovering) neighborhoods, disrupted social networks, variations in resiliency, and social and environmental (in)justice. As geographers we have the technologies and spatial skills to map, predict, and ultimately understand these landscapes. This will be a common theme throughout the book: irrespective of chapter topic, geography and geographic methods play an important role.

1: CAPTURING THE GEOGRAPHY IN PAST DISASTERS

Spatial aspects of a disaster can be portrayed in terms of physical process or human consequence. The physical processes can often be described geographically in their formation (where the weather fronts meet, or the path of the hurricane), or at least their impact points can be mapped (a tornado touchdown, the earthquake's epicentre). Figure 1:1 shows an inset from a report published after the 1906 San Francisco earthquake. The map itself displays "Distribution of Earth Movement on April 18, 1906."

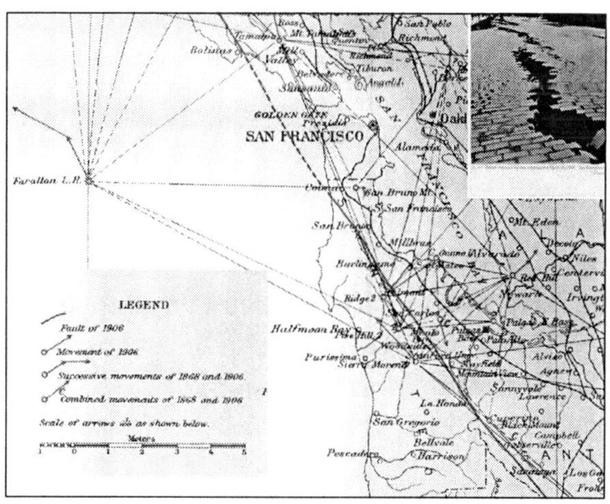

Figure 1:1. Source: Earthquake Investigation Commission. A. Hoen & Co., Baltimore: Carnegie Institution of Washington, 1908.

Figure 1:2. Source: Earthquake Investigation Commission. A. Hoen & Co., Baltimore: Carnegie Institution of Washington, 1908.

Although this image provides an interesting spatial description of the physical expression of the earthquake—where the known (at the time) fault lines are and the amount of earth movement, it doesn't capture the human impact. The same report included two further maps showing sections of the built environment that succumbed to the earthquake-related fire. Figure 1:2 displays the area of San Francisco that burned[1]. However, even these maps don't capture the human emotion and suffering associated with the burn areas. This human dimension is hinted at in the inset "disaster" postcards often produced after such events. One shows a relief camp with the flames in the background, while the other has been "used" by a witness who describes how the earthquake affected her.

Another difference between Figures 1:1 and 1:2 is the spatial scale involved, an issue that will be discussed in more detail later in this chapter. Using geographic terminology, Figure 1:2 is a larger (or finer) scale map than Figure 1:1. Arguably, we have to zoom in even further, to the scale of neighbourhoods, streets, and even individual buildings, to begin to understand how all the other complexities of a disaster interact to affect people.

Alternatively, in Figure 1:3 we zoom out to a small scale (large area) map of Bengal. As an illustration of a non-US disaster, and in keeping with the historic cartography showing primary and secondary impacts presented in the first two

1 The detail in this map is good enough that it could be combined with other spatial data, such as current information about San Francisco within a Geographic Information System. This affords researchers the possibility of "recreating" some of the spatial aspects of the disaster or visualizing the destruction that could occur if the same event happened today.

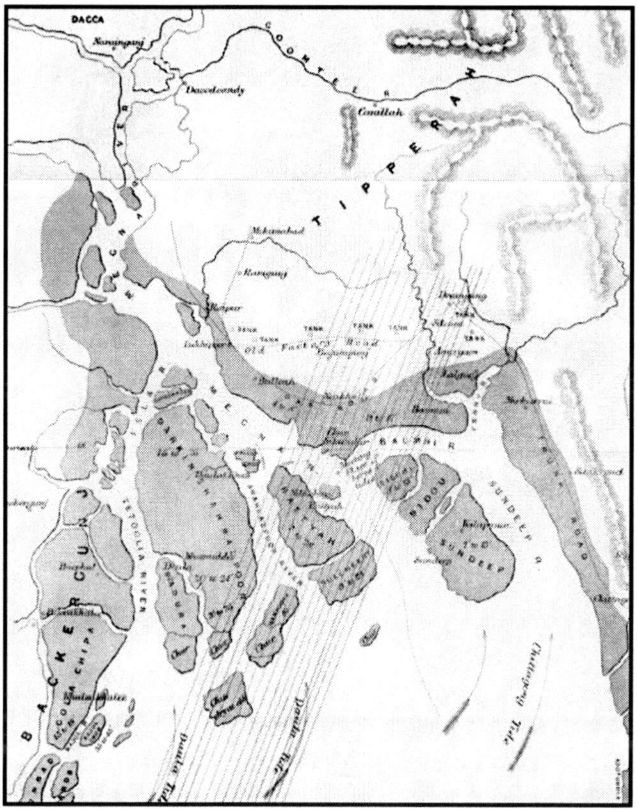

Figure 1:3. *A map from an 1877 report to both Houses of Parliament on the "Bengal Cyclone and Storm wave of the 21st October–1st November 1876; and the subsequent cholera epidemic."*

illustrations, Figure 1:3 displays the impact of a cyclone (the same as a hurricane) that struck Bengal in 1876. This map displays the geographic output of the physical event—the "*Depth of Inundation*" measured in feet, and the track of the "*vortex measured by parallel lines.*" However, the death toll for this event far exceeds that of Katrina or any other hurricane in US history[2]. Although the only example of cartography in the report displays the physical process—the spatial pattern of cholera deaths could also have been mapped as well (and frequently were in contemporary reports)[3].

Q: Have you seen a newspaper or other popular media map describing a disaster that has combined both physical and human impacts?

Aside from this one example, this book will exclusively concentrate on the United States for reasons of controlling both social and geographic space in order to make the comparisons more valid. However, the reader should be fully aware that in any year there are many disasters with severe human consequences scattered

2 The report concludes: "... it is distressing to see that the ravages of cholera, following close upon the track of the first-named disaster, have carried off no less than 75,000 lives. I gather from the present papers that the total mortality occasioned by the cyclone wave and the ensuing cholera amounted at the end of last March to about 165,000 ..."

3 Cholera is (usually) spread by people drinking or washing food in water polluted by fecal matter. It was known even at this time (1877) that boiling water helped lessen the chance of contracting disease—but the upheavals, forced migrations, and general stresses found in the wake of a natural disaster often provide the perfect environment for epidemics.

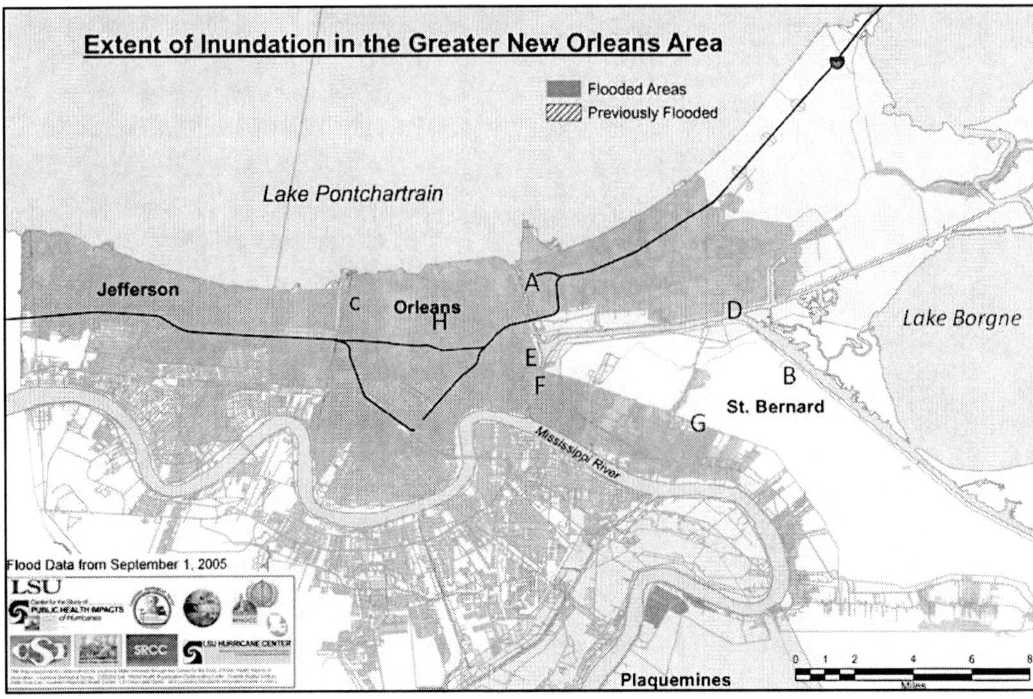

Figure 1:4. *Shows a map created in the Louisiana State Emergency Operations Center (EOC) during the response to Katrina. A further layer of letters corresponds to the sequence of events following the approach of the hurricane to the rupturing of the levee walls. The flood sequence presented below the map comes from the excellent "The Times Picayune Flash Flood" animation still accessible at http://www.nola.com/katrina/graphics/flashflood.swf. Readers are strongly advised to see this resource.* Source: Map created in the Louisiana State EOC. Specific author is unknown, but all groups who contributed data, skills or technology to this map are cited in the bottom left. Letters correspond to flood information taken from the "The Times Picayune Flash Flood."

across the globe, a great deal of which far exceed the human loss of life experienced in U.S. disasters such as Katrina[4].

These first few maps help us understand the size, process, impacted area, and consequences of the disaster. If these disasters had happened today (such as with Katrina), modern mapping technologies would have eased the collection of spatial data and made map making and information dissemination almost real time.

As many of the examples used throughout this book will involve Katrina, it is useful to briefly describe *spatially* what happened.

4 Katrina will mostly be used in this book instead of "Hurricane Katrina" and "Hurricane Rita" to capture all of the actions (including human agency) and the ongoing consequences of this disaster.

A. 4:30am—Leaking along the Industrial Canal
B. 5:00am—The levee protecting St Bernard Parish from Lake Borgne begins to fail
C. 6:30am—Water begins to leak through the 17th Street Canal into the neighbourhood of Lake View
D. 6:30am—The funnelling of water in the Intercoastal waterway breaches defences and floods East New Orleans
E. 7:30am—Upper 9th Ward, Bywater, and Treme are flooded by a Western break in the Industrial canal levee.
F. 7:45am—The Eastern levee of the Industrial Canal breaks flooding the Lower 9th Ward and beyond.
G. 8:30am—St Bernard Parish's flood defences fail from the Lake Borgne waters
H. 9:30am—The Eastern levee of the London Avenue Canal fails
I. (at location C)
9:45am—The 17th Street Canal level fails.

At this point Katrina makes land fall at Slidell on the other side of Lake Pontchartrain from New Orleans.

J. (at location H)
10:30am—The western levee of the London Avenue Canal fails

This spatial sequence of events underlies much of the description to come. As you go through the different book sections you may find it useful to refer back to this map.

2: SPATIAL CONCEPTS AND DISASTERS

The events of Katrina (just as with the San Francisco earthquake of 1906) can be described *spatially* in several ways—in terms of the physical impact, the strength, and direction of the winds, or the spatial extent of flooding. This last physical expression can be further subdivided into maximum height or temporal duration of flooding (how many days was a neighbourhood under water). Onto these physical expressions can be laid (or overlaid if we start to think in terms of different geographic layers) the amount of building damage (either measured by percentage of damage, or even a dollar amount), the location of 911 calls, or where the dead were found. There are several other variables that could be used, and some of these will be discussed throughout the book, but for now we can start to think of how these all fit together. It would seem to make sense that the strength of winds, along with flood height, might correspond with damage estimates, or even loss of life. One way to start such an investigation would be to map each layer—a common technique used by geographers and one which will be used throughout this book. In order to do this, a Geographic Information System (GIS) is needed.

The utility of a GIS is now appreciated in many disciplines other than geography; for example anthropologists, biologists, environmental scientists, historians, psychologists, and sociologists all understand that spatial patterns and processes are important. It is an especially important tool for disaster scientists because as we just mentioned, natural disasters are spatial—and as a result the emergency management cycle is also comprised of different geographic layers. This cycle consists of four interrelated and overlapping areas: planning/ preparedness, mitigation, response, and recovery. In other words, how society intersects with our scientific understanding of hazards. As this book's primary focus is on the human geography of disasters, aspects from each of these four stages will be presented throughout the book—again with emphasis placed on how geography matters. The next chapter will describe aspects of a GIS in more detail, but first we still need to introduce a few basic concepts of both the software and spatial thinking in general. Although two chapters cannot hope to match the many introductory GIS texts available[5], they will be enough to present an overview of the importance of the system to understanding human consequence in disaster science. For example, different geographic / GIS terms have already been mentioned in this chapter, such as spatial scale and spatial patterns, and these will be discussed in the context of disaster response and recovery.

For the lay person, a GIS has many similarities to Google Earth (GE) as it allows someone to work with different maps at different geographic scales. What makes a GIS more important as a tool for disaster researchers is the level of interactivity it offers. To use an analogy, imagine the difference between a low-end digital camera and a professional use Single Lens Reflex (SLR)—both can take pictures, however the special features of the SLR, such as an increased picture resolution, make it preferable to the professional. This is not being critical of the small digital camera because it serves a purpose well, and the pictures can be shared and enjoyed. Even the professional photographer may keep one in his pocket for certain situations where the need is for a quick response. Hopefully the reader will begin to understand this analogy further as we move through the chapters.

Perhaps the best way to introduce both spatial concepts and GIS is to work through a conceptual overview of a problem. In this case we will investigate how spatial data and GIS can be used to prepare for an earthquake in Los Angeles.

5 Although there are many introductory GIS books, as well as more specialist texts, one of the best to combine both the theory and application of GIS in a easy to digest format is Longley P, Goodchild MF, and D Rhind, (2005) *Geographic Information Systems and Science*, Wiley.

Imagine a pediatric hospital in downtown Los Angeles that wants to take a proactive stance in terms of earthquake response; particularly, they want an estimate of how many pediatric patients to expect in the first several days after a large earthquake. This influx of patients is termed a "surge." What spatial information can be used to guide officials preparing adequate numbers of supplies to meet their immediate post-disaster needs and how can this information be easily communicated to improve decision making in the planning process? Planning for this surge is essential for a successful response because children require smaller sizes of equipment and different types/ doses of medicines than adults. Therefore, for a hospital to effectively respond during the disaster means they must have the requisite medical supplies to serve this population until outside support arrives. As you read through this scenario, think about the concepts that are introduced in this particular case as they might apply to other pre-disaster planning needs.

There is a 72-hour rule that follows a disaster. This means that any community, irrespective of size should be able to fend for itself for approximately three days until help arrives. First, by definition a disaster will overwhelm local resources to the point that individuals and institutions should be prepared to support themselves for several days. Second, any response effort that is immediate will take some time to reach all areas where it is needed, especially in large urban environments. Finally, the Federal Emergency Management Agency (FEMA) is not tasked with response, so their assistance is also not immediate. This reality argues that individuals and organizations should have actionable plans to deal with disaster events. Though these plans can be for protection of internal personnel, such as through evacuation plans, phone trees, meeting points, etc., the focus of this scenario is on planning to serve the surrounding community. Specifically, we will use commonly available sources of population data to show how a post-earthquake surge of one vulnerable population sector, pediatric patients, can be calculated for a hospital. However, the issues raised in this case are transferable to other groups who serve the public.

What are some of the issues that need to be considered? Part of being prepared is having accurate estimates of surge potential (how many people will be needing services in the hours and days after an event). However, having these data are not useful unless they can be shared with those who will be involved in providing services. Data must be disseminated in a format that is easy to use and in a way that gives context that has meaning for the user. After all, for emergency responders, what good is a map of locations of nursing homes if it sits in a planning document that is collecting dust on a bookshelf? What good is this map if it is outdated and does not present any information on the number of people the home serves? What good is it if it is digital, perhaps as a layer in a GIS, but either no one has GIS software or they don't have the trained personnel to use it? Clearly, the answer to all of these questions is "not very useful." However, if data are regularly updated,

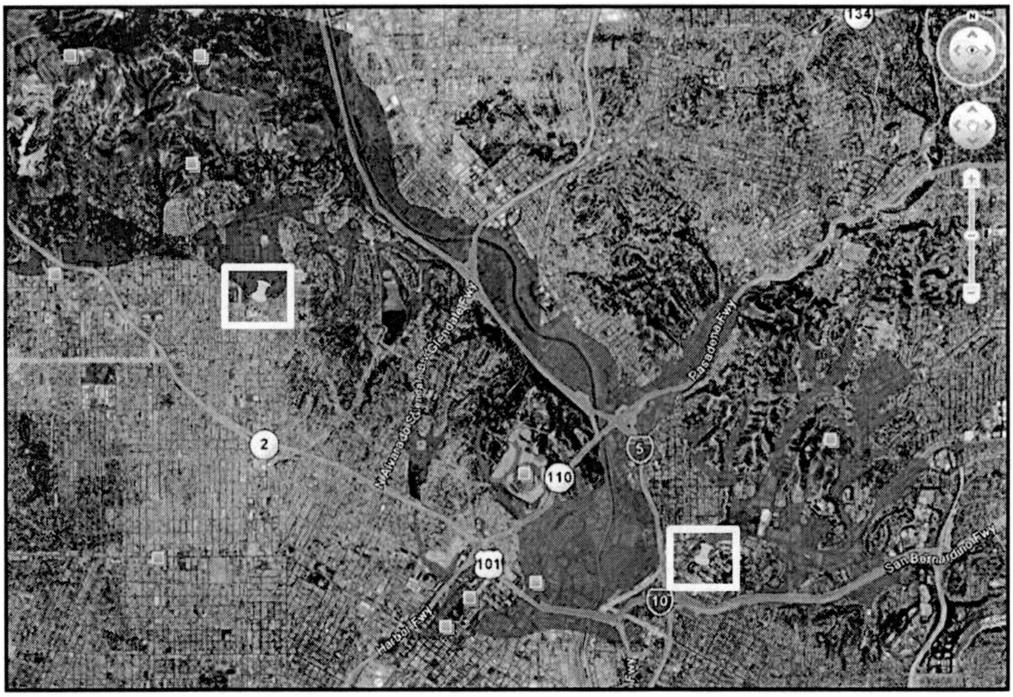

Figure 1:5. *Two hospitals are located by pins (and emphasized by boxes). These could have been located in Google Earth by using their addresses, their coordinates or even by zooming into the aerial photography. In actuality these locations were created in a GIS, as was a mask that sits on top of the map showing potential landslide and liquefaction areas.*

and are in an accessible format that can be given context (with layers such as roads, current amount of damage surrounding these places), then they have the potential to be extremely useful in planning and in response.

In general terms we need to understand basic surge potential—which means, how many patients are we likely to see coming through the doors after the earthquake? By knowing this figure we can assess where shortfalls in terms of staffing, equipment, and beds might occur. So how do we do this simple task? First, we need to know where the hospital is located. In general, this can be mapped into a GIS in one of three ways. If we know the hospital address we can match it to a street layer (we often call each individual slice of geography a layer) in much the same way Internet mapping software works. If we know the coordinate of the hospital, (maybe someone has recorded it on a Global Positioning System), then this can also be mapped. Finally, if we have high resolution aerial photography of the downtown Los Angeles area we can "heads-up digitize" the building which simply means dropping a pin or marker on the building location by clicking the mouse. All three of these tasks can also be done easily in GE.

There are several spatial concepts/issues that need to be addressed if we are to fully understand how a GIS can be used to solve a problem such as this. The first of these is the way objects interact over space (where space means a geographic area).

In this case, "objects" are people and hospitals. In effect, we need to understand how people will interact over space, how will they behave, and where will they go in relation to their hospital options? To gain this understanding, we need to consider a common geographic concept called distance decay. Distance decay suggests that the further away something is, the less importance it has, and conversely, the closer something is the more importance it has. By changing the word "importance" to "interaction," "shopping," "following a football team" or in our example *going to a hospital*, this relationship is easier to understand. All things being equal, you will go to, shop at, and support the closest team to where you live. The likelihood of you going to, shopping at, or supporting an alternative team will decrease the further away that option is. Although there are exceptions, in general the relationship works pretty well.

Q: What "exceptions" can you think of for this rule?

However, in order to maximize the accuracy of predicting how many people will come to the hospital during the earthquake, we have to think about some modifying conditions, such as, where are other "competitors" (other clinics), how large is the hospital, what is its specialty, and also, has the person been there before? Now our scenario is beginning to become a little more complicated. The size of the hospital is also important because, the bigger something is, the more likely a person is to go to it even if it is a little further away (because of more equipment, more recent techniques, and better expertise in general). This interplay between size and separating distance is common to many geographic models, and generally, if you divide the size of a place (for example the number of beds) by how far away it is[6], we end up with a "potential" value. Using this simplest of calculations, we could estimate where every earthquake victim will head for emergency care.

The second of our key spatial issues involves the way we work with data at different spatial aggregations. We can't perform the potential calculation for everybody so we have to use "spatially aggregated data"—which simply means summing all the people within a given area. It would make sense to have this aggregation by neighborhood[7], but we don't usually know the boundary of these, so instead we

6 We usually square this value as the effect of distance decay decreases at a quicker rate than just a linear relationship

7 The Los Angeles Times has a neighborhood boundary project that allows the public to view neighborhood areas in GE as well as comment on their correct spatial definition. See http://projects.latimes.com/mapping-la/api/neighborhoods/

Figure 1:6. Source: Neighborhood Boundary files from the Greater New Orleans Community Data Center. Map created using Arc Map 9.3 (ESRI) as are all others in this book.

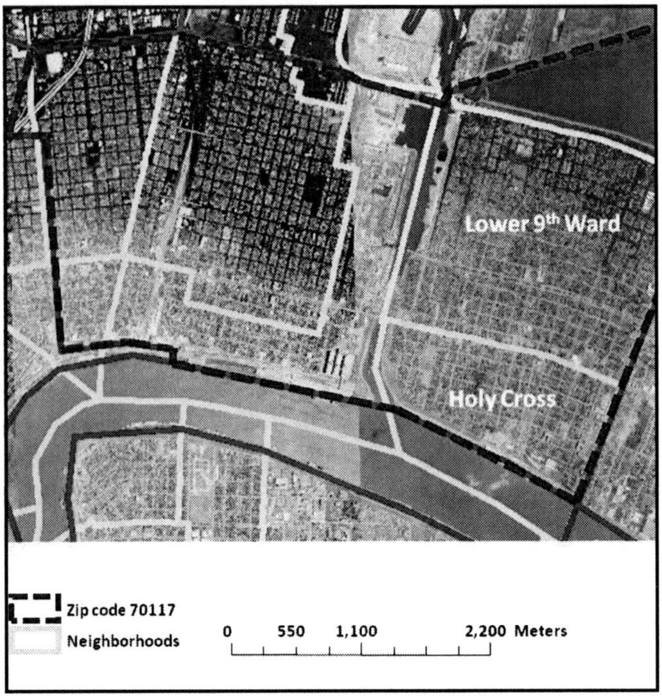

use zip codes (which is not a very useful aggregation unit[8]) or areas defined by the census, the most common being census block, block group, and tract. One of the problems with any aggregated area, but especially for large units such as a zip code, is that this doesn't match "neighborhoods" which are more representative of the processes affecting the people who live inside them. In addition, political (or postal) boundaries artificially break space and in so doing may not accurately reflect the true social (people living either side of a street) or physical (flood depth after a levee breach) landscape. Figure 1:6 shows neighborhoods of New Orleans falling inside zip code 70117. Social data made available at the zip code level, such as pre-existing health vulnerabilities, would not capture the specific living conditions or cultural landscape of the Lower 9th Ward. Similarly, any aggregate flood variable for this zip code would smooth out the high water levels found close to the levee breaks, and the relatively shallow flooding found close to the Mississippi River (seen curving at the bottom of the figure).

A further benefit of working with the census is that we can manipulate this aggregated population in different ways—we can make a map of total population, by racial subgroup, by economic status, and importantly for a pediatric hospital, by the number of children. Why is this manipulation useful? In a later chapter, we will

8 Apart from a zip code being too large and therefore encompassing too many neighborhoods, another problem is that the area is defined by postal delivery routes. It is therefore possible to have an address with a zip code number that does not sit within the zip code boundary, but rather "spurs" out along a road convenient to the postal substation. This has important implications if socioeconomic data are attached to the address by census boundaries overlaying the general zip code area. For a more detailed explanation see Grubesic, Tony H & Matisziw Timothy C On the use of ZIP codes and ZIP code tabulation areas (ZCTAs) for the spatial analysis of epidemiological data *International Journal of Health Geographics* 5:58 (2006)

discuss the importance of "social vulnerability" which means that certain groups of society are going to be at greater risk (and therefore will require extra help) because of where they live and their general inability to cope with a massive shock[9].

A third key concept related to the issue of spatial aggregation, is that of geographic scale.

Geographic scale means how much is reality modified in order for a spatial representation (a map) to be made. There is no point in making a 1 to 1 map so we have to "scale it down" so that 1 mile becomes 1 inch, or 1 km becomes 1 cm. The more we increase the proportion between our measure of reality and the measure we use on our map, the greater the geographic area that is mapped on the same size piece of paper. A 1:100 map will contain more geographic detail, but a smaller area than a 1:1000 scale map. For extensive geographic areas, we use the phrase "small scale" mapping. For small geographic areas, such as a city block, we use the phase, "large scale" mapping. Unfortunately, large scale is used in the opposite way for many other disciplines, so here we will use "fine scale" for maps of streets, city blocks or even individual buildings. For Figure 1:7 we return to the volume of maps created after the 1906 earthquake, but this time at a fine scale showing individually impacted buildings of Santa Rosa, California.

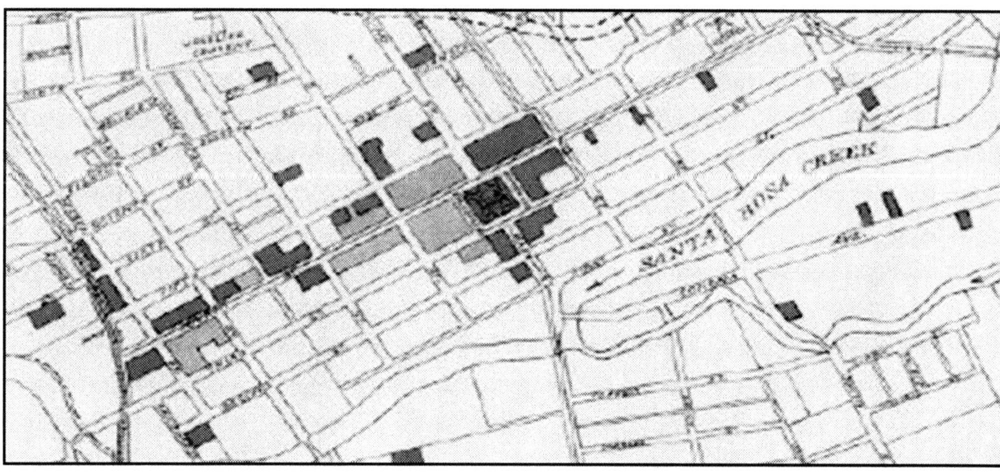

Figure 1:7 Source: Map of the city of Santa Rosa, Sonoma County, California showing the portions destroyed by the earthquake of April 18, 1906 and by the fire consequent thereto. Earthquake Investigation Commission. A. Hoen & Co., Baltimore: Carnegie Institution of Washington, 1908. The map displays actual buildings destroyed by either the earthquake or subsequent fire.

9 Although we will return to this concept later, readers are pointed to Cutter, Susan L., B. Boruff, and W.L. Shirley. 2003. "Social Vulnerability to Environmental Hazards." *Social Science Quarterly* 84:242–261.

Q: Why would a map at this scale be useful for researchers?

Now that we have introduced some basic spatial concepts that can be applied to how we obtain and manipulate geographic information, let's turn to an equally important issue of how this information is communicated. Using the example of predicting post-earthquake pediatric surge for a Los Angeles hospital, we have identified the hospital on a map, selected aggregated population data on the number of potential pediatric patients, and now we need to consider the ways we can share this information with decision-makers. After this next section we will return to the Los Angeles example and work through it in detail using the spatial concepts introduced in the chapter.

3: COMMUNICATING SPATIAL INFORMATION

Communication takes many forms for people affected by a disaster. It can be the lack of a means to communicate with others, such as friends and family. It can also be the lack of a reliable information source. Here are two "communication" tales with regards to evacuees from New Orleans. One colleague went into a shelter close to Natchitoches, Louisiana, taking two lap tops so that the people there could use the Internet to search for information. This shelter, as with many others, had no provision to relay current information. Most studies on the psychological impacts suffered during a disaster will comment that uncertainty, which can lead to "dread risk"—the negative exaggeration of any event, is at least in part caused by not knowing what is happening, and what may come. There is therefore a medical justification as well as a humanitarian imperative to inform survivors about what has happened or is happening. The second anecdote involves a couple in Morehead, Kentucky at the end of 2005. We were giving a talk about the response effort when afterwards we were approached by a couple who told me that they had originally come from St Bernard Parish—one of the hardest hit areas to the east of New Orleans. They told me how they had received phone calls from friends in their old town during the days following the storm emergency 911 calls that could not be placed to the Louisiana based EOC because the local circuits were either down or overloaded. They literally had to call Kentucky, and have their messages relayed back to Louisiana. On one hand it shows how we now live in such a small geographic space, and on the other hand, how important it is to understand the fragility of our traditional means of communication.

What is interesting is that U.S. disasters after Katrina have begun to use spatial approaches to communicate. After the California wildfires of 2007, Google Earth was used to relay daily updates as to the status of the fires, which roads had been blocked, where mandatory evacuation orders were in place—even what form of ID you would need to regain entry back into your neighborhood. This information was updated regularly by the public broadcasting station KPBS. This approach was exactly what was required for those facing the uncertainty of a disaster—even to the level of detail as to how many spaces were left in each shelter (See Figure 1:8)[10]. Google Earth has since been used in several other disaster situations, providing excellent geographic detail and a means to disseminate information quickly and easily.

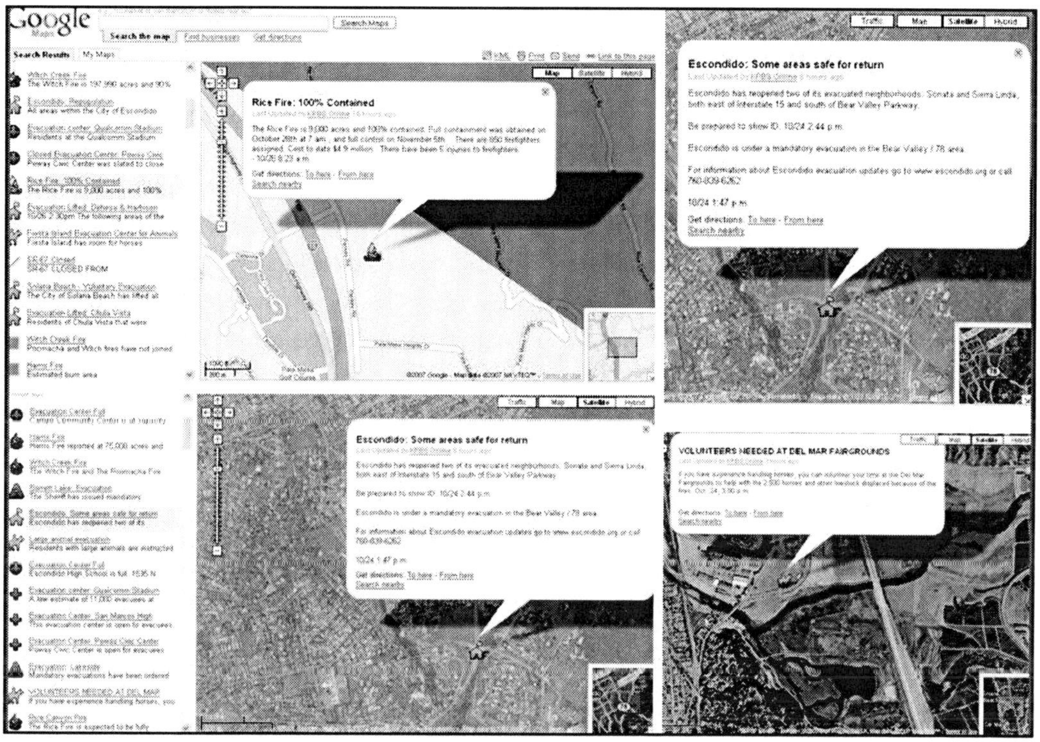

Figure 1:8. *Four snapshots from the constantly updated Google Maps page hosted by KPBS online during the San Diego wildfires of 2007. This system allowed people impacted by the fires to assess the approaching risk to their neighborhood, where shelters were located, and if they were full, and also (using standard Google Maps features) which roads were clogged. In addition, volunteers could use the system to see which goods were needed in shelters or to help animals. One nice feature was telling returnees what they would need (in terms of identification) to be allowed back into their neighborhood.*

10 Readers are pointed to the following paper Backchannels on the Front Lines: Emergent Uses of Social Media in the 2007 Southern California Wildfires by Jeannette Sutton, Leysia Palen & Irina Shklovski in the *Proceedings of the 5th International ISCRAM Conference*—Washington, DC, USA, May 2008 F. Fiedrich and B. Van de Walle, eds

Q: By using different Internet search engines—how many other Google maps/ Earth disaster sites can you find? Which of these was generated during the disaster itself as a near real time information exchange?

It should never be forgotten that a map is a communication device—just as important if not more so than the written word. The audience for the previous wildfire example was the general public, though these tools also serve major communication roles for responders. For example, during the first few days after Katrina, our World Health Organization Collaborating Center for Remote Sensing and GIS for Public Health (WHOCC) helped provide some spatial support for Red Cross damage assessment teams.

The Red Cross HQ had been set up in a large hall in Baton Rouge and the place was buzzing when we arrived. Outside in the parking lot were two high tech trucks, while larger food container rigs were also parked nearby. The area inside the hall was split into stations, many having a single computer with printer. We had been told to go to the station dealing with damage assessment. Once there we introduced ourselves and presented our first poster/map as an introductory gift. This would be an approach we would use in several other situations—presenting a useful visual aid means so much more than just "talking." The poster

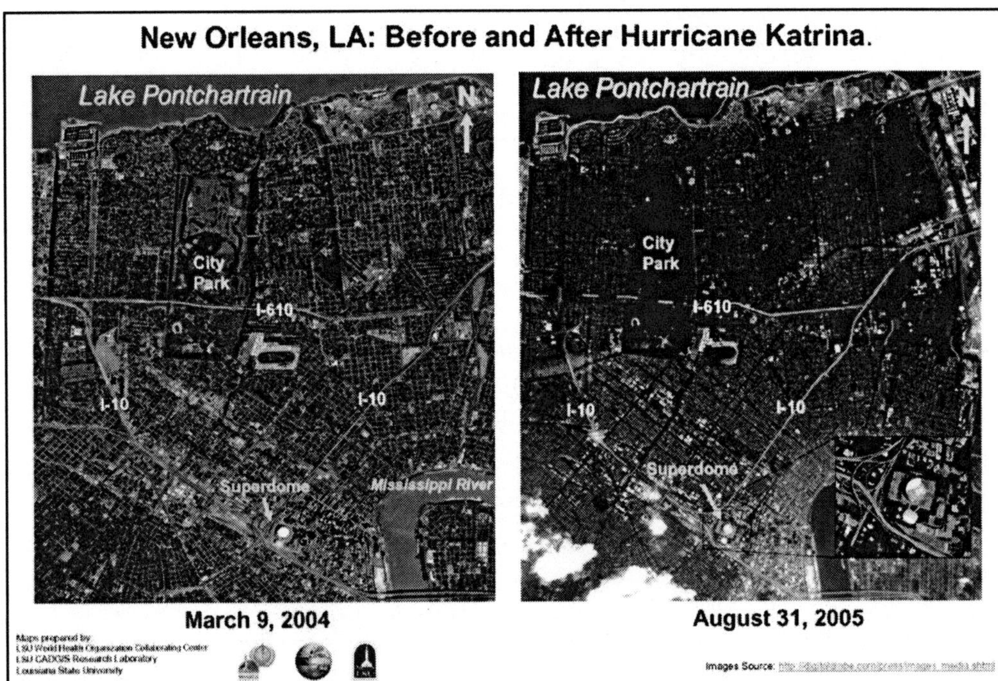

Figure 1:9. Source: Original poster created by Jason Blackburn, PhD while at the World Health Organization Collaborating Center for Remote Sensing and GIS for Public Health.

showed a before and after aerial photograph of the area around the Superdome, the flooded streets clearly visible (see Figure 1:9). One of our team members had downloaded the flood imagery and matched it with an aerial photograph from the "Atlas" geospatial warehouse housed at the CADGIS lab[11]. This immediately sparked conversation—not so much with us but between them—which is exactly the response we wanted.

Q: Figure 1:9 was created during a time of great stress and it served its purpose well. However, if you had more time, how could this poster be improved?

We were immediately asked if we could make similar poster size maps for other areas. At this point we had limited access to post hurricane imagery beyond New Orleans, but we certainly could use existing data from "Atlas." We agreed to make six posters, with underlying imagery, parish boundaries, and major roads overlaid on top. We also agreed to have them back to the group by the end of that day. These posters were to be used to help in decision making, and as a focal point for briefings/debriefings.

Q: Why do you think a poster sized map would help during disaster response briefings?

At this point it is useful to explain how Red Cross damage assessment works in a post disaster environment. This is another example where you can apply spatial thinking to a disaster-related problem. Groups of volunteers, comprised of local chapter members, as well as national volunteers flown in from all across the U.S., would go out and assess the amount of damage to residential structures. These volunteers are largely comprised of retirees as they have time and flexibility. The results of these assessments are used to determine the degree of damage to a structure so that relief payments can be metered out. The assessments are carried out in the following way, a team of two to three people drive to a previously designated neighborhood and then, without getting out of the car, they record the house number, what type of structure it is, and how much damage has occurred. These entries are recorded onto a paper sheet which is stored in a box back at the HQ. When first being told about this system I was amazed at how low-tech it was...I've often made the comment in presentations that I can only imagine that there is a warehouse somewhere, similar to the scene at the end of *Raiders of the Lost Ark*, where all these boxes of assessment sheets are stored. All this wonderful information is not

11 Readers can go to www.atlas.lsu.edu where they can download the same imagery, along with a host of other useful geospatial data, much of which was used during the response to Katrina.

only being fully utilized during the event itself but it also runs the risk of being permanently lost because no digital copy is made. Imagine how useful damage assessment data for the Northridge earthquake would be now. Although it took us several days, we eventually managed to get a Xerox copy of the sheets and geocoded[12] the damage assessments. One of the challenges faced by researchers after a disaster is how to collect data that is representative of the post disaster area and its returnee population. If, for example, you want to understand the amount of stress returnees face on returning to their homes—how do you collect that data? If you use census information to guide your data collection (for example, selecting from low, middle and high income areas), then this does not take into account the post-disaster population, not does it capture the disaster "exposure" commonly meaning the amount of damage caused by the event. If you use damage maps to help guide where you want to collect data—how do you know where to start, and how to get a representative sample of people from within this area[13]? In addition, the high stress environment of a post-disaster landscape not only includes physical impediments, such as impassable roads, but also the emotional state of returnees who do not want to be "studied" making disaster research one of the most challenging environments for academics.

The geocoded Red Cross sheets could improve this spatial sampling frame by detailing on a street-by-street basis how much damage had occurred. By always being aware of preserving spatial confidentiality, and therefore not releasing actual points on a map[14], we would be able to generalize areas of the city where the damage

12 This simply means placing addresses on a map–it will be described more fully in the next introductory chapter to GIS

13 Two post Katrina surveys conducted in October of 2005 were described in the CDC's Mortality and Morbidity Weekly Report (55), both of which employed a spatial sample frame. In the first 45 census blocks were selected, with those having fewer than 20 houses excluded. Four coordinates (lat and long) were randomly generated in each census block, and the field team navigated to these with a GPS and picked the nearest residence. Problems with using this style of survey design include a reliance on pre-Katrina population data (not the returned population), and a bias towards the less damaged areas of the city "Assessment of health-related needs after Hurricanes Katrina and Rita--Orleans and Jefferson Parishes, New Orleans area, Louisiana, October 17-22, 2005." *MMWR Morb Mortal Wkly Rep* 55(2): 38-41. In the second investigation into mold risk, census blocks with more than 20 homes were again selected, but blocks were stratified by damage into mild, moderate, severe using FEMA flood and damage maps. A GIS was used to generate a proportionate number of random coordinates (lat long) within each stratum. Field teams again used a GPS to find the random coordinate, and from this selected the house nearest or north of the waypoint. Sample limitations were similar as for the first study "Health concerns associated with mold in water-damaged homes after Hurricanes Katrina and Rita--New Orleans area, Louisiana, October 2005." *MMWR Morb Mortal Wkly Rep* 55(2): 41-4.

14 The concept of spatial confidentiality is a major concern for people using health data in a GIS environment. Unfortunately normal "rules" become more lax during a disaster response and there is a real risk of making maps that can compromise an individual at a later point. For more information

Figure 1:10. *Two A4 research maps used for collecting data on mortality locations in New Orleans. Deaths are shown as points on the map, and the hand drawn arrows and notations show field notes taken as the team visited each mortality site. The top map is the Lower 9th ward (including Holy Cross), while the bottom is the area around the London Avenue canal break*[15].

had been severe, or light, in both racially different, and economically varied neighbourhoods. This approach was actually employed by one team from LSU interested in gauging returnee stress loads. In order to help their field-data collection teams this spatial sample frame was mapped onto A4 paper—we had found that this size was perfect for field-teams using clipboards, both for search and rescue teams during the response stage, and also for our own research in New Orleans (see Figure 1:10). If we had to do this again, we would have created a GE mash-up and allowed them to spatially interact with the maps through any internet access point, even in the field if they possessed suitable mobile technology. As previously mentioned, GE is a great tool to disseminate spatial information—combining maps, images, and GIS overlays to help understand the geography of a place.

see Curtis, A., Mills, J.W., and M. Leitner (2006), Re-engineering confidential information from maps about Hurricane Katrina, *International Journal of Health Geographics* 5:44

15 Although readers might think these locations are revealing confidential information, they are actually best estimates extracted from a local newspaper.

This initial interaction with Red Cross damage assessment would lead to our adoption of a new means of spatial data collection utilized throughout this book. We were giving another presentation in Maynooth, Ireland, at the National Center for Geocomputation and one of the other presenters discussed a system whereby a digital camera (or two or three) could be synched to a GPS device for the recording of motorway conditions. This type of approach could be used to check the road surface, the state of road signs, or even the encroachment of vegetation. I immediately saw the potential for Red Cross disaster assessment. Imagine a car with two cameras, attached to either side of the vehicle, recording building damage as we drove. Instead of stopping and manually recording an address, a type of structure, and the amount of damage—all this information could be recorded on video, and synched to the right coordinate. Back at the HQ, a team working through the night could then process the video, in effect completing onscreen damage assessment sheets through a pop up window on the video output. With a little additional programming an address could also be estimated to the GPS coordinate. The benefit of the system was obvious; it would be exponentially faster and require far fewer volunteers, thus reducing the Red Cross's budget (volunteers need to be housed and fed). In addition data could be quality checked for accuracy at any point, something which obviously couldn't happen with the paper copies. Data are also recorded digitally, allowing for archiving, and distribution between multiple EOCs. Finally data could be used for other disaster research, such as HAZUS (FEMA's damage estimation program) validation.

By the end of February 2006, we had established a working relationship with the National Centre for Geocomputation (NCG), and through their generosity a single camera unit had been shipped over from Ireland. A group of us took the unit, suctioned cupped the digital video camera to the side of our car, and headed off to New Orleans. The GPS unit itself was small, no larger than a laptop battery, in fact the whole unit was extremely portable, easily fitting into a single hard-shelled case—again meaning it would be easily transportable with the Red Cross teams. We decided to shoot three one-hour films, one to cover the whole of City Park—the issue at the time was whether sections of the park should be used for FEMA trailers—and then cover as much as possible of the Lower 9th Ward.

To some degree this was complete guesswork, we had no idea how effective the GPS was, or even if it was working (there was no LCD screen—just a box with small green and red lights), we had no idea what was the right speed for either the best video image, or GPS recording, nor how the overcast day would affect GPS reception. The whole package, tapes, camera, and GPS unit, were then shipped back to Ireland. When we received the DVD's of the trip back, and after many attempts to correctly configure the browser system, we were stunned. To the left hand side of the screen, the entire pathway through the Lower 9th Ward was displayed. A red

Figure 1:11. *Spatial video collected using NCG equipment in the Lower 9th ward on 3/11/06.*

arrow indicated the direction of travel (see Figure 1:11). As we started the video, so the arrow would slowly move along the path. On the main screen, the video image was displayed. Although the video was a little blurry at times, possibly as a result of the driving speed, the amount of damage to each structure was easily visible. If Red Cross had used this data collection approach, the volunteer could have easily recorded data into a database approximating the current hard-copy collection form used by Red Cross.

The three snapshots show the same section of road close to one of the Industrial Canal's levee breaks. In the first image, the flood imagery to the left of the browser shows the water pouring through, and the barge that is thought to have punched a hole in the wall. In the accompanying spatial video image, the barge is indicated by a box. The middle image progresses along the same stretch of road. This time the flood imagery has been zoomed into so the house accompanying the spatial video image can be matched to the "map." This image shows severe damage, the house having collapsed, with the roof now on the ground. Search and rescue markings are also visible on the building. The third image adds the current *Make it Right* (MIR) building project area which corresponds to Figure 1:11. The spatial video image shows a temporary relief shelter set up for returnees.

The browser could also display multiple other cameras, though for this proof of concept we had just used one, and this meant every road had to be driven twice. A damage assessment team would affix one camera on either side of the vehicle, thus cutting the travel time in half. As a result, because of this Lower 9th Ward data run, we now have a spatial snapshot in time of what the neighborhood looked like after Katrina. The area now has totally changed; in one corner the Make it Right Foundation has begun a new home development, scattered other rebuilds can be

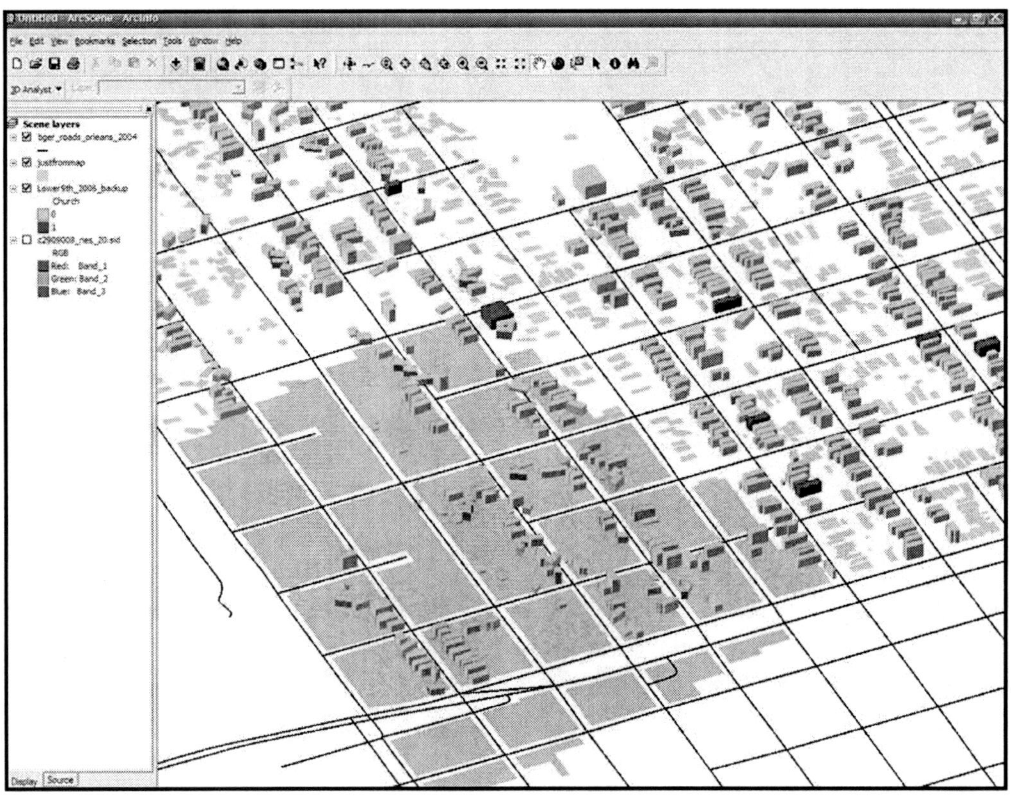

Figure 1:12. *A typical GIS window where the information from spatial video in Figure 1:11 has been extracted and turned into a 3-D map. The shaded area is the current MIR rebuilding project area.*

found between the abandoned homes and massive vegetation overgrowth. Most of the homes have been bulldozed. If you think back to the first few maps of this chapter, maps that displayed different aspects of major disasters from the past. This digital spatial surface is our modern equivalent, only now future researchers will have a video record, and because it is spatially referenced they will be able to manipulate it, analyze it, and map it in different ways[16].

Now that we have covered some of the basic concepts and approaches to thinking spatially about disasters, let's work through a real-world case that requires use of these concepts and also consideration of greater spatial complexity in the problem-solving process.

THINKING SPATIALLY ABOUT DISASTERS: AN APPLICATION

We previously identified a disaster related spatial problem—estimating the likely surge after an earthquake in Los Angeles. We also introduced a few basic spatial concepts necessary to understand and approach a solution, including distance decay, spatial aggregation, and spatial scale. We will now return to this case to apply some of the concepts we have introduced in this chapter. Four basic steps can be used to develop an accessible, all-hazards geographic resource for use in planning post-disaster surge. Step One: Identify a service area. Step Two: Identify the service population. Step Three: Improve data where possible then repeat Step Two. Step Four: Disseminate the data for use by decision makers. Notice the generality of these steps. There is no mention of a specific type of service, a specific type of population, nor of a specific type of event. Let's think through the following steps.

Step One: Identify the service area

What services need to operate in a disaster? The answer to this question ranges from local entities, such as fire departments, police, and hospitals, to state and federal agencies among others. These resources are part of a place's critical infrastructure. Knowing where these assets are located and the capacity that they provide is essential to planning and response functions. However, these data are often stored in tabular format, which does not provide a geographic context for their use.

Furthermore, to demonstrate the importance of accessibility and context, look at Figure 1:13. This image is a layer of hospital locations in Los Angeles

16 Readers should see our publication which details how three measures of damage were mapped from the Lower 9th Ward video. This publication also provides an example of a Red Cross street sheet. Curtis, A., Mills J.W., Kennedy B., Fotheringham S. and T. McCarthy (2007) Understanding the geography of post-traumatic stress: an academic justification for using a spatial video acquisition system in the response to Hurricane Katrina *Journal of Contingencies & Crisis Management* 15(4): 208–219.

Figure 1:13. *(a) A map of hospital locations in Los Angeles County, California. The GIS layer is the only data presented in the map. (b) The same map from (a) is presented with the addition of the county boundary and roads.*

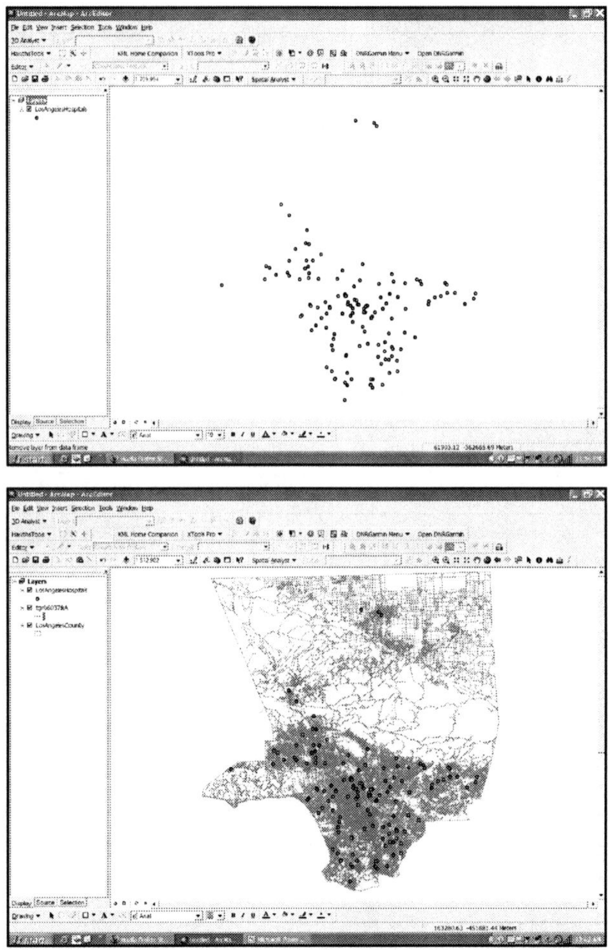

County, California. These data were extracted from a larger file of Licensed Healthcare Facilities that is publicly available from the California Spatial Information Library (CaSIL) and were originally created by the California Office of Statewide Health Planning and Development (OSHPD). Figure 1:13 (a) presents these points by themselves, while context is added with a county boundary and roads in Figure 1:13 (b).

Q: What other layers would help add context to the map?

Q: What do you think planners and emergency responders need to know about the geography of these service provision locations?

However, in paper format, this is still not a useful planning tool because not all road names and hospital names can be displayed without creating a cluttered map. Alternatively, in Figure 1:14(a), this same layer of points is presented in Google Earth and in Figure 1:14 (b) we can see how much geographic context can be gained from simply zooming into each hospital, in this case the Los Angeles County + University of Southern California hospital (LAC+USC). This example demonstrates the concepts of accessibility (a paper map or GIS layer vs. Google Earth) and context (dots on a map vs. dots linked to imagery of the surrounding area). Furthermore, this demonstrates the first component of Step One, identifying locations of critical service provision.

Q: How might data presented in Google Earth be more accessible and have more context than a paper map?

Q: How might a paper map of the same data be more appropriate for planning or response?

The second component of Step One is to select the service region. This is an area surrounding the critical service provision location. For a study considering disaster related surge potential in a hospital, the simplest form of defining a service region would be to create a catchment area outwards from the facility by a reasonable distance. From the perspective of mapping, the technique used to define this region is called buffering. In essence, a shape (buffer) is drawn around the point of interest, and in this case the catchment area is defined as the space that falls within this buffer.

Q: Buffers are commonly used in environmental health and environmental justice studies to examine the impact of hazards, such as chemical releases and traffic exhaust. How would you use buffers to study the impact of a chemical release? Or the relationship between respiratory problems and the distance a person lives from heavily trafficked roads?

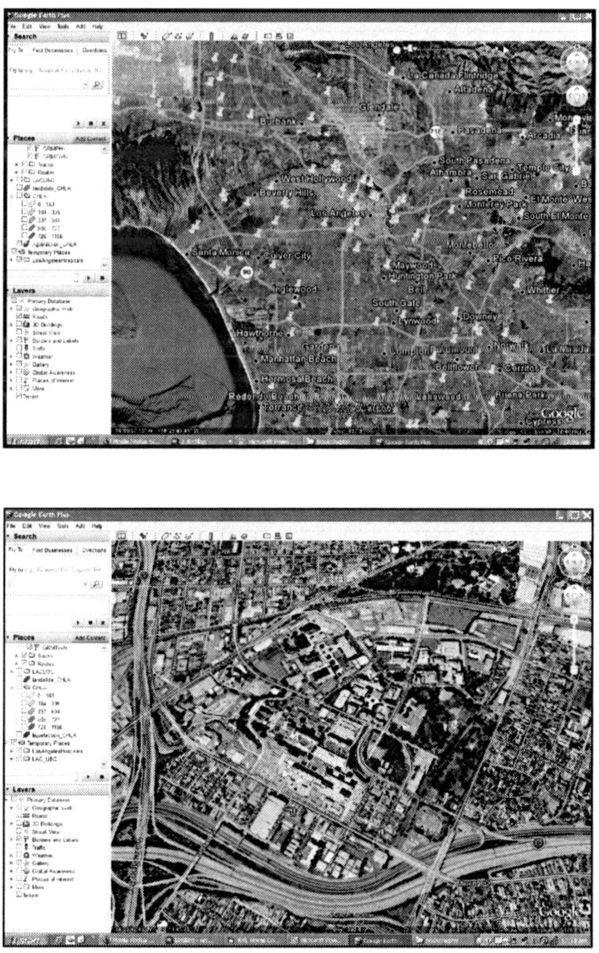

Figure 1:14. *(a) Locations of Los Angeles County hospitals. This is the same layer from Figure 1:13, but the file format has been converted so that it can be visualized in Google Earth. (b) A close-up of the area surrounding LAC+USC hospital in Google Earth.*

Buffering involves consideration of two issues: 1) buffer shape and 2) buffer size. In a GIS, buffers are commonly equidistant around a feature. However, this default approach is not necessarily appropriate for determining an area that will require specific services in a disaster. For example, using our case of post-earthquake pediatric surge at a Los Angeles hospital, we could simply generate a circular buffer at 0.5 mile, 1 mile, and 3 miles around the hospital. Though this is a reasonable method for approximating a service region, let's think about the geography underlying this selection. First, these distances (buffer sizes) could be justified by saying that they would capture population within a walkable distance to the hospital. Assuming the likelihood of damaged transportation lines (such as use of buses, local roads, highways), people would be walking (or would be carried) to reach medical services. So, in this case, the buffer size makes sense.

Q: How might buffer size vary based on the type of event (e.g., wildfire, earthquake, flood), physical geography, and urban/rural setting?

Now let's examine buffer shape. Though concentric circles around a point are commonly utilized, other shapes may be more appropriate, such as semi-concentric circles or a buffer that follows an arterial network. Concentric circles assumes an isotropic plan; that access to the center is equal in every direction. However, in reality, how often is this the case?

In a disaster, a lack of thoughtful, geographically-informed planning can have a real impact on whether or not the plan is realistic and results in a successful response. Something as simple as buffer selection can have consequences for people's lives. The stakes are this high in planning for disasters.

Step Two: Identify population in the service area

Now that we have introduced concepts of service location and the region they would likely serve in a disaster, the next step is to identify the population in this area. How many people will constitute the surge? More specifically, can we identify subgroups in this population that will need special consideration in planning disaster services? In New Orleans, after Katrina these people included the elderly, multi-generational households, people suffering from both chronic and infectious disease, the physically impaired, and the obese. Each of these issues necessitates specific plans for evacuation and medical care. However, such specificity was not utilized and therefore the most vulnerable suffered disproportionately in this disaster. What could geographic information have contributed to planning for a successful response?

Using the basic geographic technique of overlay in a GIS we can develop some simple scenarios for determining the entire population for the area, subsets of the population that display social vulnerability characteristics, and subsets of the population that live in biophysical hazard zones in addition to also being socially vulnerable. Two publically available data sources are needed to develop such geographically specific surge plans. The first is the United States Census. The second general source involves more specific biophysical risks (the source for these may be local, state, or a federal agency providing spatial data on hazard zones). In the case of the earthquake hazard in California, we can use landslide and liquefaction zones provided by the California Geologic Survey (CGS).

The Geography of Vulnerability

Planning for disaster can take many different forms. Emergency services participate in training exercises, businesses invest in disaster contingency consulting, and families develop action plans about what to do, who to contact, and where to go. In reality, planning cannot ensure complete preparation as so many different scenarios may develop in the event. This is not to say that planning is not useful, it is and if a realistic plan is practiced enough, it will pay off when needed. Though the details of a disaster may be difficult to predict, research tells us that there are some constants that will occur and therefore plans can be developed to address them. One of these constants is the disproportionate impact on vulnerable populations. Vulnerability will be addressed in detail throughout this book, so we will not dwell on the topic here except to introduce a few key points relevant to planning with geographic information. First, vulnerability can take different forms, such as health vulnerability, social vulnerability, and biophysical. Second, regardless of type, each exhibits a geography; there are spatial patterns to the locations of hurricanes, earthquakes, health burdens, and poverty, for example. We want to identify these geographies and then use them to develop improved plans for serving the people and places impacted in a disaster.

Though biophysical vulnerability is composed of measurable factors surrounding an extreme event, which in this case includes proximity to landslide zones and liquefaction zones, social vulnerability is a more complex concept. The blend of components which create social vulnerability vary from place to place. Several studies have found that factors related to these types of vulnerability, in addition to built environment characteristics[17], are linked to earthquake injuries. For example, one study found that age and disability were factors for hospitalization after the 1994 Northridge earthquake[18].

17 Peek-Asa C, Krauss JF, Borque LB, et al. Fatal and hospitalized injuries resulting from the 1994 Northridge earthquake. *International Journal of Epidemiology* 1998; 27: 459–465; Shoaf KI, Sareen HR, et al. Injuries as a result of California earthquakes in the past decade. *Disasters 1998*; 22: 218-235; Peek-Asa C, Ramirez M, et al. Seismic, structural, and individual factors associated with earthquake related injury. *Injury Prevention 2003*; 9: 62–66; Ramirez M, Peek-Asa C. Epidemiology of traumatic injuries from earthquakes. *Epidemiologic Reviews 2005*; 27: 47–55.

18 Peek-Asa C, Ramirez MR, Shoaf K, et al. GIS mapping of earthquake-related deaths and hospital admissions from the 1994 Northridge, California, earthquake. *Annals of Epidemiology 2000*; 10: 5–13.

Data Sources

Q: How can we begin to use a GIS to understand all these different layers?

The United States Census provides data on a decennial timeframe, so they are useful as baseline or approximation only, which is a limitation. However, they are publicly available and geographically complete across the United States so that these data can be used by anyone in the U.S. and at a variety of scales: block, block group, tract, though the finer the scale of data, the better. The census has both data about people (race, income, education) and also about their environment (year residential structure built, home owners/renters).

Census data are downloadable in a tabular format, but as we have already demonstrated earlier in this chapter, tables are not at all sufficient for seeing the spatial patterns in data. However, the data in these tables can be mapped in a GIS. These benefits make it a useful dataset for planning and the limitation of temporal applicability can be overcome, which will be discussed in the next section on collecting primary data.

Figure 1:15 shows the spatial pattern of only one characteristic of social vulnerability. Many more variables exist. Therefore, to gain a more comprehensive understanding of how many people may need medical services in a post-earthquake environment, these variables should be combined. Furthermore, social vulnerability varies geographically, for example the factors that make someone vulnerable on the Gulf Coast may be different in some respects to factors in Southern California. However, this is a benefit of the Census in that data about the population (including the many social vulnerability indicators) are available throughout the country. Mapping social vulnerability, then, can be modified based on what makes sense for specific places.

Hazards Zones: The other main category of publicly available data that can be used to plan for population needs in a disaster is hazard zone data. These can come in a variety of themes and are developed by several different agencies. As a result, hazard zone data are not as consistent across the country as are Census data. For example, landslide and liquefaction zones for California are created and made available by the California Geologic Survey (CGS), while flood zones are available nationally due to their oversight by the Federal Emergency Management Agency (FEMA). Finally, some local governments collect and distribute hazards zone data for their municipality. For example, in San Diego County, California, SanGIS (www.sangis.org) provides data on locations of wildfire severity zones.

Although data on social and biophysical vulnerability are useful alone, they are even more beneficial for planning post-disaster population surge when they are combined, or layered onto each other in a GIS. Applying such a process could show that, although an area contains several hazard zones, few people actually live anywhere near them (this may not matter in an earthquake, but perhaps would make a difference in a flood). Alternatively, the resulting map might show that high densities of socially vulnerable people live clustered in a hazard zone at the edges of a major city, such as is a common case for the poor living in landslide zones in Latin America.

Overlay

Now that you can access data on population and physical geography as they relate to a potential disaster event, we will discuss how to combine Step One with Step Two to identify populations in a specific service area that may be disproportionately impacted by a disaster due to the combination of social and biophysical vulnerability.

The most common means of displaying spatial information, such as numbers of elderly in a liquefaction zone, is in a graduated color map. The simplest form of these maps would be comprised of administrative boundaries, for example by census tract or blockgroup. A GIS allows the manipulation of these data into more appropriate geographies in relation to a hazard. For example, buffers can be used to create an area of impact around a key feature–either a hazard location, or facility expecting surge after a disaster. The population living or working in this buffer zone can be identified as "at-risk" for a specific event (earthquake) or have ongoing exposure (particulate matter).

However, a different GIS approach involves "overlaying" data, which in effect means stacking spatial information in the search for common intersections. For example, overlaying demographic information on a highway network might be used to show that higher levels of poverty spatially correlate with close proximity to highways, which in turn might be used to prove a thesis on environmental justice where impoverished cohorts suffer disproportionately from particulate matter exposure. For a study considering disaster related surge potential in a hospital, the simplest form of analysis would involve extending zones of impact (buffers) outwards from the facility, and extracting out (overlaying) the census derived vulnerability found within.

Cartographic analysis is comprised of looking at spatial layers to identify areas where they align or correlate, such as blockgroups that have high values for low educational attainment, high single parent households, and high numbers of families where English is not the first language. Perhaps some of these areas are also proximate to a hazard zone. A next step would be to perform a spatial analytic technique to test the apparent relationship. Then, if this area is identified as a hot spot of multiple vulnerability characteristics, it should be prioritized for updated pediatric population counts as children in these places are more likely at risk.

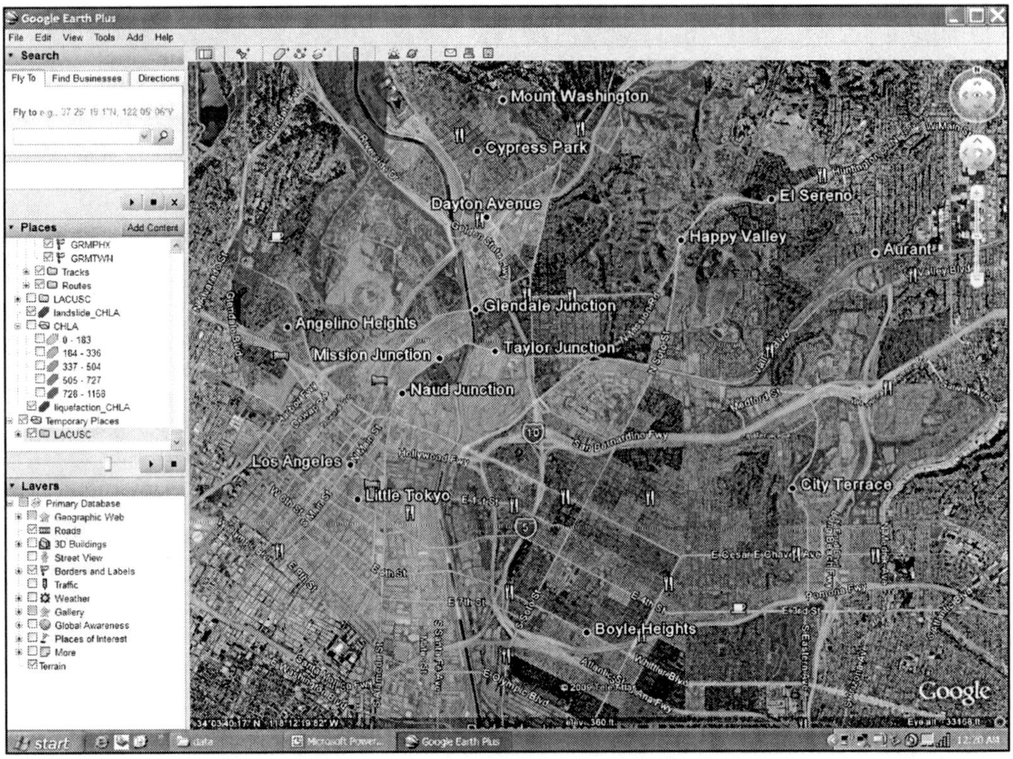

Figure 1:16. *Putting it all together: Overlaying data on service provider (hospital), immediate post-disaster service region (buffer), biophysical hazard zones (landslide and liquefaction), and a social vulnerability characteristic (median household income).*

Step Three: Improve Data, then Repeat Step Two

The basic surge prediction approach employed in this section provides a quick glance at the number of people and the types of needs that will be faced in the early days after a disaster. Therefore, this method is a suitable first step to help organizations plan their response. However, there is room for greater specificity and the approach can be dramatically improved to gain a more accurate understanding of the population needing help after a disaster. This improvement can be made in a couple of areas, such as refining the buffer (which we have already discussed), but also through considering issues of distance decay and competing destinations.

A major improvement for planning post-disaster surge predictions could be made through collecting primary data, such as more accurate information from the vulnerable population falling inside the buffer zones. A simple survey could be targeted to people located inside the service region to ascertain answers to simple questions such as, *where would you go in the event of a disaster*? Alternatively, a survey can be conducted of the neighborhood built environment to capture both physical and social characteristics.

Step Four: Disseminate the data for use by decision makers

GIS software is now relatively easily integrated with Google Earth KML, so layers and even entire maps created in a GIS can be exported into Google Earth. Of course, this is by no means the only format for distribution, but it is popular due to the familiar Google interface, ease in opening a KML file (once Google Earth is installed on a computer, just double-click and GE will open and zoom to the KML location), one file transfer (a KML is one file, whereas a GIS layer is composed of many files) via e-mail or download, and adjustable transparency so underlying aerial imagery can be used to give context to the map. As you can see from Figure 1:16, a decision-maker can look at this map, zoom in on certain areas, and estimate the numbers of surge patients and their socio-economic context. This approach provides additional planning information such as whether or not multi-lingual staff will be required or if any pre-existing health burdens can be anticipated. This is a powerful advance for understanding disaster, driven by the utility of spatial thinking and maps.

Hopefully this chapter will have shown the importance of geography in disasters—both in terms of the physical process and human dimension. You should now be familiar with a few geographic concepts, and general approaches to solving spatial problems. We have also briefly introduced spatial data collection and communication. In order to achieve all of these aspects, we need to use a GIS—and the next chapter will continue our introduction of this spatial software tool.

TWO

An Introduction to Geographic Information Systems

THE PREVIOUS CHAPTER introduced the idea that there is geography to disasters. Although many of the previously described methods utilized a GIS, we can now take a step back to describe these geospatial tools in more detail. Again, this chapter cannot replace an introductory text on GIS, but it should provide the reader with enough detail to understand the utility of these systems.

Geographic information science (GISc) combines the technologies used to collect, analyze, and visualize spatial data with the methods and issues associated with that manipulation, analysis, and visualization. A GIS is usually the software running on a PC where this occurs. Although there are several different GIS software "vendors," there are two general categories of programs based around the way spatial data are handled—*Raster GIS,* which translates the world into a grid where every pixel or cell contains spatial information; or a *Vector GIS,* where every object in space (such as a road, house, or park) has its own shape.

The spatial information output from a flood inundation model and used to create the inset map in Figure 2:1 were a grid of data. This grid contained cells with either a "1" or "0"—1 meaning that the cell or area of land would be flooded. This is an example of a raster geographic layer. A more detailed raster flood output layer might include the water depth for each cell (Figures 2:2 and 2:3).

Figure 2:1. *A flyer for a community earthquake and tsunami information meeting at Venice Beach for 2009, and the inset map displays the tsunami evacuation area.*

Q: Can you think of any other geographic layer that would benefit from this type of spatial grid representation?

Alternatively, in a vector GIS all objects are represented by a point, a line, or an area (more commonly referred to as a polygon).

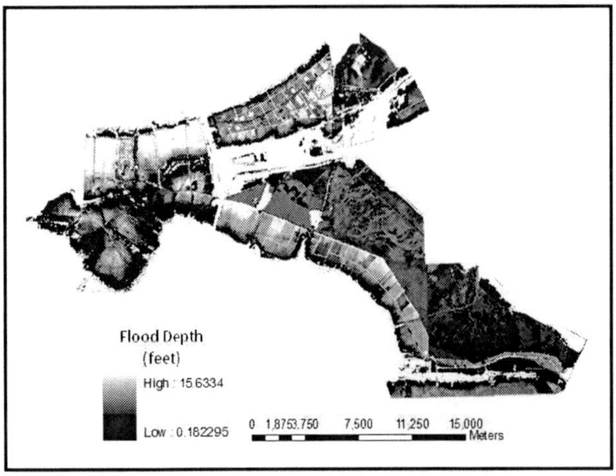

Figures 2:2 and 2:3. *These show the depth of flooding as a raster layer for all of Orleans and St. Bernard Parish, and then as a zoom into just the Lower 9th Ward area. By clicking on any cell in a GIS you can read off the flood depth at that location.*

Each of these spatial objects has other information, called attributes, attached to it. In the earlier example, the hospital would have been represented as a point (or possibly a polygon depending on map scale—think about this), and the attributes would be the number of beds, or the number of doctors etc. The attributes attached to a census tract might be the total number of people, how many children, or the proportion of people living below poverty.

As we are primarily interested in the human geography of disasters, we will concentrate on a vector GIS because the streets along which people navigate, the buildings where they live, neighborhood boundaries, and the locations of hospitals are all spatial objects. Although there are several different vector GIS packages, probably the most common software has a map window, a side bar or "table of contents" where all the different layers sit, and several manipulation, analysis, and visualization options are visible in a tool bar. In this way the software mimics other geospatial display package, such as GE.

SPATIAL DATA INPUT

Again at the risk of over simplifying, the vector GIS can be thought about as four interconnected spatial worlds—data input, data manipulation, analysis, and visualization. It is easier to understand each of these categories if we again think back to our surge potential analysis for Children's Hospital. Data input involves putting spatial data into the GIS. Luckily many geographic layers already come available in a GIS format, or at least in a file type that can easily be added. Of course, some data will still need to be captured, translated or created. For example, point data such as the location of the hospital can be matched to freely available line data (roads).

We can also add "new" data as long as it contains a spatial reference—such as a coordinate or street address. We can also use another layer within the GIS, such as a high resolution aerial photograph, similar to that available in GE, to "heads up digitize" information. For example, the hospital of our surge example could be added into the GIS as a point by clicking on the building in the photograph, or by tracing the actual outline of the building it can be created as a polygon. It is amazing how much geographic data we can create through this digitizing process. Look again at Figure 2:4. We could "digitize" building outlines, the location of flood debris, even water plumes to show direction of flow. Lines can also be added through this digitizing process.

Figure 2:4. *A typical GIS window. The left side of the map is a "Table of Contents" (TOC) that shows the different active geographic layers. In this example, we have an aerial photograph showing flooding (you can still see the levee break and flowing water in the top left), point data (mortality locations), line data (roads), and polygons (building outlines that correspond to one of our exercise neighborhoods).*

We can also add attribute data to our points, lines, and polygons. For example, if we had conducted questionnaires with local residents, either for our hospital surge analysis, or even for residents coming from the flooded area in Figure 2:4, we could add this information to the map by attaching it to a coordinate (if we were carrying a GPS), or by the respondents address.

If this book was a text to accompany an entry level GIS class, considerable time would be spent on the different methods of data input, and their potential pitfalls. However, as the purpose of this book is to concentrate on the social impact of disasters, and as most individual social data tends to be brought into a GIS through a street address, for example medical data, critical infrastructure, 911 call locations, it is important to spend a little time discussing some of the errors involved with this popular form of data input. The most common GIS approach to matching an address is to take an address and proportionally place it along a road segment bounded by two cross streets. This simply means that every road network (all city streets) consist of lines with attributes of address ranges. These ranges usually extend from one intersection to the next. By knowing this range (say 200 Main Street to 400 Main Street) we can place the actual address proportionally where we expect it to fall. If, for example, the house is 300 Main Street, we would expect it to be half-way.

Unfortunately, the age of housing, the sinuosity of the road...and in some cases just plain spatial weirdness mean that some addresses do not map that well. Therefore one important question to always ask is how much error are we talking about, and does it matter?

Figure 2:5 shows the results of geocoding addresses[1] for the *Make it Right* (MIR) foundation, a non-profit organization supported by Brad Pitt which is building a series of new homes in the Lower 9th Ward. Part of their collaboration with the Department of Geography at USC was to map who owned the different land parcels around this new construction.

Figure 2:5.

1 For the purposes of this chapter "geocoding" simply means adding spatial data—the term is often used to mean adding an address in a GIS. Another common term for this GIS function is address-matching.

Figure 2:6. *A home where someone died during Katrina (the "1D" at the bottom of the "X" marks this). In Google Street View the address is given as 654 even though the house number is 609. This represents a geocoding error—the point placed on a GIS map would be some distance from where the actual house is.*

The MIR interest was to start an information system that could easily expand to encompass a constantly changing neighborhood of construction and return. The research interest was to gauge whether this outside building initiative was acting as a rebuilding stimulus to the neighborhood. However, when the addresses were matched to the street network in a GIS, north-south street segment locations were compacted towards the southern end of the street, with some even being placed in the median of the major cross road and outside of the block face (a phrase used to describe either side of a single street segment). Parcel data—these are the legal boundaries of properties, are also displayed to give an idea of where the addresses should have been. The only way to rectify this was to physically go to each address and capture a latitude and longitude with a GPS.

Unfortunately, many GIS users would take the address as matched in the GIS as correct—and it is only exercises such as this with several addresses in close proximity to each other that the spatial bias becomes evident. Although there are several papers and books that describe the pitfalls of matching addresses in a GIS[2], one simple "test" is to use Google Street View. For any neighborhood, find a visible address (either on the house or by the curbside), and see how close the number is to that reported in the top address box. The errors found in Google Street View are likely to be similar to those found by traditional address-matching in a GIS.

2 There are several papers that discuss the errors associated with geocoding, for example Daniel W Goldberg, Craig A Knoblock, Beate Ritz, Myles G. Cockburn An effective and efficient approach for manually improving geocoded data. *International Journal of Health Geographics* 2008, 7:60

Q: Can you think of a disaster related situation where this type of spatial error will be problematic?

SPATIAL DATA MANIPULATION

The second spatial component of the GIS involves data manipulation. This can be either non spatial, spatial, or a combination of both. Data manipulation involves changing our attribute data, or using the spatial relationship between features in our GIS to create a subset. In addition, we could even create new data, for example by using the GIS to determine how far each person lives from a clinic and then adding that distance to their attribute table. Again using the surge example, the attribute table for each census tract contains information such as how many people

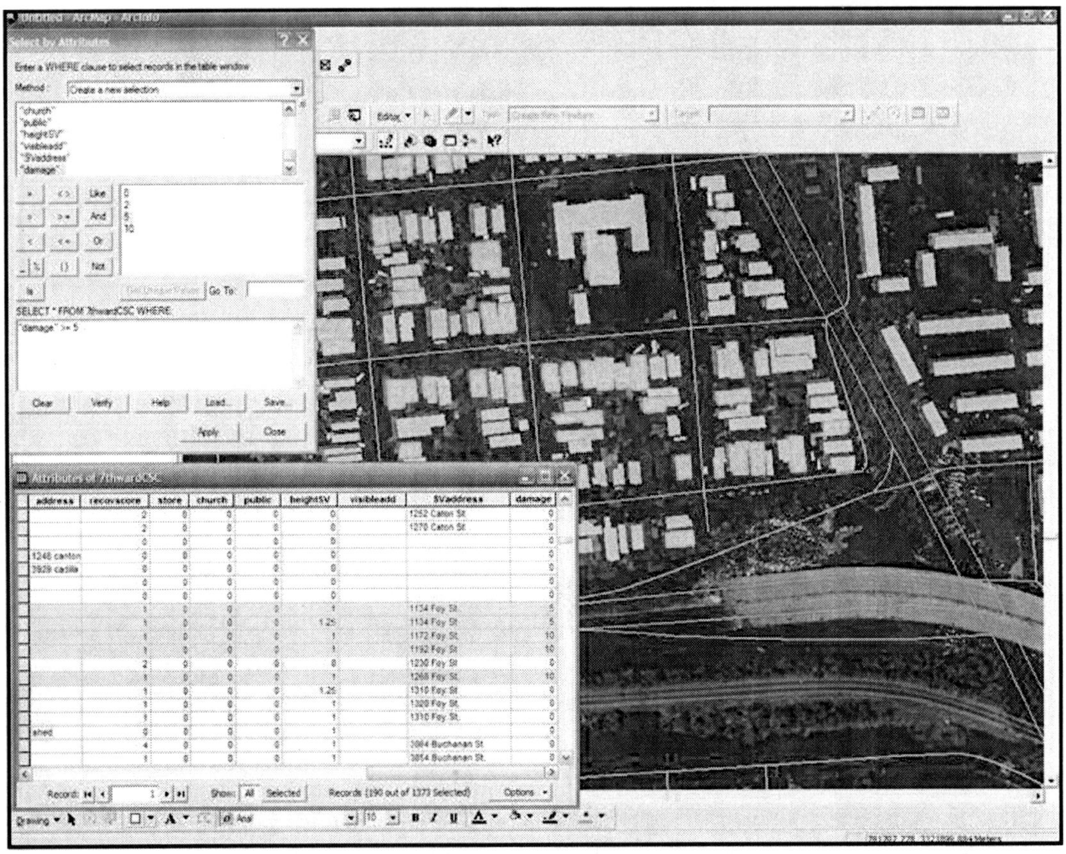

Figure 2:7. *A typical GIS window running a query on the attribute table—in this case the amount of damage for these buildings in the 7th Ward/St Bernard neighborhood. Buildings have been digitized from the underlying aerial photograph, which again shows flooding—notice how the freeway is under water. A score of 5 or above means the property has received noticeable damage.*

live there. We can perform an aspatial query on these data by extracting out only children. A spatial query would then be, find the total number of children falling inside a certain distance (the buffer) from the hospital.

Queries are extremely important in GIS use—these embrace the "information system" aspect of a GIS—sorting through large amounts of data, both spatial and aspatial, to help understand and solve a problem. For example, the North Carolina Department of Agriculture and Consumer Services has one of the most advanced GIS facilitated response systems called a Multi Hazard Threat Database (www.agr.state.nc.us/oep/MHTD/index.htm). Quoting directly from their website:

> The Multi-Hazard Threat Database (MHTD) is a decision support tool that reduces vulnerability of the state of North Carolina to natural and man made disasters. It is a secure, comprehensive database with state infrastructure with geographical information built on the North Carolina Animal Health Geographic Information System (NCAHGIS). The MHTD allows for a more rapid response with fast access to information, facilitates improved resource allocation and planning, and allows real-time management of field data.
>
> The MHTD has been utilized for a wide range of incidents and supports all state agencies in time of emergency. It provides a "one medicine" approach to surveillance, detection, diagnosis and response on the local, state and federal level. Information integration is the key to success. We work closely with our partners in NC Public Health, Department of Defense, Department of Homeland Security, Federal Bureau of Investigation, State Bureau of Investigation, State Highway Patrol, Emergency Management, and private industry.

This system provides a good example of how multiple aspatial and spatial queries of the database are needed during a response. If, for example, there is an outbreak of Foot and Mouth Disease, buffers at varying distances can be placed around the infected farm, with the locations of all Dairy, Swine and Poultry farms mapped as points (color coded by animal type) with the underlying attribute information for each farm type per buffer (how many farms, and total head/number of animals or birds) displayed as a table. In this way the hazard response team knows exactly how many farms and animals, at what different distances (which could be a proxy for infection risk) surround the suspected infection source. The table also includes the human population within the buffer areas, not as important for Foot and Mouth Disease, but certainly of concern if this was an Avian Influenza Outbreak. The next GIS facilitated aspect to the response involves quarantine and perimeter control—with various human geographic layers being added to the map, such as

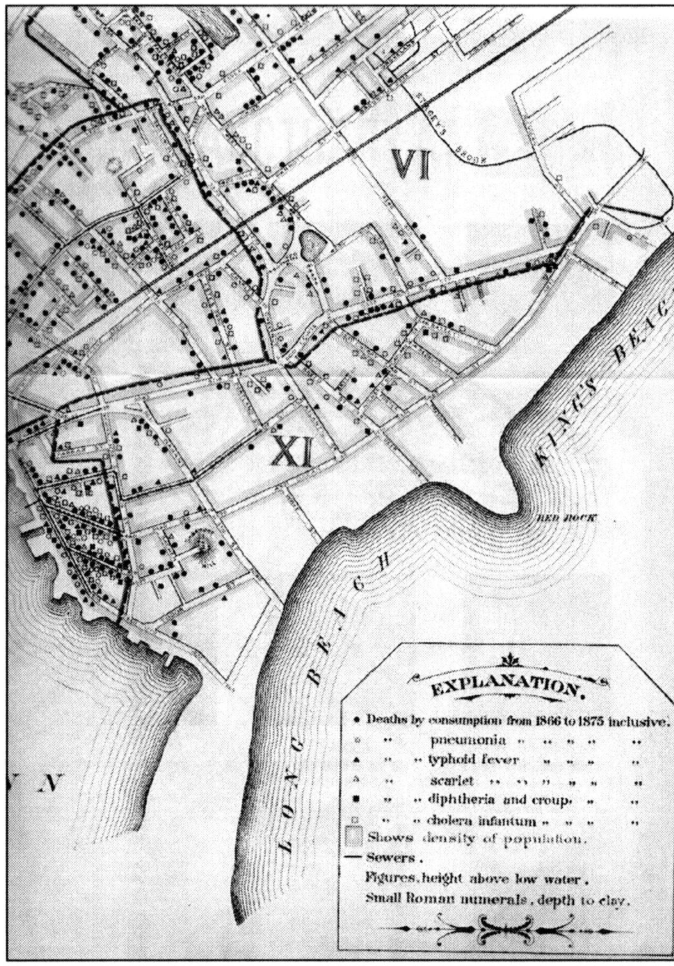

Figure 2:8. *A detail from a disease map from 1876 (Source 8th Annual Report State Board of Health for Massachusetts 1877)—different diseases are located on a map. If researchers had a GIS they could have used spatial analysis to see where each disease clustered—this being the first step in the search for reasons why disease occurred in these locations. For example, a similar disease dataset from the 1878 yellow fever epidemic in New Orleans has been analyzed to show where hotspots occur in the French Quarter.*

transportation networks, human settlements, potential staging and response areas (such as schools), and hazardous substance disposal sites, an important consideration for a Foot and Mouth outbreak where one of the greatest challenges is an effective and safe way to dispose of hundreds of thousands of animal carcasses. In addition, quarantine check points are visualized across the map where the buffer (or *cordon sanitaire* in public health speak) intersects a road. The system can also be used to access all important contact information within the buffer area, such as addresses, email, and telephone numbers of the farms so that they can be informed about the situation and hopefully this will maximize their compliance with the response). Finally, information relevant to the responders, such as farm location and important features about the surrounding terrain can be distributed in GE format, which, as we discussed earlier is easy to understand and use, and simple to distribute via email.

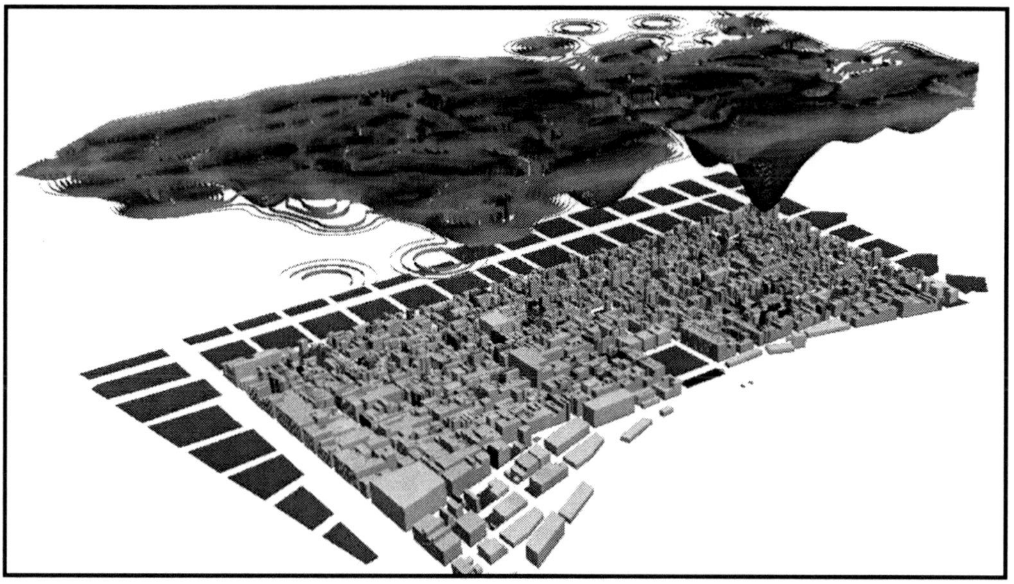

Figure 2:9. *Displays where the highest concentrations of disease were found in this yellow fever epidemic (the largest peaks extending downwards show the hotspots). The GIS also allows us to visualize the city in 3-D so that when we identify each hotspot we can start to see if there are factors about the built environment that can be linked to the amount of disease. Data source: Official Report of the Deaths from Yellow Fever as Reported by the New Orleans Board of Health (1879)*

SPATIAL DATA ANALYSIS

The third component involves the analysis of data. The use of spatial analysis is not as common in disaster science as for other academic disciplines using GIS, such as public health. This situation is beginning to change as more fine scale spatial data are collected for the post-disaster landscape. For example, disaster medicine is a developing area of research which straddles disaster science and public health/epidemiology. It is only natural for researchers in this field to want to know where and why hotspots of disaster related sickness occur.

In keeping with an "all hazards" approach to disaster preparedness and response[3], a state or city's emergency response plans will also include scenarios for a bioterror agent release, or even a newly emerging infection such as Swine

3 The Purpose of Homeland Security Directive 8: National Preparedness states: This directive establishes policies to strengthen the preparedness of the United States to prevent and respond to threatened or actual domestic terrorist attacks, major disasters, and other emergencies by requiring a national domestic all-hazards preparedness goal, establishing mechanisms for improved delivery of Federal preparedness assistance to State and local governments, and outlining actions to strengthen preparedness capabilities of Federal, State, and local entities. Where the definition is: "all-hazards

Flu. These scenarios provide the best examples of how GIS spatial analytical approaches can be used to understand a threat, and manage a developing situation.

A simple definition of spatial analysis is the identification of spatial patterns and processes within a spatial dataset. For example if there was a disease outbreak, there are various spatial analytical approaches that can be used—either to search for the origin of an epidemic, predict where the contagion may go next, match cases to an underlying environmental risk (such as proximity to a chemical plant or road), or conditions of the building or neighborhood. In addition, aspects about the person who has contracted the disease, where he/she lives, what are the behavioral characteristics of the individual, what are her/his social contacts, might be revealed by different forms of spatial analysis.

Q: Yellow fever is spread by mosquitoes—what built environment factors might we want to look for in each hotspot?

The results of these analyses can help shed light on a disease or health situation—and hopefully be used for purposes of intervention. Let us briefly consider a modern-day outbreak scenario. Imagine the release of a pathogen (for example smallpox) has occurred in a large city, but mainly in one neighborhood. The first step is to identify if there really is an outbreak or not—and although not perfect or universally applied, syndromic surveillance (which basically means any spike in reported illness, or the conditions associated with illness such as buying certain drugs) would give a first indication when mapped. If the release can be verified, and if actual cases identified, then a GIS could be used to place a quarantine around the outbreak area, much as was described in the North Carolina example. Buffers could be overlaid on a road network to find out where the quarantine should be imposed. In addition, work on social networks (basically the interactions people have with each other across space) might suggest we also watch other neighborhoods of the city within a certain distance that contain high portions of the same ethnic group[4]. Unfortunately, as was found with the recent Swine Flu outbreak, respiratory diseases are notoriously hard to spatially predict and therefore control in near-real time.

Although there are many types of spatial analytical methods, it can be argued that these fall into two broad categories—first the search for spatial patterns, and second the search for meaning or associations with those patterns. Another way to phrase this is an exploratory analysis followed by a confirmatory approach—which

preparedness" refers to preparedness for domestic terrorist attacks, major disasters, and other emergencies. See http://www.dhs.gov/xabout/laws/gc_1215444247124.shtm

4 This ethnic connection was found to be more important than geographic distance for an epidemic in Curtis, A. (2008) Three-dimensional visualization of cultural clusters in the 1878 yellow fever epidemic of New Orleans *International Journal Health Geographics* 7:47

simply means we let the data determine where anomalies occur in the first instance, and then from these results we develop hypotheses explaining why these hotspots occur with this relationship being transferable to other locations and time periods.

There is also a move by researchers using GIS in health investigations to use a "mixed methods" approach—not only relying on a traditional scientific approach of spatial analysis which involves sampling, statistical rules and "confidence" in the findings, but combining these with interviews, ethnographies, and in general trying to understand the world and its challenges at all geographic scales for the people being studied. The previously mentioned surge-interview used to find out what people will do in an earthquake is an example of a mixed methods approach if this is combined with an analysis of the underlying census surface.

In order to perform a local area spatial analysis, for example determining which parts of a neighborhood have been disproportionately damaged, or which streets are recovering quicker than others, fine scale data collection is necessary. This book will constantly refer to a spatial video system that allows for such data to be collected. As a result more sophisticated spatial analyses can now be performed on the post-disaster landscape.

Figure 2:10 displays a kernel density surface generated from building data in the Holy Cross neighborhood of New Orleans. In the top GIS window, each building has been digitized, and then by using Google Street View, each front door has been located in the building as a point. Also by looking at Google Street View, the amount of damage has been assessed for each home, and this value entered into each building's attribute table. A simple "join" process in the GIS allows these attributes to also be attached to the front door point file. A density surface is then created for these points. The output of this surface can be interpreted as darker areas mean there is a greater density of damage (in effect a combination of the number of front doors, weighted by the amount of damage for each building). In the GIS it is easy to click on any cell and read off what the damage density for that location is—roughly meaning how much damage is within a certain distance of that cell. The bottom GIS window shows the 3D sister program of the GIS. In this window the density surface has been changed to a red color ramp, which is more appropriate for a hotspot surface, and the building footprints have been "extruded" up to their actual height. In this way we get a 3D impression of where the greatest damage is, and the types of buildings falling inside the hotspot.

Q: Why would we want to use front doors and not building centroids?

Q: Why would we want to look at this surface in 3D?

Figure 2:10 Data source: Damage data for the Holy Cross neighborhood was generated by Jenny Novak Wrye, in the Department of Geography at USC.

SPATIAL DATA VISUALIZATION

The final component of the GIS is making the map. This is probably the most important aspect of a GIS because spatial data is most easily understood as a map. As we said earlier, so many aspects of a disaster are spatial, so disaster cartography has long been an important part of describing and understanding catastrophic events. Although this is not the place to synthesize the various aspects of cartography, or even what makes a good or bad map, it is useful to take a quick tour through different examples of disaster themed maps.

DISASTER CARTOGRAPHY

We would be remiss to not spend a little time talking about how we map disasters. Geographers love maps, as you are already aware from this book. Disaster maps are designed to help us understand the different aspects of a disaster, what has happened, what might happen, and even what is happening. These maps are often a descriptive visual portrayal of the sequence of events, tied by significant locations (and often times) so that the map reader understands what happened. A map of New Orleans a few days after Katrina, for example, would show the locations of the levees (and their breaks), where flooding over-washed from the lakes, and even the final depth of water throughout the city. Social data could be layered on top, such as the location of where people died. In general there are several types of sources for these maps—the news media, disaster reports, government and non-profit agencies, and books describing the events. Examples of all three types will be found throughout this book. However, the most common source of disaster cartography, even today—is the news media. Newspaper maps have always been an excellent source of information, allowing the reader to understand events, while also providing an archival source of information.

Figure 2:11 & 2:12 display three examples of old newspaper disaster cartography—each a little different in style. Figure 2:11 shows flooding throughout New England in 1936 as a series of pictographic images attached to key locations. Each scene tells a story using the map as a frame. The second map is more "scientific" in approach, capturing the path and wind speeds of the 1954 hurricane as it skirted the eastern seaboard before making landfall in Rhode Island. Figure 2:12 again displays flooding across New England in 1936, this time the map serving as a companion to a written description (indeed words supersede the importance of spatial representation in this example).

Figure 2:13, although not a map per se, shows an early attempt at disaster aerial photography. In this case the path of a deadly tornado through St Louis, Missouri in 1927 taken from a Dirigible Airship. As has already been seen in this chapter, aerial photography provides an excellent data source to capture changes to the landscape caused by a catastrophic event.

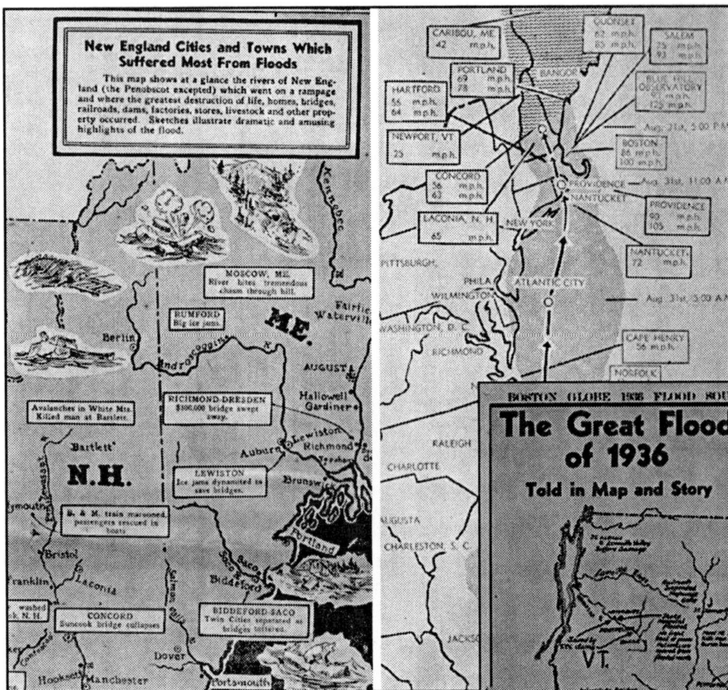

Figure 2:11. Source: Boston Sunday Post Flood Photo Section, p16, March 29, 1936 and "Hurricane path" Boston Sunday Globe Hurricane Edition, p12, September 19, 1954.

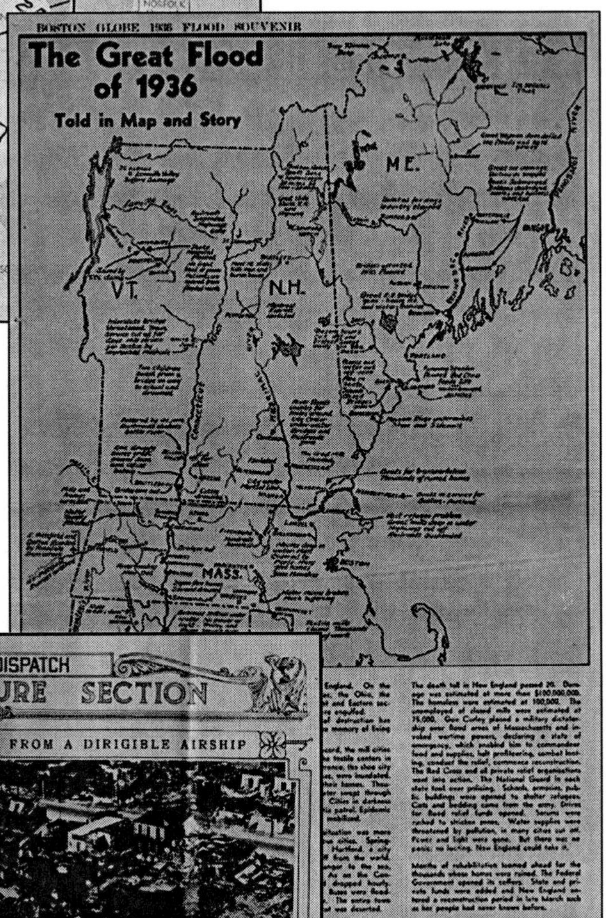

Figure 2:12 Source: The Great Flood. Boston Globe 1936 Flood Souvenir.

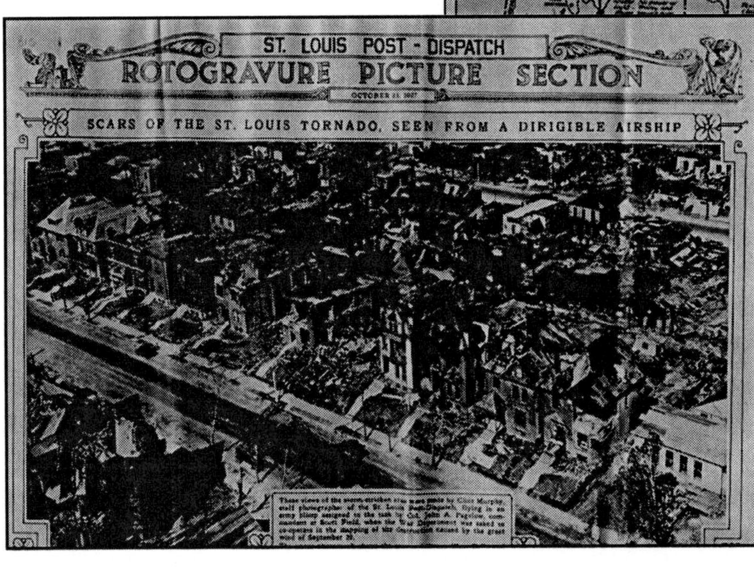

Figure 2:13. Source: *St. Louis Post-Dispatch*, October 23, 1927.

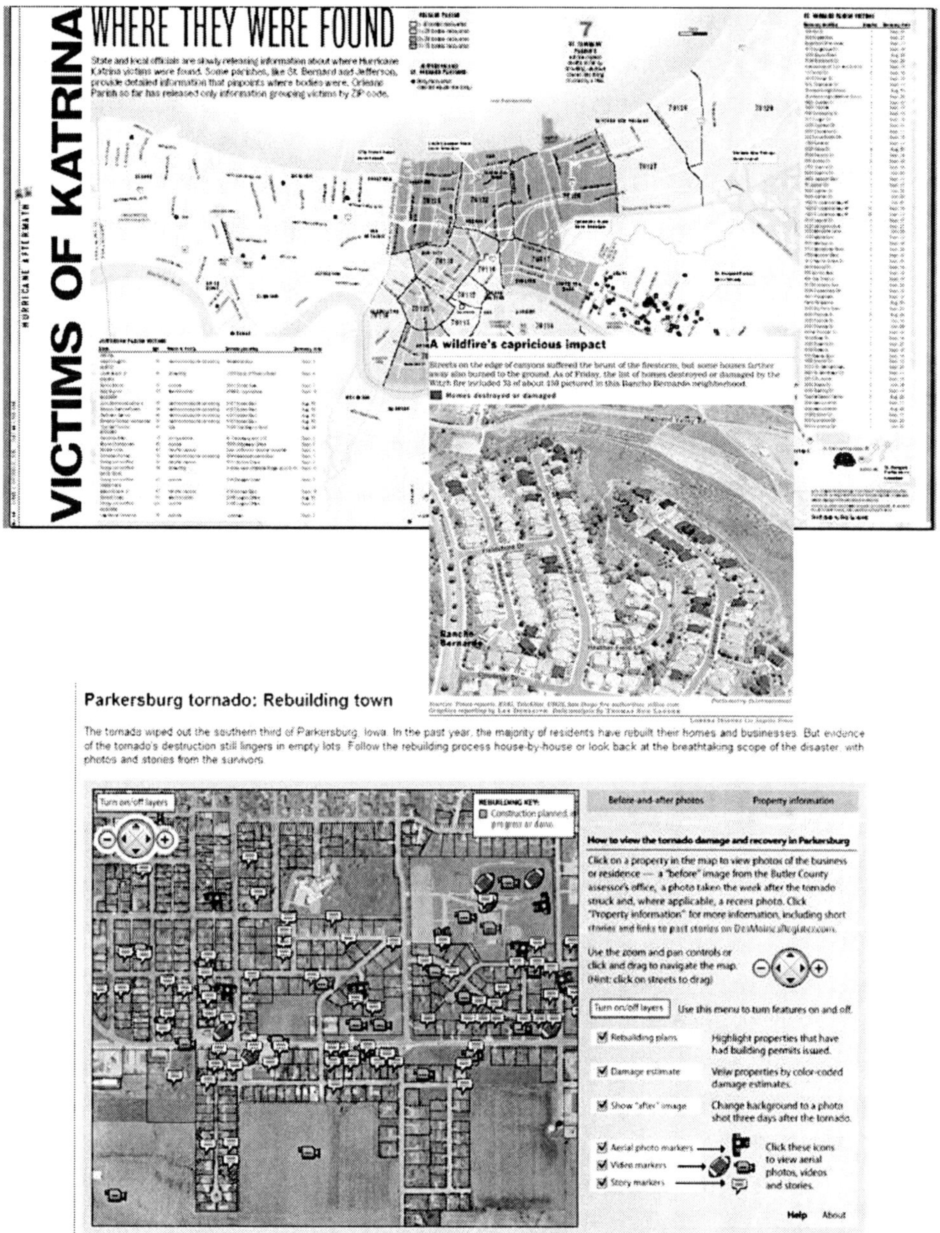

Figure 2:14. *Three examples of modern media disaster cartography, all taken from newspapers, though the bottom image is from the associated website—this map from the* Des Moines Register *(http://data.desmoinesregister.com/parkersburg/parkersburg.php) showing the recovery efforts after a tornado wiped out a large section of Parkersburg, Iowa. The top map shows the location of deaths after Hurricane Katrina, while beneath it are homes burned in the 2007 California wildfires.*
Source: *The Times Picayune*, October 24 2005, and burned homes during the San Diego wildfires of 2007 (*Los Angeles Times*, October 28th 2007).

Q: When did tornado strike the town of Parkersburg?

Newspapers can also provide an excellent resource for researchers. The large array of sources available to a newspaper, especially in combination with the near-real time reporting capabilities of companion websites, can be used to help plan journeys into disaster areas. For example, in February 2008, a series of tornadoes hit Arkansas and Tennessee. When we travelled to the area we found the tornado maps provided by the weather service to not be particularly helpful or accurate due to the coarseness of their spatial resolution. After surveying the damage in Jackson, Tennessee we used local newspaper coverage to find which other local towns had been hit (Figure 2:15). This helped us focus on the town of Huntersville which had been badly damaged. From there we drew on our knowledge of typical tornado paths, which usually is in a north east direction that allowed us to track backwards across the map to find other indicators of touchdown such as a gash in the ground or a single home which had been damaged and therefore hadn't made the news.

ONLINE MEDIA

Google Earth now makes the descriptive disaster map more accessible and interesting than ever. For example, type in "Mt St Helens" into GE and the aerial photograph of the volcano clearly displays the collapse of the north side where the landslide released the Nuee Ardante. Television news shows have also caught onto this new form of geospatial display which is both visually arresting and instantaneous, and not reliant on the skills of a cartographer. In this way CNN "flies" the watcher along the Red River in Minnesota

Saturday, February 9, 2008

The Jackson Sun

jacksonsun.com

50 cents Jackson, Tennessee

HEROES OF THE STORM

Share your story by providing the following information:
Your name, address and phone number.
Your hero's name, address and phone number.
A brief description of why the person is a 'Hero of the Storm.'
E-mail to news@jacksonsun.com. Or mail to ATTN: Heroes of the Storm, The Jackson Sun, 245 W. Lafayette St., Jackson, TN 38301.

RISING FROM THE RUBBLE

COMING SUNDAY: Read the story of the rescue at Union University from those who were there — the students and rescue workers.

SUNDAY IN THE JACKSON SUN

Huntersville mourns

Figure 2:15.
Source. The Jackson Sun, Saturday February 8, 2009.

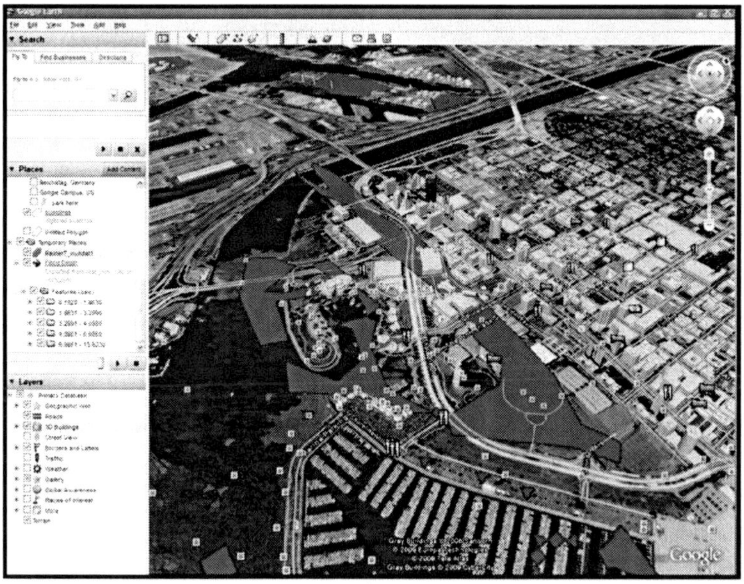

Figure 2:16.
Source: Original flood model output by Aggeliki Barberopoulou, Tsunami Research Center, University of Southern California.

(during 2009) to see which towns are vulnerable to flooding and where the levee may break.

The USGS website also provides some excellent cartographic resources, especially in association with earthquakes (www.earthquake.usgs.gov). For example, the Community Internet Intensity Maps allow you to report if you felt an earth tremor and then see the aggregate mapped results (for those of you have not felt a minor tremor, it can be hard to distinguish what is real and what is imagined). Possibly the most interesting display is a near real time GE layer that shows all recent earthquake activity for the past hour, past day and past week, along with earthquake magnitude and location of fault lines. This map provides an excellent tool to appreciate how seismically active California is. The real power of this tool, however, is in allowing the user to spatially "connect" with any place on earth which has just experienced a massive earthquake.

Maps are also an important tool for informing the public about a hazard. For example taking scientific data and presenting it in such a format that the public understands the hazards better and can use them for planning/preparedness and mitigation measures. Figure 2:1 displayed one such output as part of the Venice Beach community preparedness event, and Figure 2:16 displays a similar inundation layer this time overlaid on Long Beach, California in GE. A similar flood mask in GE was used to help inform communities along coastal Louisiana what might have happened if Katrina's path had impacted them.

Maps also provide a useful means to inform the public where evacuation and relief centers are, both before the event (Figure 2:17) and even during the disaster (Figure 2:18).

RECOVERY MAPPING

The last category of map use involves capturing the temporal and spatial changes in recovery. This is arguably the least mapped stage of the disaster cycle. There are several reasons for this, not least of which is the lack of data. Census data are no longer valid, especially if there have been massive evacuations and relocations. Although different forms of damage assessment are conducted in the days and weeks following a disaster, these are mostly a single snapshot in time—they capture some of the characteristics that may encourage or impede recovery but they cannot show the act of recovery. Even Google Street View, which for New Orleans provides an excellent impression of the post-Katrina landscape (collected mostly in August 2007), is still a snapshot. The question to ask is how much will change in a short duration? This point can be clearly seen in Figure 2:19 which shows two incremental steps along the same Google Street View path—in one the St. Bernard Housing project, where hundreds of families before Katrina is clearly visible, but by the next image it has simply disappeared.

There are several important reasons why maps of recovery are necessary. Firstly, they can help give an indication of the state of recovery for families who are still displaced and cannot journey back on a regular basis. These maps can help in the decision making process in terms of whether it is worth the effort to return. To this end, the cartography produced by the Times Picayune newspaper of New Orleans, and especially the Greater New Orleans Community Data Center are excellent sources for Katrina related geospatial data.

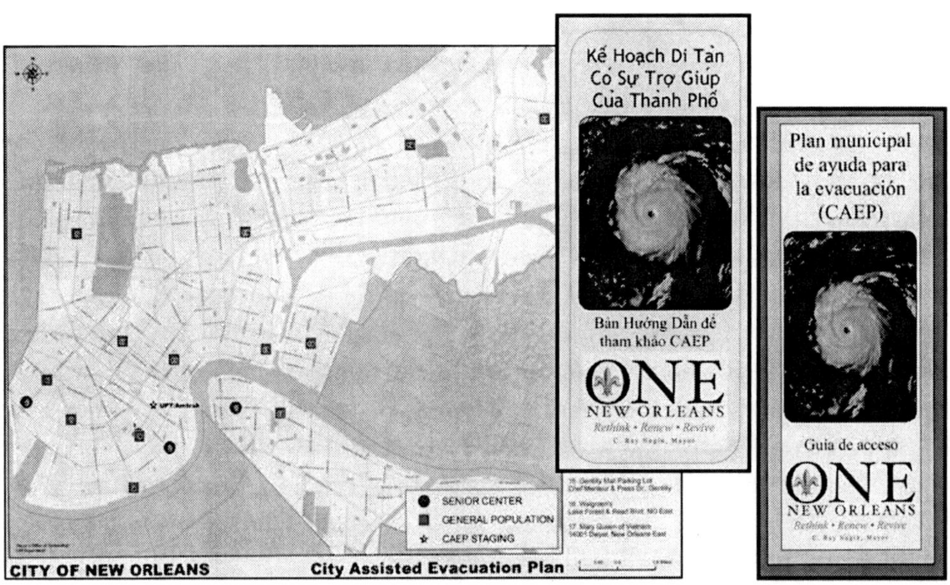

Figure 2:17. *These evacuation leaflets for New Orleans are available in English, Korean and Spanish, though the map is in English in all three.*

Figure 2:18. *This map was displayed outside the Department of Homeland security base in the French Quarter. A standard street map had been pinned to the board, with lines drawn on it showing where different relief centers were located. Pieces of paper next to the map provide more detail about these centers, such as their addressees.*

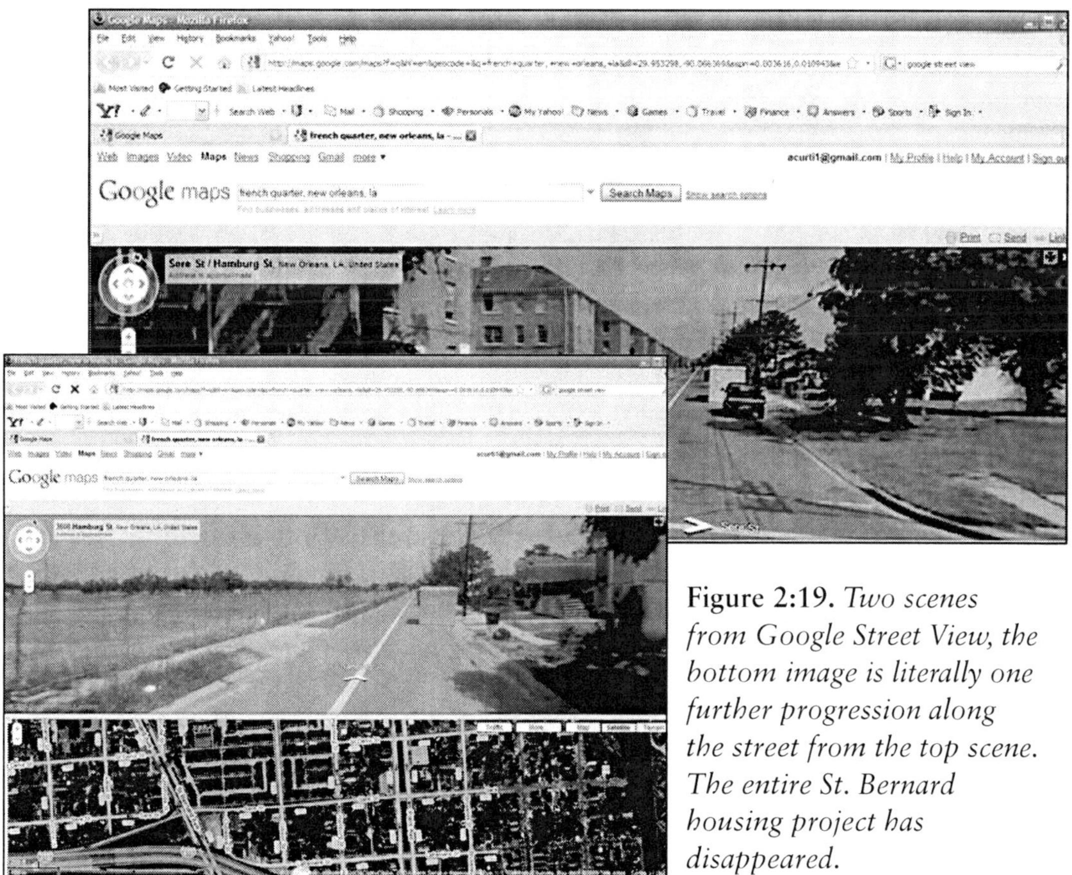

Figure 2:19. *Two scenes from Google Street View, the bottom image is literally one further progression along the street from the top scene. The entire St. Bernard housing project has disappeared.*

DISASTER MAPPING IN ACTION—GIS USE IN THE LOUISIANA EMERGENCY OPERATIONS CENTER

This chapter will conclude with a real world use of geospatial technologies during a disaster. The events of how GIS was applied in the response to Katrina have been described elsewhere[5].

A large contingent of students, faculty and staff from across LSU were drafted into the Louisiana State Emergency Operation Center to help at the GIS desk during the response to Katrina. A typical EOC consists of multiple stations at which sit a different federal agency, state group, non profit organization, or a specific tasked team (such as Red Cross damage assessment). These stations are often identified by flags, signs, or workers wearing vests with their Emergency Support Function (ESF) written across their back. On one wall are usually several screens that allow different video feeds to be displayed at the same time. The role of a GIS desk is to provide the spatial support for all of these teams in the EOC, as well as providing maps for field units, such as search and rescue. During a massive disaster response, such as with Katrina, the request for map products is never ending, some tasks being simple identification of coordinates, while others take more advanced GIS skills to manipulate geographic layers—for example, road networks and flooding information.

Among the maps made during the response to Katrina were basic navigation aids, neighborhoods where hostile activity (gunfire) had been identified, the mapping of 911 calls (geocoded elsewhere on the LSU campus), the identification of state lands (for shelter locations), shelters, critical infrastructure (especially hospitals and clinics), and levees, pumps, and breaches.

Possibly one of the most important geographic technical advances to emerge from Katrina was the use of GE. Anyone can be trained to run a GIS—it is a software application. Most people could collect spatial data—we are now all familiar with GPS. However, GE (I believe) has started a revolution in spatial appreciation. One of the most common requests at the GIS desk in the EOC was the need to place new information into or onto a map. For example, for several days U.S. Public Health Service helicopter pilots would come to the desk with locations

5 More information can be found in a variety of sources—and sometimes with descriptions that do not appear to be consistent with each other. Part of the reason for this disconnect was the massive nature of the response, the combination of different personalities involved, and a myopic reporting of each different research group. See: Curtis, A., Mills, J.W., Blackburn, J.K. and J.C. Pine (2006) Louisiana State University geographic information system support of hurricane Katrina recovery operations. *International Journal Mass Emergencies and Disasters* 24(2): 203–221; Curtis, A., Mills, J.W., Blackburn, J.K. and J.C. Pine. (2006). *Hurricane Katrina: GIS response for a major metropolitan area.* Quick Response Research Report 180. Boulder, CO: University of Colorado Natural Hazards Center. http://www.colorado.edu/hazards/qr/qr180/qr180.html. These reports differ to those described in Ivor Van Heerden and Mike Bryan's "The Storm."

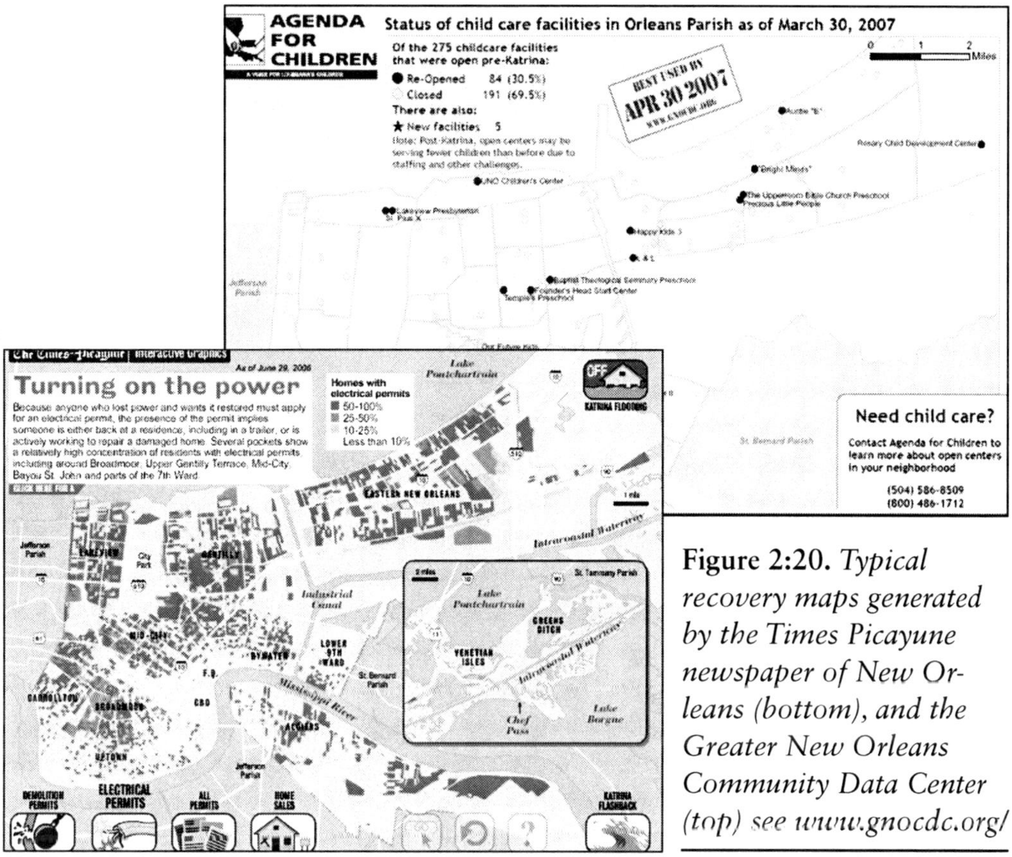

Figure 2:20. *Typical recovery maps generated by the Times Picayune newspaper of New Orleans (bottom), and the Greater New Orleans Community Data Center (top) see www.gnocdc.org/*

(usually clinics or other health care facilities) and want to know their latitude and longitude. These pilots needed to visit each facility to see if people needed to be evacuated, or if not, if the provisions were sufficient. A street address is also not particularly useful from the air. This task of geocoding an address, and extracting out a coordinate, is not hard to do in a GIS—but it still takes a little time, and it requires both software and expertise in the software. The added pressure of a continual stream of job requests, often made by different military personnel, didn't make the task any easier. It was soon realized that GE was all that was needed to accomplish many of these tasks. It was quick, it needed no prior skill set, and it was relatively fool proof in a high-stress environment. It had the additional benefit of being easily loaded on any computer with an Internet connection, and had high resolution aerial photography that allowed the helicopter pilot to actually see the clinic and make adjustments to the coordinate in terms of where was the closest probable landing zone. The geographic skill came in helping put together a suitable flight path that would cover all these clinics in the most efficient way possible—what geographers call a "travelling salesman" approach. Again, this could have been done most accurately in a GIS but probably not so quickly. Someone who works with spatial data on a regular basis can also construct such a route based on instinct.

During any disaster there are certain advances that are made due to the needs of the moment and the lack of data, techniques and technology. I often liken the situation to a battlefield hospital—most medical advances will be made in the warmth and safety of a civilian hospital. However, occasionally an advance will come from within the blood and guts of the field hospital tent, partly due to circumstances not commonly encountered elsewhere and also because these high stress environments act as a creative stimulus for some. This was true of a doctoral student, Jason Blackburn. He would play a lead role in transferring remotely sensed flood data into GE KML files. What is most remarkable about this last act is that at this time there was no easy GIS to Google transformation so he spent many hours with Google and a Homeland Security technical representative to use one of the first GE mash-ups. The first time the map was displayed, the EOC literally came to a standstill. Flood extent had been extracted from remotely sensed imagery with the help of DeWitt Braud, a remote sensing specialist at LSU. These data had been turned into a vector shape file and by overlaying it on top of the air photo of New Orleans, and then extruding (which means drawing upwards) the three dimensional buildings from down town New Orleans, you could literally see which streets were flooded—like a real-time animation of the disaster. The computer on the GIS desk was hooked to one of the overhead projectors and the map displayed on one of the three huge EOC screens. Then he flew around the city, zooming in, panning around—the usual tricks of working with GE—a transformative moment.

However this advance was not purely for cosmetic reasons as requests at the GIS desk now focused on the question of where critical infrastructure fell inside the flood area. As an example of one of these requests, two New Orleans city officials

Figure 2:21. *One of the computers running the flood mash-up in Google Earth on one of the EOC's GIS desk's three computers.*

wanted to see if a location containing toxic material was under water or not. This was because they wanted to see if the chemicals may have spilled, and secondly whether they would be able to get close to inspect them.

A further benefit of the KML was that it was easily transportable between EOCs. In any disaster there are likely to be several EOCs in operation—it is possible that each municipality has its own response center in addition to the main state EOC. Although most state, federal and associated groups all had desks in the state EOC, many of these also had a separate building at which they coordinated their own response. For example, the Office of Public Health had a small GIS support team that had been transplanted from New Orleans. Our LSU team did its best to also provide support for their operation, with the KML being a useful way of sharing and displaying flood imagery. If a catastrophic earthquake was to hit the Los Angeles area, not only would the state EOC be mobilized but so would individual city command centers (such as Los Angeles itself, Pasadena etc) and a regional

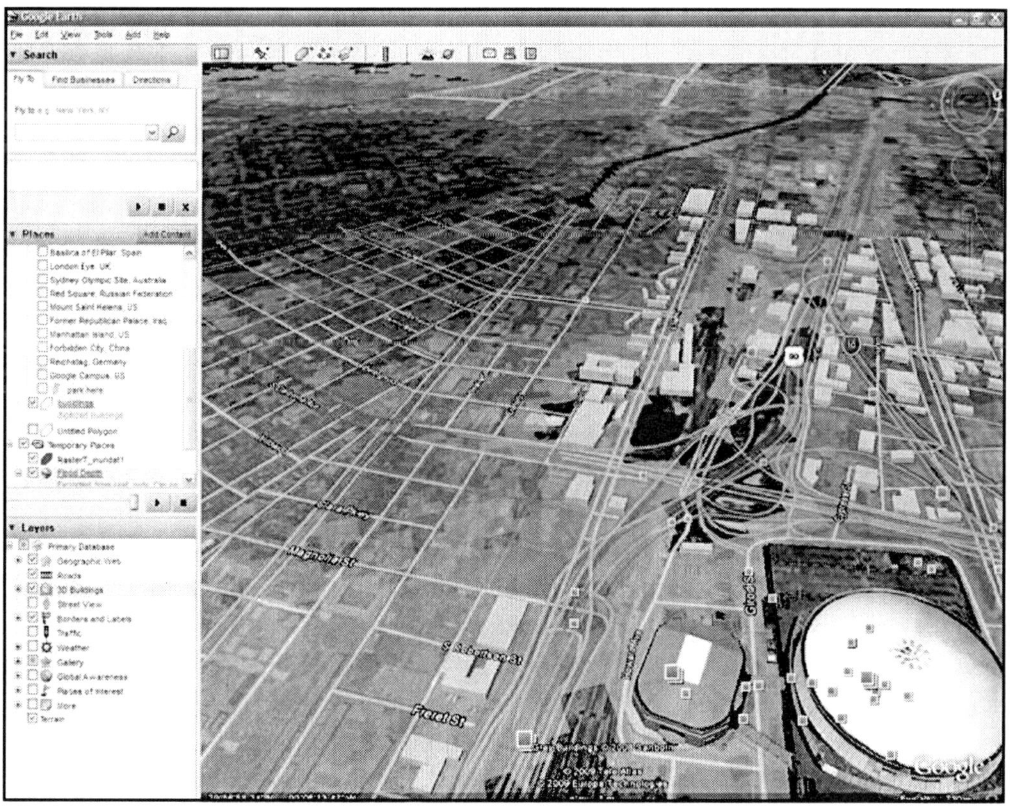

Figure 2:22. *The flood mash-up displayed over 3D buildings in Google earth; the Superdome is at the forefront of the image. Original data manipulation by Jason Blackburn and DeWitt Braud.*

EOC (REOC). In addition, Red Cross has also recently opened its own high-tech EOC in the area.

Although there were several other geospatial advances that came out of the Katrina response, the second evolution that I think is worthy of note was the development of a functional system that could cope with the sheer scale of the requests being generated. The GIS support desk in the EOC was initially and primarily staffed by the Hurricane Center from LSU and they should be applauded for the massive amount of work they accomplished during those first days. With the addition of other LSU volunteers it soon became apparent that the initial work-schema was not sufficient to handle the expanding operation. Three main problems emerged; organization, data storage, and technology. An EOC environment is a highly stressed work place—requests and orders arrive by the minute. In such an environment it can be hard to prioritize requests. The GIS desk tended to work as a three person team, usually with one overseer, one "GISer" and eventually one "Googler." Initially, the jobs that came in were recorded on paper and the supervisor attempted prioritization. However, problems emerged in terms of finding out if a job was completed, especially if the request was not of the highest priority. Even if the job was finished, and the map was printed, secondary problems ensued in terms of knowing where the plot was, where the digital copy resided on the hard-drive, and if any GIS layer had been created that could be subsequently used in other requests. This was particularly problematic as new GIS volunteers rotated in (remember this was a 24/7 operation for many days). There were several occasions where a job had to be replicated because the current team either didn't know if the map had been finished, know that a similar request had been previously made, or just didn't know where the map was. We would later find out that although the GIS desk was the prime geospatial response location in the EOC, there were other options for producing maps both within the EOC itself and at agencies such as FEMA when the Joint Field Operation (JFO)[6] became functional.

Some of us realized that this system was unworkable and needed to be rethought. After a few brain-storming sessions, an ad hoc committee decided to modify a tracking system that was in place in the Climate Center at LSU and that could be used to organize multiple flows of data between different end users. The system would work by having each "job" taken at the EOC GIS desk entered into a system that would track who was responsible to complete the task, if and when the job had been completed, where the file was stored, etc.

A second organizational nightmare involved the logistics of performing GIS work in the EOC. It soon became obvious that processing all GIS requests in the high stressed environment of the EOC was ineffective, so tasks began to be

6 This is the main hub of activity for FEMA at a disaster site.

Figure 2:23. *The Clearinghouse cooperative—the inset figure shows an example of the original file folders for all the data flowing into the clearinghouse during the response to Katrina.*

delegated out to other computer labs, mostly at LSU. This had two benefits—firstly, it removed actual "work" from the hostile environment of the GIS desk and in effect enlarged the work force. In addition, the greater resources of LSU could be utilized, especially printing and plotting. A concurrent advance that facilitated this "outsourcing" was the donation of a 20 terabyte server that could hold all data that were being generated. Although this outsourcing was more efficient, it led to even more organizational problems in terms of tracking jobs and products—again, hence the need for the tracking system.

The centralized data storage system that emerged from this chaos included two map-based querying engines developed by Intergraph which allowed a user to browse through the large library of images generated after the storm by "clicking" onto a geographic location. This data storage approach eventually became the GIS Clearinghouse Cooperative which was the spatial data site for Katrina for a considerable period after the disaster. Although other disaster-related clearinghouses had previously been developed, they were not the norm, and this foresight has benefited many Katrina related research projects[7].

The LSU involvement at the EOC GIS desk was a success. Many volunteers worked tirelessly to help out. Although there are several geospatial lessons to be learnt from the LSU involvement in the EOC, maybe the greatest one is that having a university as a back-up not only provides access to a resource of people, data and hard-ware, but also the academic mind-sets to solve problems.

This chapter has provided an overview of the geospatial techniques and technologies used when considering the human geography of a disaster. Many of the other chapters in this book will revisit the sections described here in more detail. However, you should now have a general appreciation of how a GIS works, and be able to understand its continued reference throughout the chapters to come.

7 Further information regarding the GIS Clearinghouse Cooperative can be found in the following paper, Mills, J.W., Curtis, A., Pine, J., Kennedy, B., Jones, F., Ramani, R. and D. Bausch. (2008) The clearinghouse concept: a model for geospatial data centralization and dissemination in a disaster, *Disasters* 32:3 467–479

THREE

Disaster Connections—The Johnstown Flood: Part One

THE FOLLOWING TWO chapters make direct comparisons between one of the worst catastrophes in U.S. history, the Johnstown Flood of 1889 and Hurricane Katrina. The purpose of this approach is three fold. Firstly, it gives the reader familiarity with a major disaster in U.S. history as told through a contemporary account. Secondly, through comparing these events the similarities between these disasters are emphasized, and thirdly, the "geographic eye" is emphasized in the different accounts to again stress the importance of space. The first of these chapters will concentrate on the disaster itself, including descriptions of the horrific aftermath, and the response of the Board of Public Health. The second chapter concentrates on the immediate clean-up, including both the removal of debris and bodies.

One of the reasons for stepping back to the 1800s, apart from exposing readers to a disaster which they may not be familiar with, is to illustrate the similarities among natural disasters of all types and between these two in particular. This extends to the actual event, the threat to life, and then the clean-up and recovery. These similarities also provide a strong justification for archiving all contemporary disaster data so that we can continue to make these comparisons, and hopefully learn for the next time. A challenge we have for the reader is to find another disaster from history, ideally before 1940 and use original sources such as reports, newspapers or contemporary accounts, or even from reading one of the many popular histories available[1], and see how many similarities can be found with the

1 There are many books readers can choose from, though my recommendations with regard to hurricanes and flooding in particular are "Isaac's Storm: A man, a Time, and the Deadliest Hurricane in History" by Erik Larson about the Galveston hurricane of 1900 and which provides an interesting companion to the storm that struck the same city in 2008; "Black Cloud: The Great Hurricane of 1928" by Eliot Kleinberg which includes the social injustice associated with the loss of life around

description presented here. The Johnstown flood will primarily be presented using the original language contained in the Pennsylvania Report of the State Board of Health for 1889. Into this account will be the comparisons and photographs drawn from Katrina. The use of the original language serves three purposes; firstly, it portrays the events with greater clarity (and emotion) than would our own reinterpretation. Secondly, it allows the reader to see what was important for disaster response during 1889 again without the risk of reinterpretation, and lastly, it showcases the eloquent (though a strange word to use for this event) language used in the late 1800s[2]. All descriptions from the previously mentioned Board of Health report will be indented. One thing to be aware of before starting is that in 1889 there was no real federal response and certainly no FEMA (though it was the first domestic non-war related response by the American Red Cross). Many of the activities initially performed by government agencies in the EOC during Katrina, and then afterward by FEMA became the responsibility of the Board of Health as (understandably) externalities to society were often gauged in terms of their threat to public health and welfare.

DESCRIPTION OF THE JOHNSTOWN FLOOD

> Dr. Benjamin Lee, Secretary, State Board of Health.
>
> Dear Sir, I have the honor to submit this, my final report, of the work performed by the State Board of Health forces in the Conemaugh (V) valley since the disaster of May 31, 1889.
>
> On two occasions I have submitted hastily written reports of the work accomplished by the Sanitary Corps. This report is designed to cover more fully all the difficulties that have confronted the corps, to give a more detailed account of the ground work of meeting the requirements of the hour, the magnitude of which they recognized at the outset. No other Board of Health has ever been confronted by a task of such herculean

Lake Okeechobee; "The Storm of the Century: the Labor Day Hurricane of 1935" by Willie Drye, one of the most powerful storms to strike the United States with winds carrying sand grains that could literally remove flesh from bone; "Rising Tide: The Great Mississippi Flood of 1927 and how it Changed America" by John M. Barry which describes the flooding of Mississippi and Louisiana, again including elements of social injustice with regards to the purposeful flooding of St Bernard Parish to save New Orleans.

2 Readers are encouraged to read newspapers from the 1800s as the descriptions of events are wonderful, often flamboyant with an air of adventure and excitement missing with today's reporting—whether describing a bear devouring a man, or the steady progress of cholera across Europe to the United States.

> proportions; how well they met and performed that task is evidenced by the fact that no epidemic followed the disaster.
>
> In presenting this, a retrospect of Pennsylvania's greatest calamity, I have only to say that it is simply given as a preface, a general outline of the cause and extent of the nuisance which you called on the governor of the state to abate. I do not essay the impossible task of picturing the scenes, or telling of the thousands of horrors that accompanied the devastation. Records of the dead in a dozen morgues tell how densely the shadow of death enveloped the valley.
>
> Specialists in meteorology claim that three storms centers met in a dire embrace upon the summits of the Allegheny Mountains and were held in check on the east by a wall of high barometric pressure. We know that the storm stood as if anchored above the crests of the Alleghenies, and for thirty-six hours discharged an amount of rainfall unprecedented in the history of the country—about seven inches of water falling in fourteen hours. The headwaters of the north and south forks of the Conemaugh were in this gigantic atmospheric disturbance. The greater part of the watershed of the forty square miles, contiguous to these two mountain streams, drains toward the west in to the Conemaugh valley. Stony creek, rising in the south and flowing northwardly along the base of the main range of the Alleghenies, empties into the Conemaugh at Johnstown. The combined floods from these three mountain streams submerged Johnstown on the morning of that ill fated Friday, five feet deeper than ever before known.

Figure 3.1 and 3.2 provide two interesting examples of disaster cartography. Figure 3.1 originates from a book describing the events published in the same year of the disaster—a feat which has carried on to today, where there is always a race for the first account or commentary of any disaster. The second map is a large fold-out accompanying the Pennsylvania Report of the State Board of Health for 1997 Vol. 1 (one of several details is shown here). On first glance the map appears to be similar to the first, showing the valley with additional detail sketches common to this period. However, on closer inspection, many of the sketches actually portray aspects of the flood. Working from east to west we come to Bridge No 6 which in parentheses is "gone." The relief "Camp Conemaugh" is north of Conemaugh. As we move along the river, past the "Ruins of Roundhouse" we can see that several of the buildings are drawn askew, as though moved by flood water. This is most evident in the sketch of Johnstown itself that shows a destroyed town with only the largest buildings remaining. Passing through town we come to the Stone Bridge, which will be referenced throughout the following description, and the mountain of debris that accumulated there. Close by are the tents of another relief camp.

Figure 3.1. *"Map of the Deluged Conemaugh District" from the book, History of the Johnstown Flood, by Willis Fletcher Johnson, published by the Edgewood Publishing Co., 1889.*

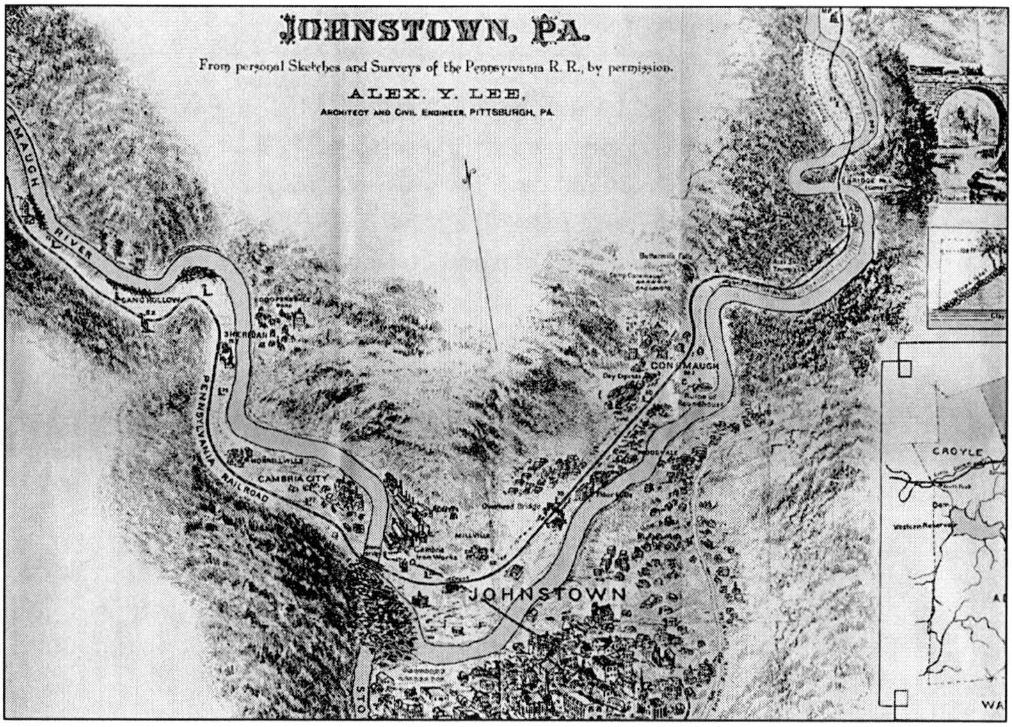

Figure 3:2. Source: Pennsylvania Report of the State Board of Health for 1997, Vol. 1.

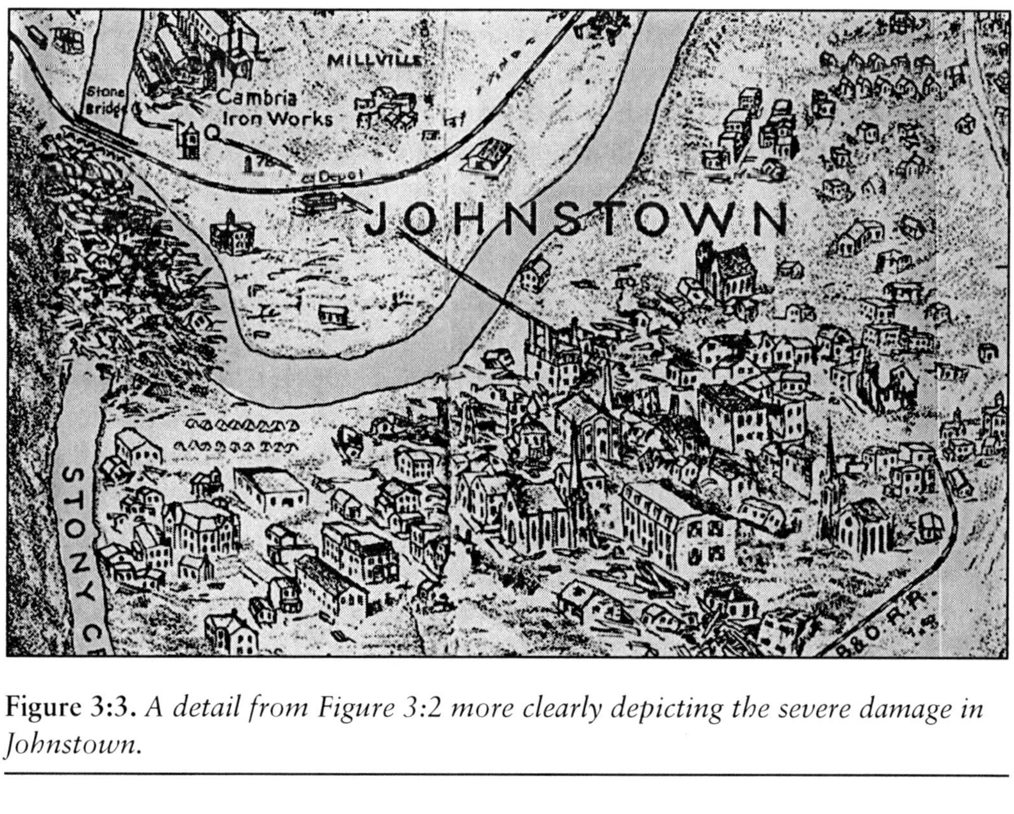

Figure 3:3. *A detail from Figure 3:2 more clearly depicting the severe damage in Johnstown.*

The Board of Health included the following sentence: "People were imprisoned in the tops of their houses and entirely cut off from aid." This became an all too common image following Katrina as well with people waiting desperately to be rescued. As researchers we had access to the original 911 calls generated in the days following Katrina, and these captured this desperation:

"In 2 and a 1/2 storey home with rising water"

"Owner has gun and will not let others on roof"

"Six kids trapped in attic"

"Four people trapped in car with rising water and power lines on car"

"Her and four kids on 2nd floor and water rising"

"Two trapped, one is an infant. There are two sharks in the water."

Even in the weeks following Katrina, evidence of how the residents of New Orleans became roof and attic dwellers can be seen across the post-disaster landscape. Some buildings had messages requesting rescue or supplies spray painted across their roofs[3]. Other homes had holes punched through their roof to allow escape. One of the more common tell-tale signs of search and rescue, apart from the spray painted crosses, were the holes punched into the attic so teams could check if anyone was left inside. In a later chapter, the psychological impact on children of being forced onto their roofs is revealed in their art, the traditional home (a square topped with a triangle roof) being replaced by a triangle alone (only the roof) as will be discussed in greater detail in chapter eight. In our conversations with people who had weathered the storm, we were often told of the fight to get as high as possible, either in their own building, down the road to the only two story home on the street, or even to nearby overpasses. Many of these stories described the speed in which water filled the streets, often only stopping at or just below roof height. One lasting impression we had from the deserted city were the rows of cars parked along the overpasses in order to escape flooding.

The four images in 3:4 show different aspects of human vulnerability and floods. In 3:4a a homemade escape rope hangs from the attic window while 3:4b shows

3 As of the writing of this book, a home with a spray painted message across its roof is still evident in the Holy Cross neighborhood, and can be seen in Google Earth/Street View on Dauphine St, New Orleans, LA.

Figure 3:4.
Source: 3:4a photograph taken in Mid City, New Orleans, on 10/12/05. 3:4b is from the Boston Globe 1936 Flood souvenir. 3:4c photograph taken on 10/29/05 around the London Avenue canal break (eastern side). 3:4d is from the spatial video data collection run of the Lower 9th ward, 3/11/06. Photographs by Curtis.

a similar scene but from just under 70 years earlier in Massachusetts. Figure 3:4c shows an escape hole smashed through the roof of a home close to the break in the London Avenue canal; the car in the foreground is almost buried from the sand that washed out of this break. The final image is again taken from the spatial video, which even seven months later shows one of the many abandoned boats found in the neighbourhood.

> Nine miles up the valley and among ridges of the Alleghenies, nearly six hundred feet above Johnstown, was a reservoir called Conemaugh lake, a body of water over two miles long, nearly three-quarters of a mile in width and from ten to fifty feet in depth. At 3.15 in the afternoon, when the water in Johnstown had reached the second floor of many buildings and had already receded an inch or two, the water in the lake overflowed the dam. In a few moments the overflow washed out a large section of the dam, and scarcely fifteen minutes later the lake was empty, the whole valley a mass of seething waters and the crowning calamity of the century, for this country at least, was being enacted.

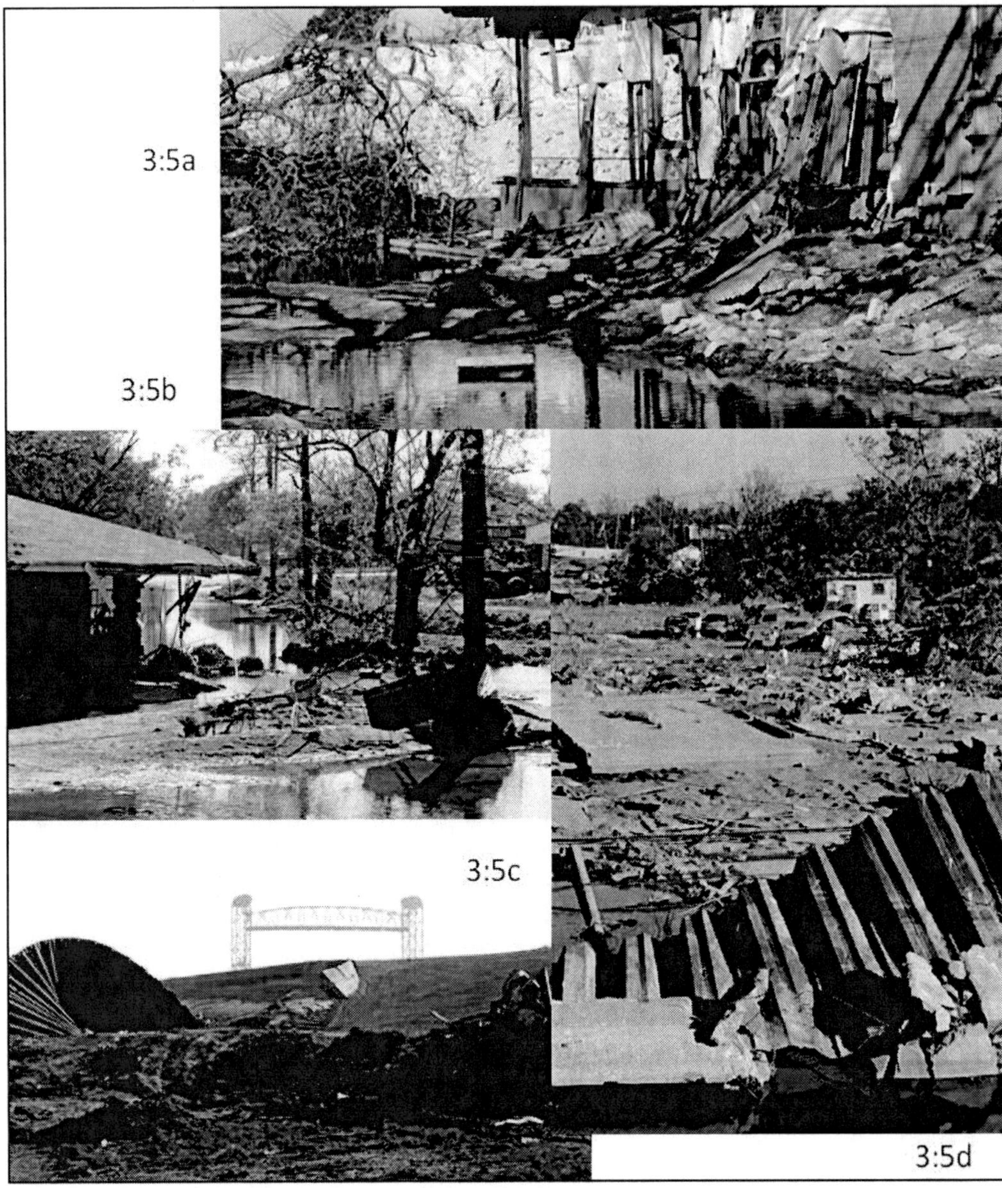

Figure 3:5. *Image 3:5a shows the damage to one of the buildings closest to the 17th Street canal break, the levee wall is seen in the background. 3:5b is looking away from the eastern break of the London Avenue Canal levee break, the force of water having obliterated the side of the house in its path. 3:5c & d are the northern break of the Industrial Canal, 3:5c looks through the break (the buckled corrugated iron is the actual levee wall) towards the bridge that spans the canal. 3:5d looks out from the break, a house once stood directly in the floodwaters path, and now all that remains is the concrete base. There is still a considerable amount of water in the streets at all three levee breaches.*
Source: All photographs taken by Andrew Curtis on 10/29/05.

There is a striking parallel between this description and Katrina. People were expecting the hurricane. It was obviously a terrible experience for those who didn't evacuate, and for many the wind damage was accompanied by over-wash flooding from Lake Pontchartrain and Lake Borgne. However, no one expected the subsequent flooding as multiple breaks in the levee system drowned the city. Just as with Johnstown, the initial natural disaster was followed and made far worse by a failure in human-engineered structures.

> The immense body of water formed, with its accompanying debris, a resistless wave the full width of the valley, sweeping the pine-clad hills on each side, and stripping them to the naked rock. It denuded the valley for a distance of twelve miles and left it a total ruin, a river bed filled with massive rocks and stones many feet deep. In fact, it was so clean that the medical inspectors, in making their survey, found nothing of an unsanitary character to report. The mountain of water came with terrific force, holding to its embrace, engines, cars, bridges boulders, houses, trees, and massive machinery which ground to fragments everything in its path. All were caught up and rolled over and over to be left, a heap of twisted iron and splintered wood, half buried in the sand and mud. It was a moving mountain of shattered homes and ruined industries. Where before stood immense brick and stone buildings not even a trace could be found.

The force of water breaking through the levee walls had the same effect in New Orleans, smashing through homes around the 17th Street and London Avenue Canals, and literally washing away the homes by the breaks in the Industrial Canal in the Lower 9th Ward. Fig 3:5 shows the damage at ground level.

Figure 3:6 displays a map of building footprints from the Lower 9th Ward showing the "sweeping" pattern of homes away from both breaks of the Industrial Canal, but especially the hole by the previously described barge. These building footprints can be used in a GIS to map the flow pattern of the flood waters. By drawing a line from the original address to where

Figure 3:6.

the building ended, several calculations and visualizations could be made. How long were the line segments (a proxy for home displacement and therefore flood strength), how does the orientation of these lines change (which would be a proxy for flood direction). It would even be interesting to run a density surface of these lines[4]. Of course, one challenge is being able to match the flood moved home to its original address. Another challenge, is that many buildings are severely damaged, meaning it would be hard to find their original address, and digitize their shape. Possibly this exercise would work better as we move further away from the major break, where building movement was not so great and we could be more confident of making matches. Although this wouldn't show the worst flooding and damage, it would still capture how the water fanned out through the rest of the Lower 9th Ward.

> The torrent of water, when it reached Johnstown, about forty-five minutes after it left the lake, was at least twenty feet high. It struck Johnstown and its contiguous boroughs, giving its thirty thousand inhabitants but a minute's warning. The wave, pushing before it all the material it had gathered in a twelve mile rush, plunged through the town without regard to the river's channel. Its force carried it some distance up the Stonycreek valley; it struck the mountain side and rebounded, and at the stone bridge of the Pennsylvania railroad it piled the debris more than forty feet high. The piling up of wreckage over so many acres of ground formed a dam that even this terrible power could not move, and the waters were held in check until the railroad embankment gave way. Then with an awful rush they carried destruction and death to Cambria City. This check in the flood's rapid course caused the waters to surge back and forward, forming whirlpools and eddies—a maelstrom of seething waters and a wilderness of grinding, crashing buildings. All this was done in less time than can be conceived. Professor Sanders, who stood on the highest point of the Morrell Institute building and had an unobstructed view of Johnstown and the waters, said: "The destruction of the city was almost instantaneous; the eye could not follow the racing waters. At some points the flood ran like a high sea, throwing logs and debris high above the water, at other points the pressure exerted forced the water to rise vertically in a body. There was an awful absence of human cries. The onslaught was so sudden that the people were unable to shriek their despair. A cloud of dust arose from the city and for a moment was suspended above the waters,

4 A common analytical technique in a GIS—the density of points, or in this case, lines, for a given area is calculated. Those areas with high densities (a great number of points/lines) can be considered hotspots.

Figure 3:7. Source: Frank Leslie's Illustrated newspaper, for the week ending June 22 1889, picture entitled "General view of Johnstown from Green Hill." The fires across the city are clearly depicted.

> and as in terror I whirled around to face the stone bridge, I saw the water returning, Johnstown the center of a maelstrom."

Toward the eastern edge of the Lower 9th Ward, much of the debris pushed by the levee break also came to rest. The damage was greatest closest to the levee break, where the water poured through, which was as expected. The further the buildings were away from the break so the damage lessened (though still being considerable)[5]. However, at the eastern edge all the debris came to rest in not a dissimilar way to that described at the Stone Bridge.

> The horror of the scene that followed was graphically portrayed by the Johnstown Tribune, in its first issue after the flood, and the following extract is submitted in place of an attempted description. "In a moment Johnstown was tumbling all over itself; houses at one end nodding to houses at the other. And on sped the wreck in a swirl, the angry water, baffled for a moment, running up hill with the town and the helpless multitude on its back, the flood shaking with rage and dropping here and there a portion of its burden, crushing, grinding, pulverizing all. The back, with great frame buildings floating along like ocean steamers, upper decks crowded, hands clinging to every support that could be reached, and so on to the great stone bridge, where the houses, piled mountain high, took

5 This is an example of that core geographic concept, distance decay—the further apart two bodies are, the less action exists between them, in this case being damaged by flood waters pouring through a break.

Restricted License (Research & Educational Use only) Survey - holycross_1_18_08
File Edit Help
Display Speed
Slow Fast
Cam-1 On
Cam-2 On
Cam-3 On

<< Figure 3:8. *The top photograph (Curtis) was taken on 10/12/05 in the Uptown area of New Orleans. The entire block had burned to the ground the only thing left standing being the brick columns of the houses, the chimney stacks, and decorative ironwork. The bottom image is a spatial video image (the process described in Chapter 1) from 1/18/07 for the Holy Cross neighborhood in New Orleans. This is the output from a three camera system (the two different thumbnails at the bottom show the left and front facing camera views). The main camera reveals the black remnants of a home that had literally burned to the ground. In the background another home with a blue FEMA tarp has black scorch marks on its roof. What is also interesting about this location is that this scene of devastation lies directly opposite a child's playground (see in the left thumbnail). For any families that have returned, if they use this playground then this is the scene surrounding their children. Remember, this is almost 18 months after Katrina.*

> fire and burned with all the fury of the hell you read about—cremation in your own home, perhaps a mile from its foundation; dear ones slowly consumed before your eyes, and the same fate your own a moment after.

The co-relationship of flood and fire may not be immediately obvious, however in the days and weeks following Katrina several fires erupted in the city. On one day we were flying with the Civil Air Patrol above New Orleans and we could see a huge column of smoke clearly visible from somewhere in the Uptown area. During our walks through the devastated city in October, several neighborhoods had burned buildings. The most notable was a whole block of buildings that had burned, only the brick front columns to the entranceway remaining along with scorched metal facades, and brick chimneys standing like sentinels (see Figure 3:8). Although I cannot be sure, I am convinced that burned block corresponded with our smoke plume seen from the air. During the next few years, as we catalogued the neighborhood recovery process, we again would see evidence of abandoned buildings which had caught fire and burned down. However, these events pale in comparison with the fire that erupted in the city of Grand Forks in North Dakota during the Red River flood of 1997[6]. A combination of firefighters having to expend energy on rescuing people who had not heeded the mandatory evacuation order, and the flood waters which both hindered access and reduced water pressure to fuel the hoses contributed to the severity of the fires. The conflagration was eventually fought with the aid of a plane dropping red retardant, a more common sight for residents of California who often witness this in association with wild fires. Why is there this

6 The events of the Red River flood are described in Ashley Shelby's "Red River Rising: The Anatomy of a Flood and the Survival of an American City."

connection between flooding and fire? Three general reasons for the modern link are the impact of electricity and water (given as a reason for the start of the Grand Forks fire), the lack of fire services to fight the fires (or impediments to their effective deployment), and lastly, possibly as a way to claim insurance for a property not carrying flood insurance. It is with earthquakes, however, that the full menace of fire occurs, sometimes causing more deaths and general destruction than the quake itself. The Great Kanto earthquake that struck the Tokyo area is one such example, with cooking fires being thrown against paper walls eventually resulting in a fire that literally boiled some alive as they took refuge in small water bodies[7]. Of course the most famous North American example is the fire that ravaged San Francisco (and mapped in the first chapter) after the 1906 earthquake as the residents sat in chairs and on hillsides watching the disaster as though at the theatre.

> The water receded rapidly until it reached the level of the break in the railroad embankment, but all night the site of the town was filled with a vast lake of water. Shrieks for mercy were unavailing; human skill was powerless to save hundreds who were imprisoned in the wreckage. The burning pile lit up the turbid waters of the dismal lake beneath which slept thousands cold in death. Few eyes closed that night except in death, for those who escaped to the hill-sides kept vigilant watch through the long dreary hours of the night, as did others with only a foothold on some partially wrecked building, shivered, in dripping garments, in the cold night air. Many died from shock, exposure and wounds through that long age of night. Death! Silence! Eternity! What names of terror to the beings who longed and prayed for deliverance in the darkness, and amid the din of crumbling walls.
>
> At length morning came, and the muddy waters had sufficiently subsided to allow the work of rescuing the perishing to be commenced, but amid scenes of desolation that cannot be pictured. There was wretchedness on every side. Horrors piled on horrors! Wrecked hearts, bankruptcy, orphanage, widowhood, childlessness, obliterated homes! Wreck, ruin, and deposit of slimy mud, in some places five feet deep, covered the site of the town.
>
> At an early hour the survivors began to realize their position. The flood of terror had been succeeded by the calm that precedes a storm of insanity. After some hours this gave place to a desire to obtain food and shelter,

7 It might be a challenge to track this book down, but the descriptions of the earthquake are both gruesome and compelling; Peter Hadfield's "Sixty Seconds that will Change the World: The Coming Tokyo Earthquake."

a place of safety; to find friends, loved ones. They realized that they were face to face with fate and future. They must do something. Action severed the cord of madness and brought the living to their senses. But where to go, what to do, no one could tell. Many wandered aimlessly around. Over and under the wreck they went, in and out of the ruins, seeking yet dreading, to find a friend. Some could be seen standing stunned and tearless by the disfigured clay that once was instinct with a life closely linked with theirs. Some survivors, for days and days, walked and talked among the ruins of their former homes without shedding a tear or dropping a word of sympathy. Some, for weeks, travelled up and down and over the wreckage, seeking the members of a lost family amid scenes so harrowing, that it is a marvel that anyone could look upon them and escape insanity.

Figure 3:9. *3:9a shows a temporary morgue set up in the school house, the coffins stacked outside. 3:9b gives an indication of the amount of debris the relief teams faced. 3:9c illustrates survivors consulting a "wall of information" for news of friends and family members. 3:9d shows the beginning of the relief aid that poured into the stricken communities.*
Source: Frank Leslie's Illustrated newspaper, for the week ending June 22 1889.

This description provides the first indication of the impact these events would have on the mental health of those who survived, though the public health concern as to issues of stress, methods of coping, and disaster related psychopathology were really not part of the public health world at this time. Now it is arguably the greatest public health concern following a disaster in the United States and as such a later chapter will be dedicated to it.

> There were left, however, strong hands and brave hearts, and these commenced the work of rescuing the injured, the aged, the weak, and removing the dead.
>
> In a walk, the morning following the disaster, scarcely a step could be taken without discovering some dead body. In one place sixteen bodies were found together, crushed and mangled beyond recognition.
>
> In a walk, from the lower to the upper end of Main (S)street, over fifty dead bodies were counted. The debris was filled with the dead bodies of human beings and animals. As soon as possible, morgues and hospitals were opened, and the exposed dead and injured removed. This work was attended with many difficulties. They had to be carried long distances and over almost impassible wreckage.
>
> In the afternoon people began to reach Johnstown from the country and to bring much needed assistance—a silver lining to the dark cloud of death, that touched the hearts of the wretched survivors.

It should be remembered that at this time there was no real coordinated federal response, just the efforts of the State Board of Health and volunteers coming to the disaster. Just as with many Katrina survivors, help was days away, many having to spend that first uncertain night not knowing if they were safe, if they would be rescued, or if friends and loved ones were safe. By the next day, and again we found this with many Katrina survivors, people would begin to take control of their own actions, walking through flooded streets to the Superdome (a shelter known to everyone in New Orleans), or even beginning the long wade out of town along the Interstate. One survivor we talked to said he had rescued a child from a care center, but after spending one night in the Superdome, decided it was safer for his family to wade along I10 towards Baton Rouge.

> Sunday, the second day of the disaster, the sky was cloudless and the sun's rays warmed up the shivering community. The outside world commenced to extend its sympathy, to invade the town, to help extricate and bury the dead and feed and clothe the twenty thousand survivors. In their greatest hour of need, when the valley was still covered with water, and when its

Figure 3:10. *The distribution of pins gives a good indication of how far-a-field the responders came from. Of course this resulted in challenges in terms of being able to provide good directions to people unfamiliar with New Orleans. Photographs by Curtis.* Source: This map was in the federal "Homeland security" response camp set up in the French Quarter on Decatur.

suffering people were literally shut in, the world opened her heart and unloosed her purse-strings to succor the living and perform the last sad rites for the dead. Words will never convey an idea of the noble work done by the various relief committees upon their arrival at Johnstown.

On this, the Sabbath day, Johnstown was the scene of confusion; thousands of strangers crowded the ruined city; every avenue of relief swarmed with generous and noble workers. There was a tremendous demand on human energies. The dead everywhere; hundreds in the morgues—thousands to be recovered. Then demand for coffins, undertakers, men to excavate the graves and teams to carry the dead, was most urgent. But the grey mists had scarcely arisen from the hills in the morning until hundreds of funerals were coursing their green sides. Without a friend to drop a tear, hundreds have been lowered into unmarked graves, and in this life their

> fate to friends will never be known. All day the work went on and still, the morgues were full, more than four hundred bodies were recovered. The corps of seventy-five undertakers toiled from sunrise to sunset, and day after day, to consign the army of the dead to its last resting place.

With any disaster volunteers will flood in. Some genuinely want to be involved in any way possible, even if all they can offer is physical labor. Other benefactors have skill sets that are needed—emergency services, especially doctors and nurses from other areas, being the obvious example. There are still several active web pages of different search and rescue units from other states who flew as a team into the disaster area[8]. However, geographers (and especially GIS specialists) are needed as well, to help understand the disaster-transformed landscape, provide mapping support, and possibly even field-based geospatial data collection[9].

The description of family members after the Johnstown flood wandering around in their search for lost loved ones (and depicted in 3:9c) also occurred after Katrina as people went for weeks in the shelters not knowing where the rest of their family were. The strongest media images of such walls full of the faces of lost loved ones are associated with the September 11th terrorist attacks and the Indian Ocean Tsunami of 2005.

> The immense pile of wreckage at the great stone bridge continued to burn, to rage over the tangled graves of—only God can tell how many. Mothers and fathers hung around the morgues for days and weeks searching for a missing face. Some haunted them for months, going from one to another and never missing a day. While some were finally rewarded, the weary searching of others was in vain.

DESCRIPTION OF THE JOHNSTOWN RESPONSE

> Enough has been said to convey, in some measure, the conditions and urgent needs that confronted the officers of the State Board of Health on the morning of June 3rd. No other set of men in all the flooded district had a more difficult, laborious and painful task before them.

8 See the Nevada Task Force One 1 website at http://www.nvtf1.org/katrina/CCFD_Katrina.htm The same team had also responded to the World Trade Center, as well as Hurricanes Gustav, Ike, and Rita.

9 One such volunteer force is GISCorps, a volunteer group run through URISA. Their "missions" in terms of disaster response, and even recovery can be seen at; http://www.giscorps.org/index.php?option=com_content&task=view&id=22&Itemid=59

> In giving a history of the operations of the sanitary corps, we labor under some disadvantages from not having full data from which to compile it. Considering the confusion that reigned in the valley during the awful two weeks succeeding the flood, it is remarkable that any organization was effected and records kept. We regret that we cannot at this time include a complete record of the disinfectants received and used. Large quantities were donated, and we would like to give the donors credit. In many cases disinfecting places without any account being rendered to our department. Some work was performed by volunteers and they left without rendering to the office a report of their services or of those who worked under their direction. Few of the inspectors rendered an itemized report of the work performed in their districts. This history, therefore, must be to some extent, a general one.

The issue of data preservation during a disaster remains. This has already been mentioned as a continuing problem but it is worth re-emphasizing the importance of good record keeping. It is understandable that during the height of response, when people are being taken from roof tops, and immediate health care needs are being seen to in the shelters, or in the care given to animals at National Veterinary Response Team shelters[10] (or in the case of Johnstown, the care for the injured and the burial of the dead) that more emphasis is placed on helping, and less on the recording of the act. For non-academics this may seem understandable, however, the only way we can learn from a disaster so that future events are handled more effectively, which in turn may mean the saving of countless lives and reducing suffering, we must have an accurate record of what happened—including successes and failures. One way to maximize the likelihood of such data being stored is to have an effective response plan in place, with data archiving being part of that plan.

HISTORY OF THE SANITARY WORK

> Early the next morning (June 4th) a consultation was held and it was decided that the situation was a dangerous one that the immense amount of decomposing matter must be removed or its effects counteracted at

10 This group, which is part of the U.S. Department of Health & Human Services, comprises of private citizens who receive federal employee status during a disaster. Expertise within the group includes not only Veterinarians, but also Animal Health Technicians, Pharmacists, Epidemiologists, and Communications Specialists, to name a few. During a disaster the team "provide assessments, technical assistance, public health and veterinary services under the guidance of state and/or local authorities"—for more information see their website at http://www.hhs.gov/aspr/opeo/ndms/teams/vmat.html.

Figure 3:11. *Examples of different disaster industry vehicles scene around New Orleans in October 2005.*[11] Photographs by Curtis.

once, that the Board of Health should at once secure quarters and put a force of men at work. In carrying out this order we encountered many difficulties.

Wherever bundling were left the material was piled fifteen or twenty feet high. In some localities, covering a number of acres, buildings were so jammed together that a passage was affected with great difficulty. In fact, the greater portion of the town site was one mass of debris, thrown together in such an inextricable network as to almost forbid the contemplation of its removal. The survivors, a homeless, half crazed mass of humanity, seemed perfectly helpless and unable to give the Board the least assistance. All departments of their local governments were paralyzed. They were without protection from the lawless hordes that were pouring in from every section of the country. Chaos reigned supreme. Thousands of people, train after train of supplies, relief committees and gangs of volunteer workmen were constantly arriving. All formed a confused mass, everybody trying to find someone else.

There is an interesting choice of words in the previous passage about the "lawless hordes"—does this suggest that looting occurred?

> The day was spent in an almost fruitless effort to secure residents for clerks, time keepers and laborers. We were compelled to seek men wherever we could find them and it was not until later in the afternoon that we succeeded in securing a small force. These were at once put to work destroying dead animals, which were already becoming very offensive.
>
> At this time no arrangement had been made to pay the men. There were no facilities for feeding and lodging a force. To provide food in a devastated valley for the twenty thousand survivors, the visitors and the rapidly gathering forces, making in all about thirty thousand, was the question of the hour. We were embarrassed with difficulties at every turn. There was no use in trying to employ men unless we could at least furnish them with food and shelter. Another consultation was held and it was resolved to employ a force of carpenters and at once commence the erection of a camp and furnish the men with lodging and board to effect as soon as possible an organization of forces. This organization was not fully completed until the 8th and that only by working night and day.

Even today the logistics of supporting the large number of disaster professionals and volunteers who fly into a disaster region are extremely challenging. We have already mentioned the challenges faced by Red Cross in terms of housing their volunteer staff in Baton Rouge—a city that had already been swelled beyond capacity with the mandatory evacuation order of New Orleans. What readers might not be as familiar with is how reimbursement works in a post-disaster situation. This might seem rather callous in terms of people who volunteer expecting to be paid—and many do not. However, a large portion of responders do receive reimbursement based on MOUs (memorandums of understanding—they also go by other acronyms). These contracts basically state that in the event of a disaster, a pre-determined service between a responder team and (for example) FEMA is agreed at a set price. This means a job can be done without the need for wasting time on making sure workers are paid, or haggling over prices in a demand-intense situation.

On a related note, unless you have never been in a disaster, you might not be aware of the large "disaster industry" that exists. Within a few days of Katrina,

11 The profits made by corporations in times of war and disaster has stirred considerable debate over the recent few years. There are several interesting documentaries and books on the subject, possibly the most well known being *The Shock Doctrine: The Rise of Disaster Capitalism* by Naomi Klein.

Baton Rouge began to fill with trucks of various sizes. The roles of these professionals varied, from general clean-up, including managing liquids (such as sewage) to "Trauma Scene Cleaning" which is a polite way of saying bodies and body parts. We had wondered where these trucks were stored when disasters don't occur, but sometimes they are hidden in plain sight. Anyone driving on the 110 heading north into downtown Los Angeles can see Red Cross supply trailers parked outside the convention center—other Red Cross trailers are scattered throughout Los Angeles County at strategic points (the importance of geography again coming to the fore—placing these trucks so they can effectively serve the most amount of people in a disaster).

The Board of Health report indicates that disaster professionals or at least businesses with the flexibility to work disaster environments existed even during the Johnstown Flood.

> Articles of Agreement with McClean & Co
>
> Articles of agreement made and concluded this________day of July 1889, between C McLean & Co., of Philadelphia, Pa., parties of the first part, and General D.H. Hastings, Adjutant General, for, and in behalf of the State of Pennsylvania, part of the second part:
>
> Witnesseth. The parties of the first part hereby agree and bind themselves to all the work which the Board of Health of the State of Pennsylvania may require at and in Johnstown and the adjacent boroughs, in a prompt and thorough manner, and to employ such number of men and teams as the supervising engineer may prescribe from time to time, and to employ said force at such places and in such manor as said engineer may direct.

FOUR

The Johnstown Flood— The Days and Weeks After: Part Two

JUST AS WITH the previous chapter which inserted Katrina comparisons into an account of the Johnstown Flood of 1889, this chapter follows a similar pattern, though now the description moves more to the aftermath, the clean up, and the location of refugee camps. Just as with the previous chapter, these comparisons will be inserted throughout the Johnstown description. However, the chapter will start with a description of our flight over New Orleans and some of the more devastated bayou areas.

FLIGHT OVER NEW ORLEANS ON SEPTEMBER 17TH, 2005.

This trip was our first chance to see New Orleans as a disaster landscape, rather than an abstract situation we had been interacting with through the EOC, or watching on television. Sections of the city around the University of New Orleans were still under water, while the effect of the wind on roofs, especially the race course and the Yacht club were all too evident. We would eventually see the confused medley of masts and boats from the ground, but from on high the impression was one of a giants discarded plaything. The rest of New Orleans was hazy; large naval boats including an aircraft carrier were docked along the French Quarter, and at least one thick plume of smoke rose into the sky from Uptown. Our observation of the city was cut short by a large military transport rising rapidly from an airbase somewhere close to the city, causing our pilot to bank and dive to avoid a collision. I suppose it is surprising that there weren't more collisions of this type over the city given the amount of aircraft, especially helicopters that filled the airspace. The plane's radio was full of "chatter" from the different aircraft, which included the media, military, FEMA, and civilian relief efforts flying over the city.

Quite understandably there was a huge military presence in both Baton Rouge and New Orleans (see Figure 4:1). The images A to C show some of the many military hummers that patrolled the city, A and C around Holy Cross on October 21st (actually the bridge in the image crosses the Industrial canal), and B crossing water gushing from a hydrant in the 8th ward on October 17th. Military patrols would

Figure 4:1.
Photographs by Curtis

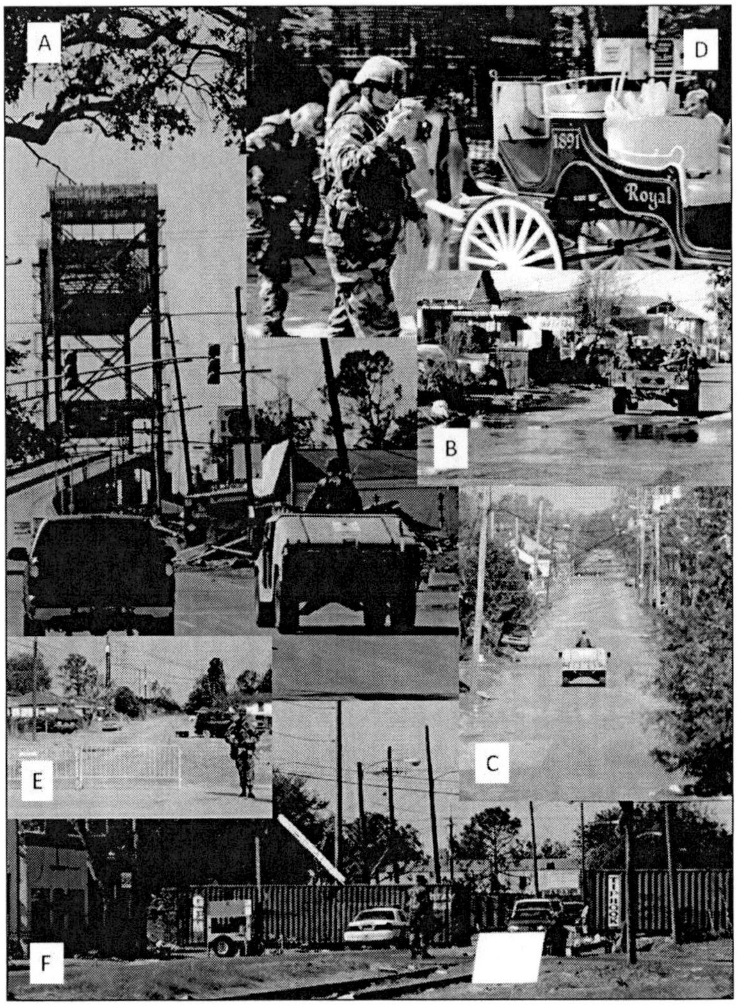

often stop us and ask for identification, and the National Guard would be asked back in to some of the more devastated neighborhoods many months later in order to control the emerging crime wave. There were also barriers in place—for example "E" shows a guard stopping people entering the Lower 9th ward on October 26th. Although the rest of the city was "open" this section remained closed for some time. We were told unofficially that the area had not been fully swept for bodies. Mortalities were found in other areas of the city even after this date. More than once we walked by a freshly painted "body removal" date, knowing we had walked by the same house with a corpse inside only days before. Figure "F" shows a barrier erected between the Lower 9th ward and St Bernard Parish. There are two major thoroughfares between the parishes; the one pictured here was blockaded by containers while the other had cars piled on top of each other. Picture "D" shows a typical scene in the French Quarter, tourists being replaced by soldiers who still wanted to do "tourist things."

After our near miss with the military transport plane, the pilot thought it was safer to fly away from the city and head out along the bayou, away from where the destruction had mostly been caused by levee failures, and instead where the damage was by storm surge and powerful winds. At this point I would have to say that one of the most important technological improvements for field researchers in recent

Figure 4:2. *The sports arena at the University of New Orleans. The neighborhoods in the front of Lakefront Arena are still under water.*
Photographs by Curtis.

years has been the digital camera. Now we can record any scene from multiple perspectives, without care of the cost of developing and printing film, setting our cameras on a high resolution that allows us to revisit the scene at a later time and extract further detail from the image. This is important as the senses are often overwhelmed in the wake of a disaster. Unlike other field work where one may happen upon a topic or subject, and then take the time to consider and record it, in a disaster images, sounds, even smells hit with machine gun rapidity. It is almost impossible to remember every scene, and certainly some things get missed—the written message on a roof, an escape-hole punched through an attic, images on the periphery of a scene that the eye misses on the first round. Indeed, I remember commenting during the flight that New Orleans didn't seem that badly damaged. In fact, I had not seen any flooding within the city—it took later careful examination of the photographs around the University of New Orleans campus near Lake Pontchartrain to see that the streets were still under water.

Certain scenes from this flight stuck with us. Firstly, as we flew along the bayou the sheer savagery of the storm was more than evident, the homes being smashed into kindling—not just missing roofs, or jumbled boats—these buildings no longer existed. Matchsticks! Secondly, in one of the bayou towns a large white church was all that remained, its stark white broad base and tapered steeple rising from the flood waters, a thick band of muddier water flowing from behind it and wrapping around. The third impression, was the fragility of the bayou, homes hugged the narrow inlets of water that came in from the marsh, each building being less than a

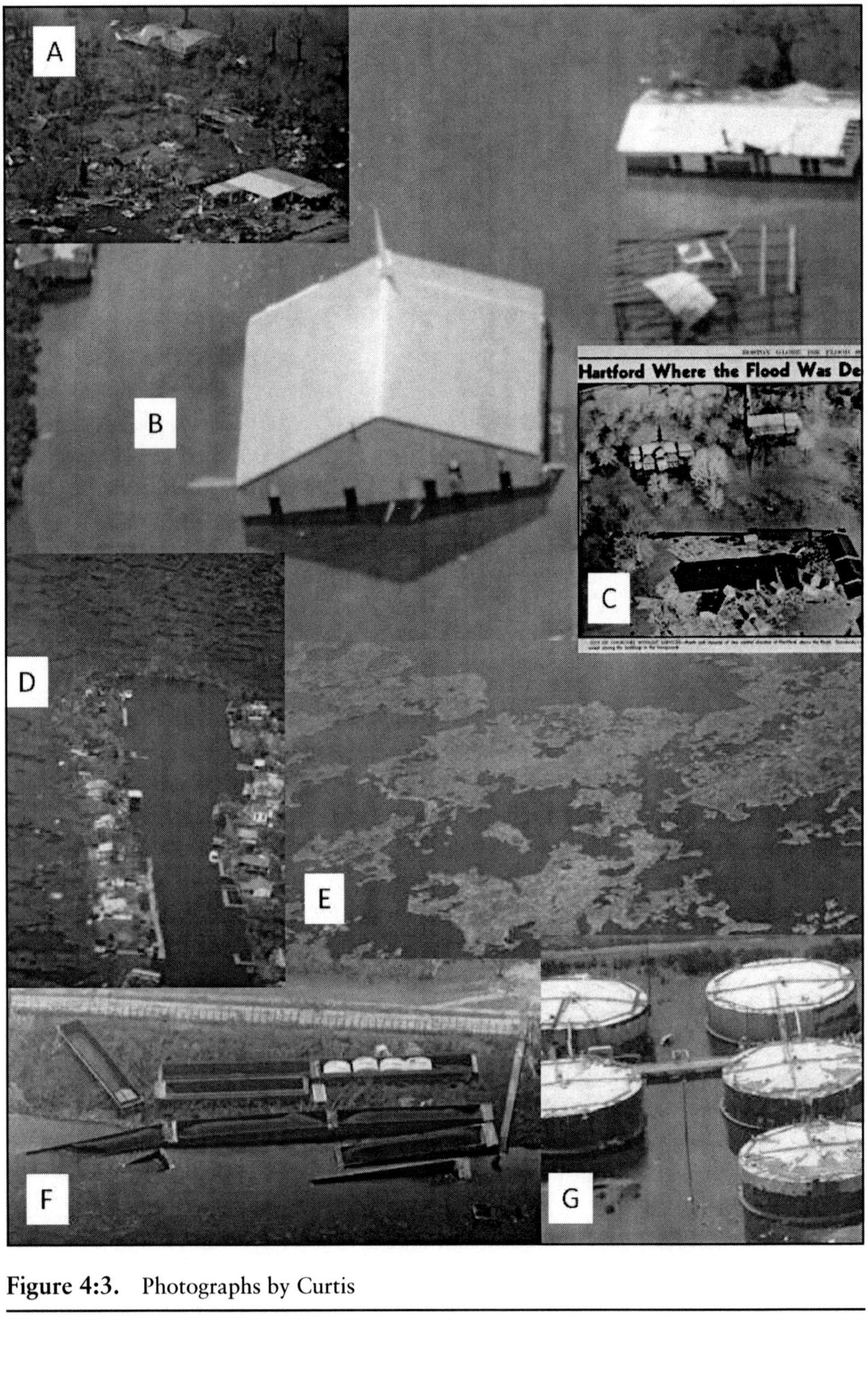

Figure 4:3. Photographs by Curtis

few feet from both water's edge and water's height[1]. Finally, this trip gave me three visuals of the storm's environmental impact. Coal barges lay beached on the levee, some scattered down to the water's edge, and others half sunk—their contents spilling out. Huge oil holding tanks were battered and shedding metal, their bases under water. And earlier, while flying over New Orleans, at one outlet from the city into Lake Pontchartrain, a dark black plume of contaminated water pushed out into the lake.

Figure 4:3 shows various photographs taken on the flight with the Civil Air Patrol. "A" shows the type of damage found along the bayou, houses smashed to pieces, washed into each other, and still sitting in flood water. Images "B" and "C" show the repetitive nature of disasters, both capturing churches poking through floodwater, one in 2005, the other in 1936. Images "D" and "C" show the fragility of the bayou. In "D" fishing camps line one inlet, beyond this, what was land is now a pocked mixture of vegetation and water. This is even more clearly shown in "E" where water now occupies 50% of the picture—this would have been all land a decade before. One of the reasons for this are the channels cut to provide access to the laying and maintaining of oil pipe—unfortunately a side effect is the hastening of wetland loss. Images "F" and "G" provide two examples of the environmental impact associated with Katrina—"F" shows coal barges run aground on the levee, while some have sunk in the channel spilling their load. "G" pictures oil or gas tanks, sitting in flood water, their roofs and sides badly damaged.

Who knows how long it will take to fully assess the environmental impact of Katrina. The role that petrochemical processing has had on the loss of both culture and wetlands in the region is discussed by Mike Tidwell in his book, Bayou Farewell, and wasn't lost on some of the residents comments scrawled across their own homes "Murphy Oil and our elected did this—at night people cried." Most of us thought the water in New Orleans was a toxic broth of pollutants, sewage, spilled chemicals and who knows what else. On one of our trips to the London Avenue Canal break with some FEMA colleagues, when helping someone push their truck from the thick sand drift that had washed from the levee wall (supplemented by the sand dropped by helicopter to stem the floodwater) I slipped and cut my hand, and spent the rest of the trip wondering exactly when it would fall off. My only disinfectant (of sorts) was liquid hand sanitizer, obtained from one of the tables scattered through the city where volunteers helped returnees with some basic supplies, and which I proceeded to liberally push into the wound. We

1 Readers are strongly encouraged to read Mike Tidwell's "Bayou Farewell: The Rich Life and Tragic Death of Louisiana's Cajun Coast" which provides an overview of the culture of the Bayou, especially the interaction between big industry (drilling for oil), shrimp fishing, and wetland loss. The topic of wetland loss is also integral in understanding why the Louisiana Gulf Coast, including New Orleans, is now more vulnerable to hurricanes.

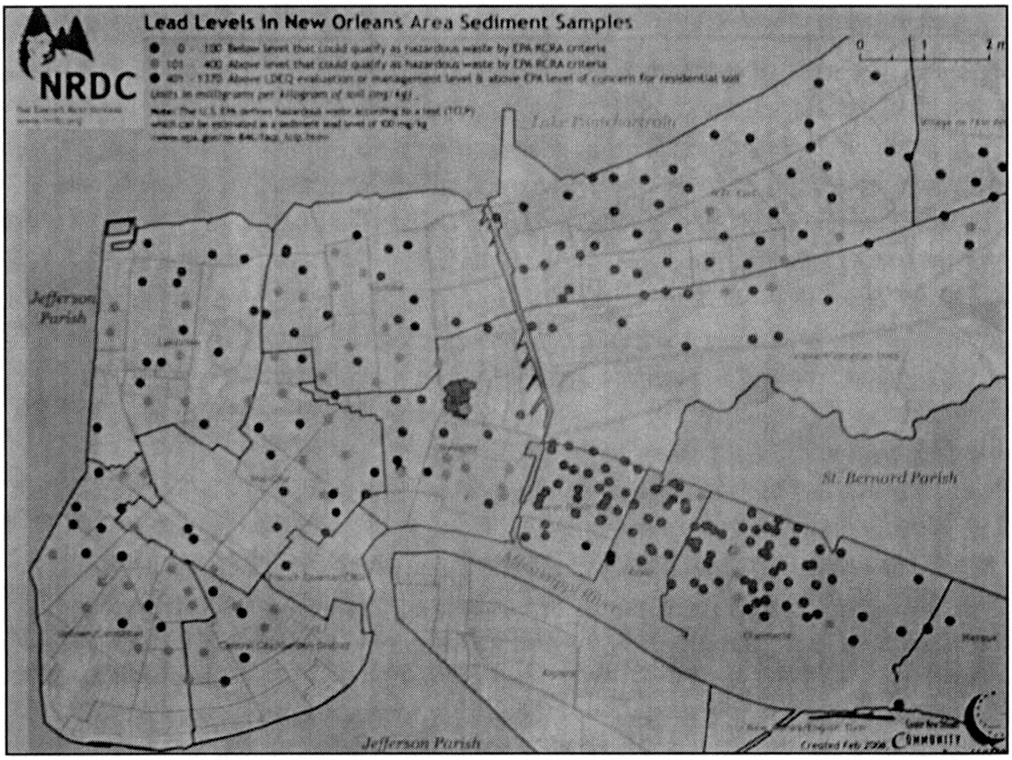

Figure 4:4. *One of several environmental impact maps pinned to a wall and photographed by Curtis at the GERI / New Orleans Health Disparities Initiative Meeting held in downtown New Orleans, June 12th 2006. The other maps included Diesel, Arsenic and Benzo(a)pyrene levels in sediment samples.*

would later be told that tests of the water revealed relatively little to be concerned about—something which has been backed up in the literature with initial papers of concern petering out into few producing hard data[2].

By the time of our flight the activities in New Orleans had moved from search and rescue to clean-up. The immediate threat to life had passed and now the process of allowing the city to function had begun. This would include draining neighborhoods, clearing streets, checking on health hazards (especially chemical spills), debris movement and then removal and the start of home gutting. All through this process the search for bodies would continue. In many

2 An example of both "concerns" and then actual analyses are: Danny D. Reible, Charles N. Haas, John H. Pardue See Toxic and Contaminant Concerns Generated by Hurricane Katrina. *Journal of Environmental Engineering*, Volume 132, Issue 6, pp. 565-566 (June 2006); **S.M. Harmon and D.E. Wyat** Evaluation of post-Katrina flooded soils for contaminants and toxicity to the soil invertebrates *Eisenia fetida and Caenorhabditis elegans Chemosphere* Volume 70, Issue 10, February 2008, Pages 1857-1864.

ways, these were the same actions as described in the following section about the Johnstown Flood.

> By the 5th of June fully six thousand men were at work in the valley. System and organization commenced taking the place of loose, ill directed and spasmodic work. A city of tents and camps had sprung up to shelter the homeless. A large force was engaged in clearing away the great masses of wreckage and opening the streets. Here and there little patches of what were streets had been cleared. But so solidly packed was the mass of ruins at the stone bridge, that dynamite, derrick and human labor made little impression. Disinfectants were coming in by the car load: our hastily gathered forces were traveling back and forth over the wreck, seeking out the dead animals and burning them, and scattering disinfectants over all the masses of decomposing organic matter. Horses, cows, dogs, cats and large quantities of meat and provisions were imbedded in the wreckage and mud. The work of extricating and removing them to a safe place to burn entailed hours of not only disagreeable but excessively severe labor. The corps of sanitary inspectors went into the thick of the debris, and in a little while the bodies of fifty dead animals, mostly horses, were removed and burned.
>
> The disinfecting crew went daily over and into the wreckage, in and out of ruined buildings and wherever they found a

FRANK LESLIE'S ILLUSTRATED NEWSPAPER

Figure 4:5. *Johnstown Flood—Beginning the clean-up,* Source: Frank Leslie's Illustrated newspaper, for the week ending June 22, 1889.

smell of decaying matter sought out the cause and disinfected it, reporting the same to the head office, when a gang was at once sent out to destroy it. During the first week an immense amount of decaying material was destroyed and large quantities of disinfectants used. But amid confusion, urgent demands and want of proper organization, no record of the amount of work performed could be kept. There is no way to compute the amount of animals destroyed by the sanitary crops. One foreman says that his force cremated nearly three hundred horses and that the total number of animals destroyed aggregate sixteen hundred. Mr. J.W. Parker, who with a gang patrolled the rivers up to Johnstown, reports that they burned over a hundred and fifty carcasses and about one hundred drift piles.

Figure 4:6. Photographs by Curtis.

Frequently a search and rescue "X" would be accompanied by a message. Some of these were to direct subsequent teams, or even animal rescue to a home where a pet had been found. Sometimes the message simply read, "bad smell" indicating that (probably) something had died and was still in the wreckage. Unlike with the Johnstown example, it appeared from our experiences that the animals that had died, mostly cats and dogs, were left to rot and be discovered by their owners.

Figure 4:6 displays some of the gruesome discoveries waiting for their owners. Image "A," taken in the Lower 9th ward on October 21st, 2005 shows a typical message left by human search and rescue for the animal teams that followed. The way to interpret this "X"[3] is the search and rescue team notation to the left of the "X" (CA 5 = California Task Force 5 Urban Search and Rescue Team, and 1/8 = 8th Marine Regiment), the top is the date of the search, the right side is anything of note found in the premises (such as "cat"), and the bottom is whether any living or dead were found at the scene. Therefore, in Figure 4:6 "D," a photograph taken in the Lower 9th ward on October 21st, 2005 describes a "bad smell" as possibly meaning a body was inside. One of the most difficult challenges to face is the loss of animal life after a disaster. Image "E" from the flooding of 1936 depicts a typical outcome of large scale flooding, while "B" (and "C"—the spray painted message above the dog) is from the St. Roch neighborhood of New Orleans, on October 19th. There were several such animal corpses, or at least signs telling of dead animals inside a home during our walks through New Orleans. It appears as though most of these animals were left for the owners to deal with upon return. It is hard to imagine the additional psychological damage this would have had on the returnee.

> As soon as possible two circulars of information were published and distributed. The first assured the community that the Board was prepared to meet all the needs of the inhabitants of a sanitary nature promptly. It instructed them as to the best means of maintaining the present condition of general good health. The second was directed to heads of families and recommended that women and children should leave the place for a few weeks, suggested that those living in over crowded house would be better off in tents placed on the hillsides, and urged the imperative importance of cleaning the cellars of their houses of all mud, and refuse, and of strewing them freely with disinfectants which would be furnished, without cost, by the Board. Nine health bulletins were published, from time to time, giving an exact statement of the condition of the public health, to allay

3 A standard search and rescue marking originating, or at least popularized, from the Mexico earthquake of 1985.

> any fears of an epidemic of disease. Circulars were also distributed telling the people how to clean up and to properly live in this emergency. A large amount of official pamphlets pertaining to the public health and giving information on the care of infants, and the treatment of infectious and contagious diseases, have been distributed at various periods. More than thirty thousand pages of printed matter have been distributed.

Distributing information through printed material was also a common practice after Katrina. With the disruption to traditional media, hand-outs, posters and signs stuck into road medians became common sights across the city. Hand-outs, sometimes available from the information tables scattered throughout the city or wedged into a buildings door included contact information for assistance from official sources such as the City of New Orleans, or FEMA—information regarding professional or volunteer help (especially from church groups), or the result of surveys assessing how much damage was "officially" attached to each property. Posters ranged from piecemeal construction informing returnees about different assistance centers, banners spread across road medians advertising services such as "mental help" or the location of a Red Cross shelter, to small posters stapled to

Figure 4:7. Photographs by Curtis.

commercial premises prohibiting any reopening until a full health inspection had occurred. Road medians soon became full of plastic signs of the type usually used to promote a political candidate, but these advertising services of local contractors, the most common being "home gutting" and "mold remediation." Even some official sources used this approach to inform returnees of available services, such as the reopening of Children's Hospital.

Figure 4:7 shows a small selection of the signs which appeared on just about every surface after Katrina. "A" was an Acorn sign advocating against bulldozing the Lower 9th ward, "B" a door hanger left by an insurance adjuster, "C" housing information taped to a door by HANO (Housing Authority of New Orleans), and "D" which shows a typical street median, full of various contractor information, and most interestingly, informing local residents that Children's Hospital was again open.

Q: What is Acorn?

> The people were prompt to act on any advice given by the Board, or its sanitary corps. They labored zealously to remove from their premises every vestige of the slime that the cruel flood buried them under.

Many of the returnees to New Orleans and the other coastal parishes were faced with immense clean-up tasks. We heard several stories of homes full of mud, some with dead fish and snakes inside, and how objects from inside the house had floated only to settle in strange configurations, such as a couch now resting on a sink top.

The first major flood I had encountered was along the Ohio River in 1997. I had taken some students to the towns of Maysville in Kentucky, and then Aberdeen in Ohio. This was also my first real insight into the social injustice that often accompanies a disaster and the strength of community that bands together afterwards in the reestablishment of the neighborhood. On the Kentucky side of the river is a levee wall designed to protect the community of Maysville. Unfortunately, the poorer town of Aberdeen on the opposite side could afford no such protection. This meant when the river flooded (which all rivers will do...it is part of their natural life cycle), instead of 50% of the waters over-spilling on either side, the Ohio bank received a disproportionate volume of water. As a result the homes on the Ohio side were flooded to their roof tops. I was lucky enough to talk to one of the residents who informed me that this was a poverty trap, as these homes were the only ones the people could afford—they were cheap because they would sporadically flood, and yet no flood insurance was available. Therefore the people who lived here had to balance being able to afford a home against the certainty of losing possessions. The same individual told me that similar floods had occurred in the 1950s, and he had remembered how the entire community had mobilized with brushes, moving from home to home to brush away the mud as the last of the flood water receded so as not to let it dry in situ.

> A number of prominent physicians from other cities came as volunteers and did everything they could to assist in pushing forward the sanitary work. They went back and forth over the entire length of the devastated district, directing the work and performing many arduous tasks with untiring zeal and energy.
>
> ...The flooded valley, extending on the west from Coopersdale to east Conemaugh on the east, was divided into twelve districts, and a deputy medical inspector placed in charge and held responsible for the sanitary conditions of the same. The organization continued the same for about ten days, when the volunteers were compelled to relinquish their duties. In leaving they took with them the respect and gratitude of all who were benefited by their labors.
>
> The work of the deputy medical inspectors was to see and direct the work in their particular district, report daily to the executive office the amount of work accomplished, and to offer any suggestions which they thought would facilitate the work in their districts. Each inspector was expected to make a daily round of his territory, reporting to this office everything discovered of an unsanitary character and directing the removal of all putrefying and decaying vegetable and animals matter: to establish headquarters where disinfectants, information and directions would reach him; to open sub stations for the distribution of disinfectants and circulars; and to put up placards at different points where disinfectants were distributed free of cost. Each inspector was furnished with gangs of men, with foreman over each, and under his direct command.
>
> The history of the manner of working in one district will, in general, be the history of the operations in all. In some instances the demands were more urgent and required a large force for months to remove the immense masses of wreckage and material that harbored filth and was a dangerous source of disease. Some districts were swept with such violence to render the locality habitable.
>
> The character of the work necessitated the formation of gangs for doing certain kinds of labor. There were gangs with carts who gathered up the garbage and destroyed it; gangs of men who cleaned cellars and buildings of their flood deposits, where people were compelled to remain; a disinfecting gang to distribute the disinfectants and to use them whenever a body or carcass was being extricated; to disinfect the dumps,

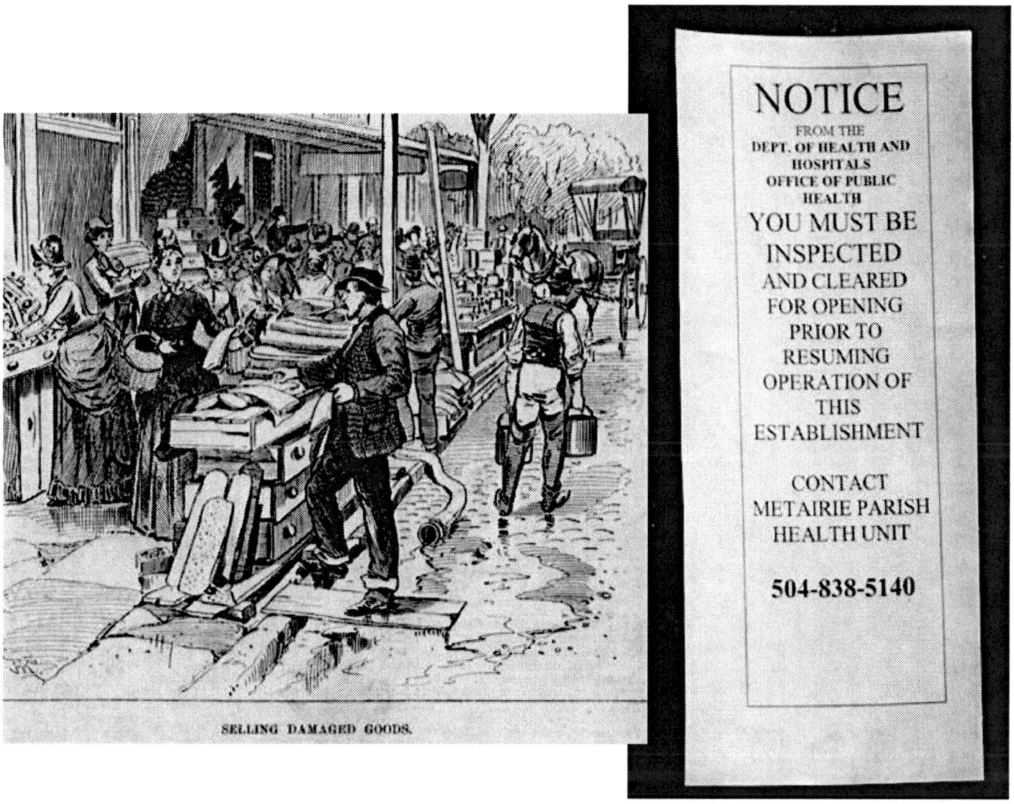

Figure 4:8. *A scene of selling flood damaged goods taken from Frank Leslie's Illustrated newspaper, for the week ending June 22, 1889, while next to it a Katrina equivalent of local restaurants having a prohibited to open sign nailed to the door.*

> cellars, and whatever or wherever there was thought to be a source of disease; a gang to gather old clothing, bedding and all matter of offensive character and burn it. This offensive material covered the whole valley to some extent, was found especially where there were piles of drift. Nearly everything that the water touched it ruined, and much that the people had undertaken to save was in a few days discarded and thrown out with the general wreckage. Each day's work averaged twenty-five wagon loads.

The streets were full of debris after Katrina; owners and landlords removed everything from the lower floors of the flooded buildings to the curb side. Some of these possessions had been flood-damaged; others were apparently unscathed and looked valuable. For example, in Figure 4:9 musical instruments and original paintings were stacked outside. Although it is impossible to know for sure, we were told that many landlords simply evicted tenants by October 1st (this being the second "first of the month" for which rent had been missed—especially if the family had been evacuated to shelters in Texas or relocated even further afield). The landlords

Figure 4:9. *What was left outside the house, from original art and musical instruments in perfect condition, to obviously flood damaged household objects, sometimes stacked as high as the roof. The bizarre and disturbing; both the mounted deer's head (from Arabi, St. Bernard Parish), and the teddy bear (from the Lower 9th Ward) are lying on the cracked cobblestone-like mud left behind by the floodwaters.* Photographs by Curtis.)

thought they had every right to reclaim their property. What is interesting is that for the month of October there appeared to be little looting/scavenging of these flood-caused discards. For example, we walked over many compact discs, some still in plastic store-wraps—the discs inside would have been fine. But we never once thought about picking something up—it was a strange feeling we liken to not wanting to disturb the dead.

> The foreman of laborers reported every evening to inspectors in charge and they in turn submitted a report to the chief medical inspector. The foreman of the disinfecting gang reported to the clerk in charge of disinfectants. All the inspectors and clerks submitted each night, a history of the day's work and called attention to such matters as necessitated orders from the Executive. The reports in connection with the large numbers

> of complaints reviewed every day were carefully read over and a plan of operations for the coming day decided upon. Orders were then made to cover every department of the work and instructions transmitted to the men in charge.

A similar process occurred in the EOC during the Katrina response. The reports would return from the various search and rescue groups, each of whom would have their own territory, and from these reports the next-day's action items would be decided.

> Hundreds of complaints came to the office every day, those needing immediate attention were at once referred to the inspector in charge of the district and an inspection ordered. Those complaints usually called our attention to some existing nuisance, a request to clean a cellar, to inspect a building, a yard, or street, while others notified us of the finding of a body or a carcass.

Figure 4:10. *Various "incentives" not to take anything. The most interesting of these was the warning not to take the caravan. Due to the universal destruction of property any mobile home suddenly became a valued possession. Also of interest are the messages from the folk artist "Dr. Bob" who told the outside world he had "3 hungry, very mean dogs," and that he would shoot looters, while on the other side of the refrigerator he uses the catch phrase he is famous for, looters "be nice" & leave.* Photographs by Curtis

> There was a great need of public water closets to accommodate the thousands that swarmed in the devastated districts. A force of carpenters was detailed for the purpose of erecting such conveniences and the work was continued until seventy-five public privies had been constructed. A special corps of men under a competent foreman was selected to disinfect these daily and to see that they were kept in proper condition. They also disinfected twenty-five closets erected by other forces. As these closets served their time and purpose they were removed, the vaults excavated and disinfected, and refilled with clean earth. At the time the state work closed there were in existence about twenty public closets.

The same was true after Katrina; many portable toilets appeared throughout the neighborhoods for returnee use, as well as to serve the needs of personnel involved in the clean-up.

> A very large proportion of the survivors sought refuge among the residents on the higher ground skirting the flooded portion of the valley. These houses became very much crowded with the wretched people who carried with them from the general ruin flood-stained articles, and there was danger that these places would soon become permeated with the effluvia, and prove a source of epidemic. After fully considering the dangers, and designing to know the condition of the people and their sanitary surroundings, it was thought best to have a canvass made of the whole district from Coopersdale to South Fork. This work was placed in the hands of the inspectors, extra inspectors being appointed to cover the ground not already included. These inspectors were instructed to proceed without delay to make a sanitary survey of their districts, reaching every house without exception; the survey to be recorded in blanks furnished for the purpose. They were to ascertain the name of the owner or tenant, number of street, number of rooms, number of families, males, females, and children, and whether there had been any typhoid fever, diphtheria, or scarlet fever in the house within six months: to report the sanitary condition of cellars, living rooms, yard, privy, stable and surroundings; to make note of all facts bearing upon the health and comfort of the people; to abate all nuisances as they found them, and to carefully inform the people how to dispose of garbage and to use disinfectants.

Figure 4:11. *Two maps resulting from similar surveys conducted after Katrina by the Rapid Population Estimate Project.* Source: New Orleans Emergency Operations Center.

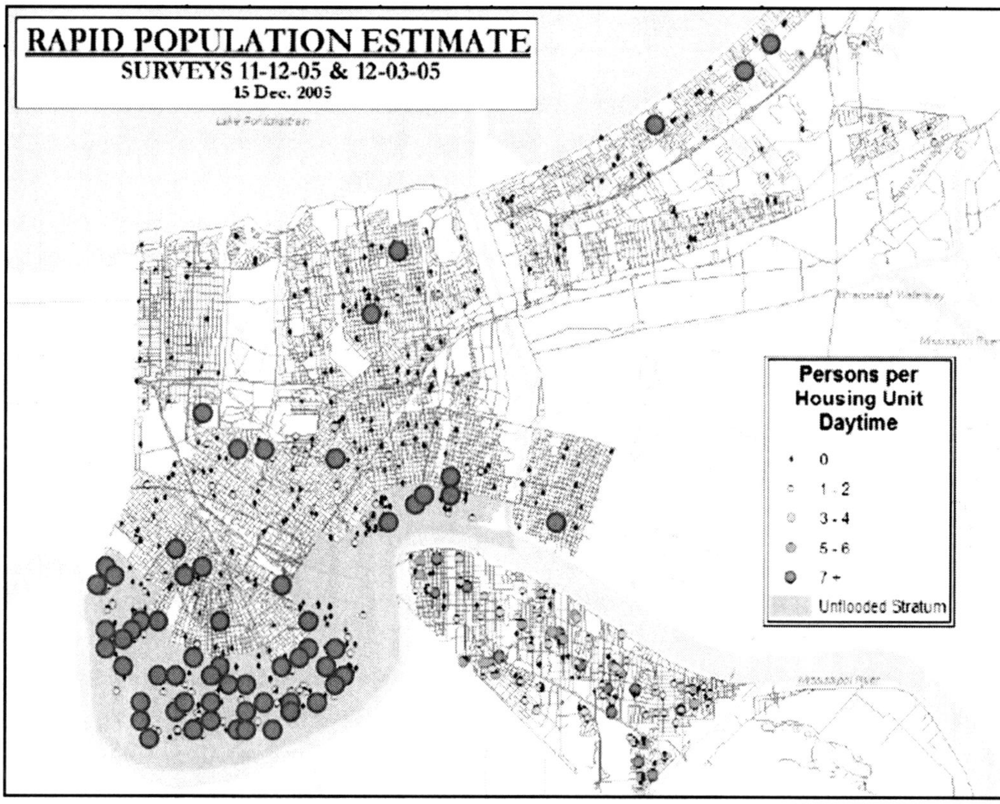
RAPID POPULATION ESTIMATE
SURVEYS 11-12-05 & 12-03-05
15 Dec. 2005
Persons per Housing Unit Daytime
0
1 - 2
3 - 4
5 - 6
7 +
Unflooded Stratum

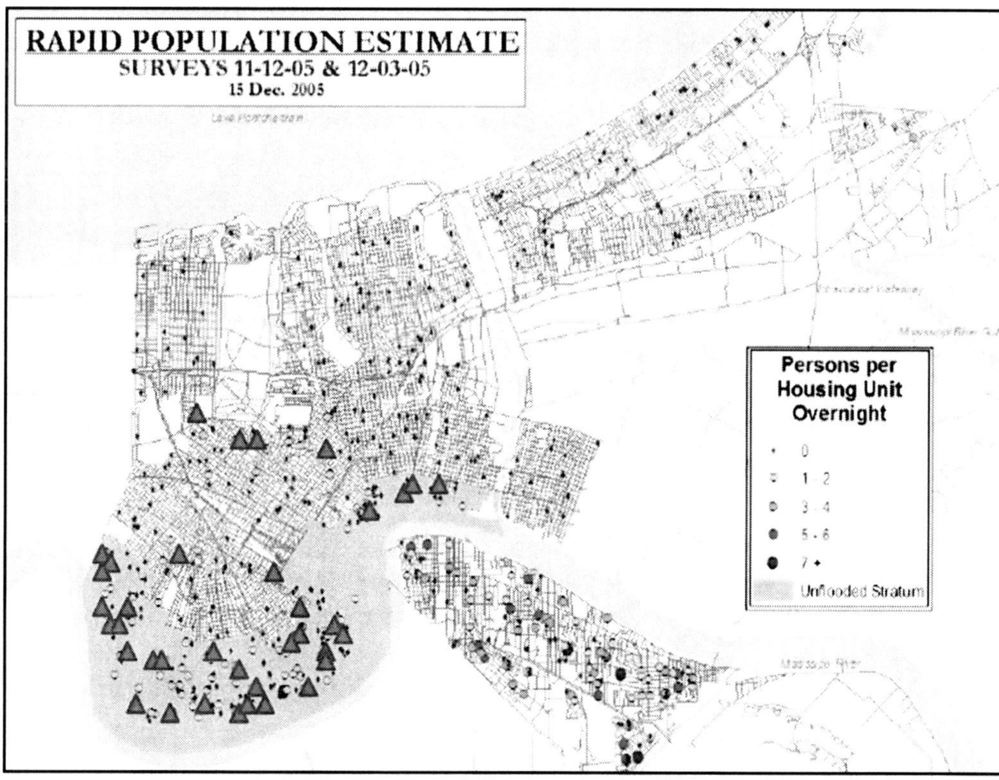
RAPID POPULATION ESTIMATE
SURVEYS 11-12-05 & 12-03-05
15 Dec. 2005
Persons per Housing Unit Overnight
0
1 - 2
3 - 4
5 - 6
7 +
Unflooded Stratum

The undertaking of such a task is as immense now after a disaster as it would have been after the Johnstown flood. However, the compilation of data into an information system (and hopefully a GIS) would make any such survey far easier to analyze today. The different attribute fields described in the previous section would have made an excellent GIS with maps being created of population density, state of the home and disease type. Sometimes, and more by luck than judgment, a researcher stumbles across these original historic surveys, either in primary document form or as a detailed report summarized after the event.

With the majority of the New Orleans population (supposedly) having evacuated, one of the biggest challenges was producing accurate estimates of returnee numbers. There was also speculation about whether Mayor Nagin could actually maintain that title as the voting population was far more than decimated (which actually means the reduction of a population by 10%). One of the only comparable situations in U.S. history was the flight of the population from Memphis during the 1878 yellow fever epidemic.

In the population survey maps (Figure 4:11), the symbology has been modified above the Mississippi River to make the patterns easier to discern in black and white. The day time returned population (probably working on their property) of more than 3 people per unit is displayed as large circles, while the equivalent night time population are drawn with triangles. The area of the map shaded to show the unflooded section of the city coincides with the highest concentration of both day and nighttime returned population. What is evident from the sections of the city, more to the north and east, is the presence of what we call "pioneers." These people are returnees who begin to rebuild even though they are isolated. None of these can stay the night because at this point utilities had not been restored. In the population report a comparison is made between these spatial patterns and the damage survey map seen in Figure 4:12.

Although Figure 4:12 does not work as well in pure black and white, it is still possible to see the concentration of severely damaged property which corresponds to the levee break areas, especially the Lower 9th Ward, though interestingly there are also many properties scattered throughout the unflooded section. These were damaged either through hurricane wind action, or possibly from some of the subsequent fires.

> The work was commenced on the 5th of June and completed in about ten days. In this canvass 2,687 houses were visited containing 13,712 rooms and a population of 18,602.
>
> A large amount of valuable information was secured and used to excellent advantage in future operations. In the territory covered by the canvassers were many houses that had been vacated, being unsafe, or

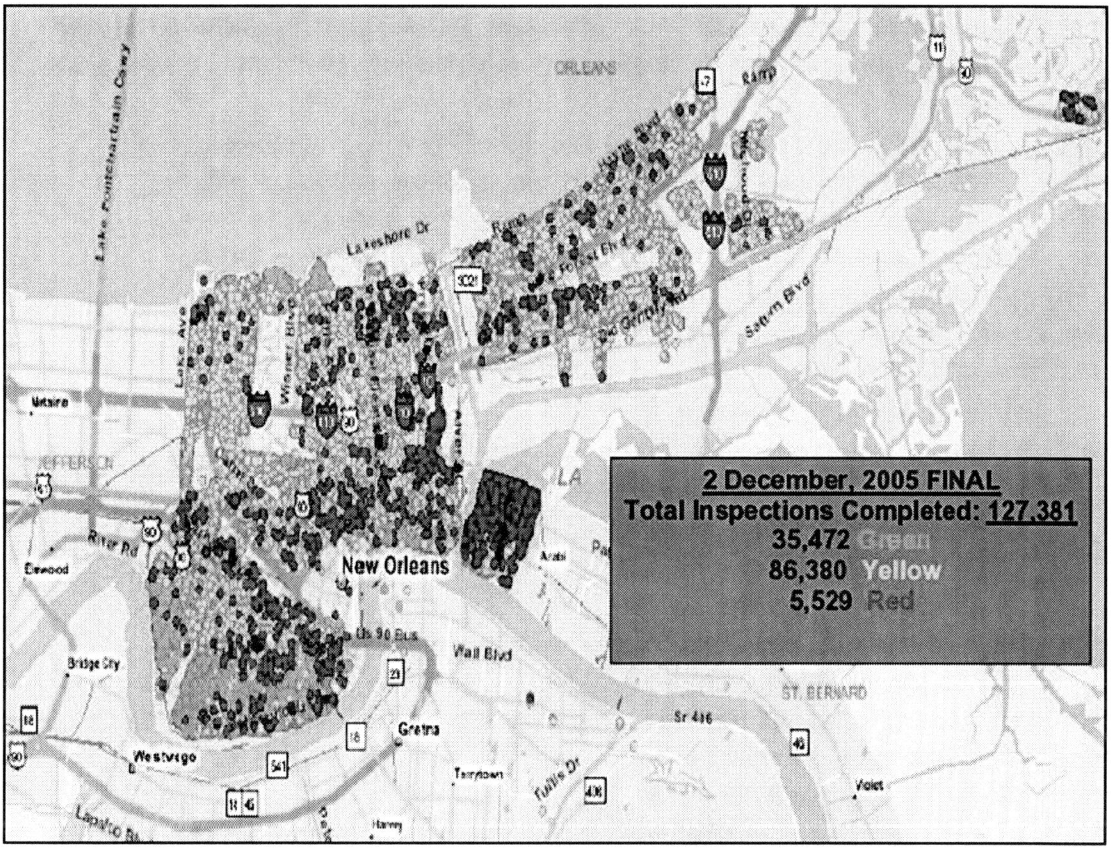

Figure 4:12 Source: New Orleans Emergency Operations Center.

filled with mud and surrounded by piles of wreckage. In making use of the survey blanks only those houses were reported where people were living.

While the average to a room is one and three-tenth, dwellings were often found that were overcrowded to a dangerous extent. The highest number found to a room was eight, but three and four to a room was quite a common thing.

The population, of the canvassed districts, before the flood was 26,326, the survey shows a decrease of 7,724. These figures do not represent the number lost in the flood, for during the ensuing two weeks a part of the population had sought shelter and food outside these limits.

A brief sketch is now given of the various districts appointed to the deputy medical inspectors for the purpose of giving a general idea of the condition of each and the amount of work necessary to put them in a habitable condition.

Figure 4:13 accompanies the following geographical description in the Pennsylvania Report of the State Board of Health for 1889. The key below goes with the map:

1. Johnstown, First, Second, Third and Fourth Wards, population 5,363
2. Kernville, Fifth and Sixth wards of Johnstown, population 3,452
3. Hornerstown, Seventh ward of Johnston, population 2,417
4. Conemaugh borough, First and Second wards, population 3,971
5. Prospect, population 810
6. Woodvale, population 1,247
7. Franklin, population 672
8. East Conemaugh, population 1,074
9. Mineral Point, population estimated 150
10. South Fork, population estimated 700
11. Cambria City, population 2,902
12. Morrellville, population 2,201
13. Midville, population 2,680
14. Mibnersvill, Second ward of Millville, population included
15. Coopersdale, population 573
16. Rosedale, population 311
17. Grubbtown, population 497
18. Moxham, population 293
19. Walnut Grove, population 375
20. Daisytown, population estimated 175
21. Niveneh, population, estimated 300
22. Sheridan, population estimated 40
23. Brownstown, population 287

> At East Conemaugh the valley is narrow and the force of the water was most terrific. The water picked up a four tracked railroad and miles of track covered with trains, freight and passengers, with shops and stone-house, scattering them over miles of territory.
>
> On July 6th after a month's incessant labor, often by night as well as day, resting only on the Lord's day, we have but just succeeded in clearing the streets of that portion of town which was not swept away, of the tangled mass of debris consisting of houses, passenger coaches, freight cars, locomotives, furniture, trees, bridges, corpses and carcasses—which encumbered them up to the very roofs of the houses, and once more converting them into thoroughfares. Fortunately, we have been able up to the present time to maintain a very fair condition of public health, but the

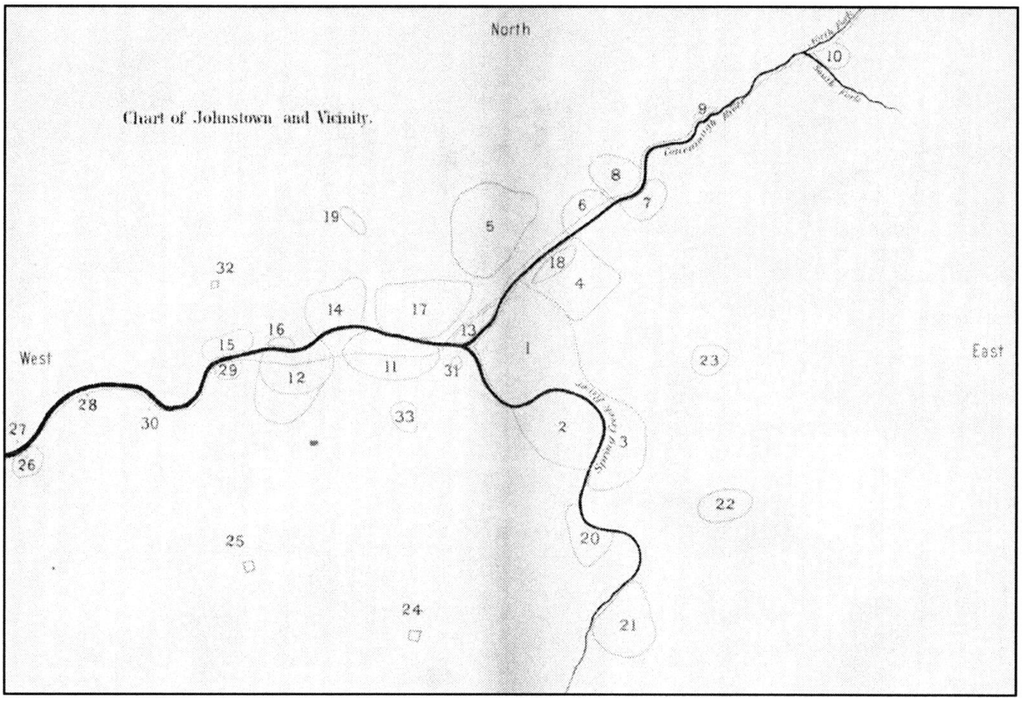

Figure 4:13

> fervid heats are yet to come and it will require unremitting care to prevent some serious outbreak of disease.

The post-Katrina landscape held several similar strange sights due to the receding floodwater; a refrigerator on a rooftop, a dumpster on a rooftop, a powerboat balanced upside down on a fence, a home that had floated into the middle of the road, while others came to rest on top of cars.

> In Woodvale borough…the population decreased from 1,347 to 345. A large proportion of the numbers constituting the decrease were drowned. A deposit of three or four feet of clean sand covered the site. Imagination can hardly conceive that this was once the most beautiful suburb of Johnstown. The people brought to bear all the influence they could to have their streets opened, cellars cleaned and the river dyked.
>
> Johnstown borough formed the center of our operations and was by far the most important field of labor. Here were piled up the buildings and the dead, not only of Johnstown, but the greater part of what had been gathered by the flood in its sweep of the valley for twelve miles. A description of this tangled mess piled from twenty to thirty feet high cannot be given. It was too great in extent to be comprehended. The population

Figure 4:14. *Illustrates the different changes to the landscape caused by the floodwaters after Katrina. In "A" a dumpster has floated and then settled on the roof of a house (St. Bernard Parish, October 26th). In "B" and "C" homes have floated from their foundations, in "B" (St. Bernard Parish, October 26th) coming to rest in the road and subsequently being cut in two by heavy machinery clearing the road, while in "C" (Lower 9th Ward, October 29th) the unusual sight of a house sitting on top of a car.* Photographs by Curtis apart from "D" by Mills of Curtis

>> **Figure 4:15.** *Different stages of Katrina's clean-up. In "A" a train filled with debris heads out of town, while "C" shows a massive debris dump temporarily located in a street median—several trucks dumping their loads into a mass that is already thirty feet high. A temporarily erected guard station sits in front of the dump (several of these guard posts were erected throughout the city). At a finer scale, "B" shows one of these truck-teams at work, notice the large cuddly-toy frog that has been fixed on top of the truck. This appears to be a Mardi Gras object salvaged from the debris. In "D" at a finer scale yet, environmental teams would walk the streets and identify with little red flags any debris of environmental concern this case small cans of chemicals and a small generator.*
Photographs by Curtis

A
B

C
D

> before the flood amounted to 5,363. Two weeks later only 645 people were to be found and these were living in 92 houses.
>
> Nearly the whole of this district was buried under the debris, and the greater part of the state work was expended here in opening miles of streets to create thoroughfares for the transportation of enormous masses of wreckage and filth, in opening and removing the huge and densely packed drift at the stone bridge, in cleaning the river of accumulations to admit a free passage of sewage material, in excavating the filth from more than twelve hundred cellars and in cleaning as many yards of the filth laden mud. Considerably more than a thousand carcasses of dead animals were extricated and destroyed and over two thousand human bodies recovered. The removal of this deposit constituted a piece of work whose magnitude cannot well be appreciated, and which was marked daily, as it progressed, by the saddening discoveries of the remains of victims of the flood. Four months of time, thousands of men and hundreds of teams were required to accomplish the task.

Bulldozers were used to clear the debris from the streets. More than once we saw a house cut into two pieces, not by the action of the hurricane, but by some heavy machinery which had obliterated the section that had drifted into the road. The task of removing this debris was immense. Apart from the trains we saw laden with debris heading out of New Orleans, there were also huge debris dumps in the median of roads. Truck after truck would drive and dump mostly wood onto mounds that easily exceeded thirty feet. A contact of ours who worked for FEMA later told me that the more toxic waste, such as refrigerators (and there were thousands of these, taped up and left by the curb with the food inside spoiled and baked in the intense heat) were taken to a dumping site in the swamps only accessible by boats—again, it was surreal the way he described this mountain of refrigerators rising from the water.

> The work of reducing the great drift at the stone bridge required nearly six week's work. Wedged down in the river bed were houses, immense logs, timbers, hundreds of miles of wire and the roots of trees, which formed almost a continuous net for some distance above the bridge, while cars and other large objects weighted down the debris from above. There were four gangs of men employed: those who with hooks pushed the small drift into the current; the axemen, who reduced the objects which were too large; the hoisters, whose engines hauled the greater obstructions upon the river bank to be burned, and the dynamiters. Seventeen thousand pounds of dynamite were used: as much as five hundred and forty

pounds being discharged in single blasts. The work was accomplished by General Hasting's force. The whole force working to exhume and clean up the ruined city numbered about six thousand men. At the cessation of our labors those whole force numbered about five hundred. During the four months, much of the larger portion of the forces was at work in Johnstown borough.

Kernville lies for one mile on the south side of Johnstown and of the Stonycreek River. Before the flood it was compactly built up with substantial dwelling houses which sheltered four thousand people. While this section of the city was outside the terrific torrent, it was traversed by the great return current which was deflected backward from the stone bridge, carrying with it an immense burden of debris from all parts of the flooded district and depositing it among the wrecked houses of Kernville. Twenty squares were under water to the depth of about nineteen feet. During the week preceding the establishment of the district officers, hundreds of laborers were employed clearing and burning the debris. One of the first acts of the office was to open a morgue. At this morgue one hundred and eighty two bodies were prepared for burial.

Large quantities of disinfectants were taken away by the people and used, and still larger quantities were distributed by the office, from house to house at regular intervals. A gang was employed in sprinkling disinfectants over the entire district. Fifteen supply stations were established, and from these they worked so that they covered the entire district twice every week. A gang was detailed to gather up all filthy clothing, bedding etc. Each day's work averaged twelve wagon loads. A number of men were kept on the dumps to disinfect them. People cleaning their own cellars and buildings deposited the filth on the street and alleys and it became necessary to remove these deposits and to clean out exceptionally filthy cellars. This work became a very important part of our operations.

Upwards of fifty dead animals were destroyed, fifteen stores, meat markets and slaughter-houses cleaned, disinfected and rendered sanitary. Disinfectants were scattered regularly, and several squares cleaned that had not been touched by the contractors of the state.

Cambria city lies below Johnstown on the Conemaugh River and suffered greatly in loss of life and property. About one fourth was almost completely wrecked and buried under debris. The population before the flood was 2,902 and the decrease amounted to 689.

The United States Engineer Corps from Willet's Point, New York harbor, arrived here on the 8th. They erected five temporary bridges and two pontoons. This gave us means of transportation to the districts and greatly

> aided us in the distribution of disinfectants and in making daily inspections. As the streets were opened and means of communication improved, the force was reduced, the districts enlarged and inspectors discontinued. The sanitary forces which had directed their efforts to recovering dead bodies, destroying animal and vegetable matter, and building and disinfecting public water closets, now gave their attention to the condition of the cellars, yards, alleys and streets. Hundreds of people were living over cellars and rooms filled with every imaginable kind of filth and were unable to clean them. Hundreds of cellars where buildings had been, now open and filled with a filth deposit lined the streets—their owners absent, many of them never to return. In fact, very few property owners had sufficient available resources to pay for any work. To protect the public health it became necessary to condemn hundreds of cellars and have them cleaned.

A similar situation occurred in New Orleans—though here many of the houses were condemned and then left abandoned, some still remaining that way three years after the storm.

Dr Geo W Wagoner, Deputy Medical Inspector for the Kernville District provided a more detailed description of the events in his district.

> Before the flood, it was compactly built up with substantial dwelling houses which sheltered 4000 prosperous and happy people.

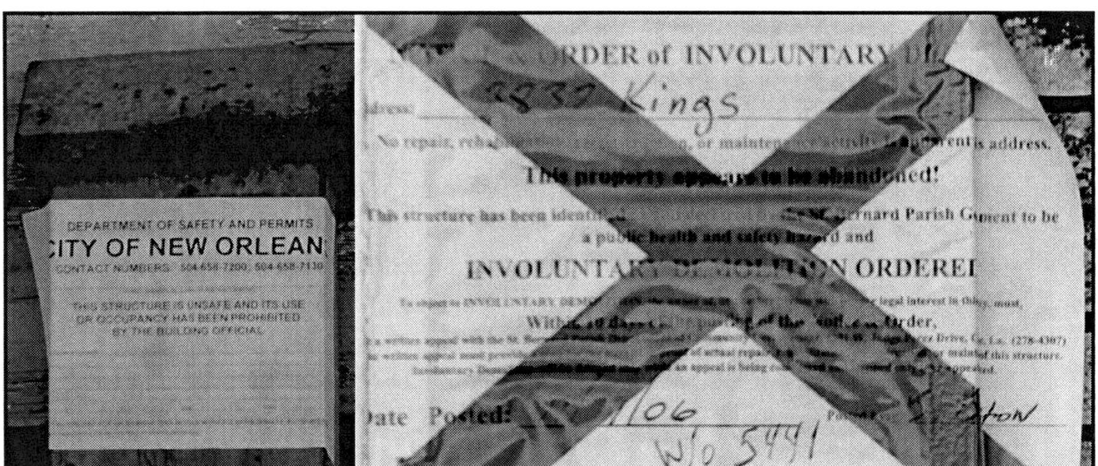

Figure 4:16. *Two notices condemning different buildings, one from New Orleans on October 14th, 2005, the other in St. Bernard Parish, the photograph taken on December 23rd 2006.* Photographs by Curtis

> While this section of the valley was outside the terrific current that swept down the Conemaugh River, it was traversed by the great return current which was deflected backward by the Pennsylvania railroad stone bridge. This return current carried with it an immense burden of debris from all parts of the flooded district and deposited it among the wrecked houses of Kernville. Twenty squares of the south side were under water. On Morris Street, at Joseph Morgan Jr's residence, the water was 19 ft 3in deep. After the waters subsided this extensive district was covered with a compact mass of debris, under which lay scores of dead, in the slum and filth that fell from the burdened waters.
>
> Many bodies were taken out of the wreck and prepared for burial at the corner of South and Napoleon streets. No morgue had been established, and these last sad duties were performed in the open air and in full view of the excited people. One of the first acts of the office was the opening of a morgue in a damaged house, where the dead were cared for in a decent and orderly manner away from the gaze of the curious. One hundred and eighty two bodies were prepared for burial in this district.

The media was full of stories about how the dead of New Orleans were treated. Many were wrapped in sheets and left by the side of the road. There are even photographs showing children playing besides the corpses. This traumatic sequence would also remerge in the art the children would create as seen in Figure 8:1.

> The force of laborers assigned to me traveled back and forth over the wrecks, seeking out the dead animals and burning them, and scattering disinfectants over all the masses of decomposing organic matter. These gentlemen—being volunteers—were soon compelled to relinquish their duties, but in leaving they took with them the respect and gratitude of all who were benefited by their labors.

This became a problem in New Orleans after Katrina. Although there were a large number of volunteers who came for the initial rescue operations, many would soon have to leave because of other responsibilities, such as a regular job. This caused a serious problem for animal rescue who were left without the necessary labor to feed all the abandoned pets. The following is an email I received on 01/21/2006 .

> THE STARVING ANIMALS OF HURRICANE KATRINA NEED YOU NOW
> Today in New Orleans, we had ONLY FIVE PEOPLE to feed the

THOUSANDS of
Katrina animals who remain on the streets. These animals are trying to stay alive long enough for their people to return or to be rescued.
Animal Rescue New Orleans has mapped OVER
2,800 Food/Water locations that need to be refilled!!
MOST Food/Water stations are not refilled for over a WEEK because we don't
have enough people to feed the animals.
It is desperate down there.
The animals are starving again.
Two more emaciated dogs came in yesterday ...
PLEASE GO FEED THE STARVING ANIMALS.
If you can't go yourself, then help spread the word that we need more volunteers.
Crosspost far and wide.
THIS IS URGENT.
If you go down to help support ARNO's Food/Water program, don't let ANYTHING or ANYONE stand in your way. Please don't stay at the shelter—
GO OUT AND FEED THE ANIMALS THAT HAVE NOT YET BEEN RESCUED. The animals that are NOT in the shelter need you MORE than the animals in the shelter.
The shelter animals are fed every day—but the thousands of animals on the street may eat ONLY ONCE A WEEK!!
PLEASE GO FEED THE STARVING ANIMALS.
They need you. The animals need you to go work hard to get them fed—to work a 12-14 hour day—to refill over 100 Food/Water stations a day.
It's the only way we can keep them alive.
We used to keep them well-fed.
Now we are struggling just to keep them ALIVE.
PLEASE GO FEED THE STARVING ANIMALS that are living under homes ...
For more information on how to help the thousands of Katrina animals, please visit the Animal Rescue New Orleans website:
THANK YOU for anything you can do to help the innocent victims of Katrina.

A very large proportion of the survivors sought refuge among the residents on the high ground above Napoleon and South streets. These houses became very much crowded, and as all of the wretched people brought with them flood-stained articles from the general ruin, these places soon became permeated with the effluvia of the soiled goods. Canvassers were sent to each house to ascertain the number of people occupying it, the condition of their health, the condition of the house and its surroundings, and all facts bearing upon the health and comfort of the people. Directions were given in regard to the disposal of garbage, cleanliness about the house and yards, and all were invited to call at the office for disinfectants, where instructions were given as to their use. Fifteen supply stations were established, and from these they worked so that they covered the entire district twice each week.

I had the occasion to protest against the contractors depositing the mud and filth taken from the streets, on the bank of the river, to narrow its channel, obstruct the flow of the water and clog up the outlets of the sewers. I advised that this material be deposited on a low, marshy place near the hill where the river had little or no bank, and where it would be useful in protecting the district from being flooded in times of high water. My efforts in this direction, during the time the greatest quantity of mud was removed, were without avail although repeated several times.

The people soon commenced to clean out their cellars and to deposit the filth on the streets and alleys. At first it did not seem to be the policy of the Board to remove these masses but to leave them for the contractors. When the contractor's men had worked over a place they would not return to clean up any new deposits, and as the people cleaned their cellars when they saw fit, it was not long before the streets had numerous piles of the most offensive mud scattered over them. It was my judgment that the Board should not have waited for the impoverished people to clean out their own cellars, but having an anxious care for the people's health under such calamitous circumstances, should have undertaken this important work at once. It was not long, however until our force was allowed to remove the filth when deposited on the streets, and in some cases to clean out exceptionally filthy cellars.

At this time, although germ theory had displaced miasma as a source of explaining how disease spread[4], there was still a lingering concern about the health

4 Prior to germ theory, disease contagion was associated with the environmental decay of matter and the noxious gasses it produced. It made sense at the time to stay clear of the dirtiest and poorest areas of the city, as here in the filth the "bad airs" could lead to disease. Also, close to marshes, the smell of gases and decay could lead to sickness. Of course even today disease and poverty are often spatially associated, and the marshes were also home to mosquitoes—the landscape attachment to disease was therefore correct, just for the wrong reasons.

dangers posed by the deposit of so much polluted mud. Even so, I refer the reader back to the earlier comment I made about the general concern of the toxic mud left after Katrina.

The Board of health Report also contains several letters, communications and informational notes made by Benjamin Lee, MD., Secretary of the State Board of Health. Several of these again had close comparisons with Katrina. This no more poignant than in the selection of temporary housing and where it could be placed.

> 19—Decision with Regard to Portable Houses June 20th by Benjamin Lee, Secretary to Adjutant General Hastings
>
> Dear General: I have this day examined two styles of portable houses which are being put up in this place. The smaller one is much too small, flimsy in material and construction, and evidently only intended as a hunter's summer shelter. I should condemn it as a habitation for a family. The other is more substantial; has a solid floor and a shingle roof, and would stand the weather. Its dimensions are 16X24 feet. A small family could find accommodation in one of these, but families of any considerable size would require two of them in communication, in order to avoid dangerous over-crowding.
>
> The State Board of Health earnestly advises that all women and children in Johnstown and neighboring boroughs, and especially those who are living in overcrowded houses, or in houses which are damp and wet, or which are located on very low ground, should leave the place for a few weeks if they can possibly do so. All those who have lost their homes would be much more comfortable in tents than in overcrowded houses. Tent life is both healthful and pleasant.
>
> The board therefore recommends that camps be established on the hillsides in dry locations and near good water supplies. Families could live comfortably all summer, and they would opportunity to collect new home supplies to furnish their houses when renovated or rebuilt.

At a later point more specific geographic detail was provided as to where safe temporary housing could be located.

> The areas in Kernville on which the Board considers it unsafe to place portable houses have been revised, and are now as follows:

- The square on the east side of Morris Street, between Haynes and Dibert streets

- The lots fronting on the east side of Morris Street, from Dibert Street to the residence of Joseph Kuntz Esq.
- The portion of the square on the west side of Morris street from the alley along lower end of Swank's Pottery down to Dibert Street, including lots fronting on Morris, Dibert, and Napoleon streets.
- The squares on the west side of Morris Street to the hill, and from Dibert Street to Haynes Street.
- The square from Haynes street to the river and from Napoleon street to the hill.
- The lots fronting on Haynes Street between Morris and Napoleon streets and the residence of Joseph Mogan.
- All other parts of the South Side are now open for habitation.
- The restriction is removed from the entire Seventh ward.

In today's world these details would have been mapped out using a GIS. Indeed one of the tasks in the EOC, and then later during the weeks that followed, was the

Figure 4:17. *Two examples of trailers, one flamboyantly decorated as the owner awaits his new Make it Right home in the Lower 9th Ward (August 2008), while the second from the NCG spatial video browser displays a typical street scene of two trailers squeezed into any available gap around the home (January 2008).* Photographs by Curtis

identification of open land owned by the State or City on which the white FEMA trailers could be erected. These would end up on three general land types—large expanses containing hundreds of trailers—the most famous of these in Baton Rouge, Renaissance Village (which provides the backdrop for the chapter on children's art therapy), on smaller lots, especially parking lots where four to six trailers could be squeezed and in the front or side of a home, wooden steps leading up to the front door, a long white pipe bringing utilities inside, and sometimes a roof and skirting added to give the impression of a complete home. Although these trailers were necessary as temporary residences, they would soon become health hazards with the potential to create a negative health legacy.

In an interesting additional comment regarding how people should rebuild to avert similar disasters, Lee advocates for families to move away from the flood plains. In New Orleans, the same strategy involved FEMA advocating rebuilding at specific heights to also avoid future flooding. For many home owners this meant raising a structure fifteen or more feet. After Katrina you would sometimes see a single home being raised high above their neighbors. One contractor explained how this was done, either through burrowing beneath the home and using multiple jacks, or sliding a tongue from a truck beneath the home and then lifting. Obviously, the home owner needs funds to be able to raise a home, though grants are available. There is another reason why a home owner might be reluctant to do this though. The act of being the only home raised above your neighbors, especially in the more modern developments where the entire neighborhood is constructed in a similar fashion, may deter the act.

Letter to James McMillan Esq., Chairman of the Finance Committee from Benjamin Lee, July 26, 1889.

> My own judgment, after a careful survey of the whole flooded region from South Fork dam to Morrelville, is that homes should henceforth be placed upon the sides and summits of the hills, and the river bottoms be reserved for business and manufacturing purposes. A system of street railroads and inclined planes will make this perfectly feasible, and the time occupied in going from home to place of work or business, need not be half as long as that similarly spent by the citizens of New York, Philadelphia, or any other large city. I cannot express myself too strongly in urging the good people of Johnstown, whose welfare is so near my heart, to pause before erecting residences again in these exposed locations.

A final sequence of communications between Benjamin Lee and the President's office provides another parallel to Katrina. The political arguments between the

Figure 4:18. *A house being raised on metal jacks.* Photographs by Curtis

three levels of government regarding who was to blame have been well documented[5]. President Bush, Governor Blanco and Mayor Nagin of New Orleans, and the levee board have all been criticized—each being ridiculed by home owner's signs erected throughout Orleans and St Bernard Parish. Although not as vitriolic, one can still sense the frustration between Lee and the Federal Government as documented in the Board of Health Report

Letter to President Harrison from Benjamin Lee, July 27, 1889.

> Dear Sir: The extreme urgency of official duties in connection with the disaster which has befallen this town, has prevented my earlier acknowledgement of your kindly expressed interest in my efforts to prevent an epidemic, as a result of the flood. Your suggestion of temporary deportation of women and children was most timely, and was acted upon as far

5 Two of the more popular books to document the events of the disaster, including the political dimensions are *Breach of Faith: Hurricane Katrina and the Near Death of a Great American City* by Jed Horne, and *The Great Deluge: Hurricane Katrina, New Orleans, and the Mississippi Gulf Coast* by Douglas Brinkley

as possible. I take the liberty of enclosing my health bulletins up to the present time, which will show you that the efforts of our Board have not been unsuccessful. A problem now confronts me of grave importance, which, owing to the provisions of the State constitution, I am powerless to meet. The Conemaugh and Stony creek rivers have been so filled with sand and debris, and at the same time, their banks have been destroyed to such an extent, that the whole plateau on which Woodvale and Johnstown stood, is in danger of inundation from very moderate rainfall. The temporary structures which the homeless survivors are erecting may at any moment be washed away. Congress will be applied to, and I presume successfully, to dredge and dyke the streams. But in the meantime, immediate protection is needed. It occurred to me that you might perhaps see your way clear to assign me a portion of the contingent fund placed at your disposal, to be used for this purpose. This would enable me to tide over the danger, until such a time as more permanent improvements can be made.

The response from E.W. Halford, July 30:

The fund to which you allude is expended under the suggestion of the Surgeon General, Dr. Hamilton, to whom your letter will be referred. The President, however, is of the opinion that no funds would be available under that special appropriation, for the purpose you indicate.

And then from Supervising Surgeon General J.N. Hamilton on August 2nd, 1889.

Sir: In regard to the placing of a portion of the epidemic appropriation at your disposal, and in reply, to invite your attention to the enclosed circular.

At the same time, I have to request that you will kindly specify the particular diseases from which an epidemic is threatened, and the kind of supplies that will be required, in case of a request being made by the Governor.

The circular includes:

The President of the United States is herby authorized, in case of a threatened or actual epidemic, to use a sum not exceeding one hundred thousand dollars, out of any money in the treasury not otherwise appropriated, in aid of State and local boards, or otherwise in his discretion, in preventing and suppressing the spread of same.

Benjamin Lee's response on August 5th

> In reply to your inquiry as to what particular diseases I fear an epidemic of, in the case of a portion of the site of Johnstown being again under water, I would say typhoid fever, malarial fevers of a low grade and that form of pneumonia which is usually called typhoid pneumonia. Typhoid fever is already manifesting itself in the valley, and I am confident that it would commit serious ravages should even a very moderate flood again submerge these towns.

And then from Supervising Surgeon General J.N. Hamilton on August 13th,1889.

> Sir: Referring to the matter of the epidemic fund and of its use in dyking the Conemaugh, I have to say that it is not contemplated that the appropriation should be used for such purposes.

FIVE

Wildfires in the Wildland–Urban Interface: The Case of San Diego County, California

THE PREVIOUS CHAPTERS in this book have introduced the reader to the concept of the spatial nature of disasters, and how we use different geospatial technologies, especially GIS, to approach disaster research. We have also used descriptive sources to capture different stages of a disaster, from the actual event through stabilization and then clean-up. This chapter combines elements of all of the four first chapters; understanding the human geography and spatial aspects, using GIS approaches and disaster cartography, and drawing on narratives (from newspaper accounts) to present a growing hazard in Southern California—wildfires.

Southern California has a long history of wildfire occurrence, however it has not been until relatively recently that these events have impacted humans and therefore been described as disasters[1]. Although wildfires can occur naturally, for example through lightning, many are caused by human action. As cities sprawl, activity in the wildland-urban interface (WUI) grows, making it more likely that 1) human action will cause a wildfire and 2) that humans will be impacted by a wildfire, regardless of the ignition source. Look at the series of maps in Figure 5:1. These show the fire burn history of San Diego County, California from 1910 to 2007. Then look at Figure 5:2, which displays wildfire severity zones along with addresses (both residential and commercial) from SanGIS.

Q: What spatial patterns do you see? What is the spatial relationship between population density and potential wildfires? What are the implications of these patterns?

DATA SOURCES

San Diego County, California is an especially good place to study the human geography of wildfires with the aid of a GIS. This area is home to several excellent sources of geospatial data useful for wildfire investigation. The Department of

1 Cook, Steve. 1997. Wildfire adapted ecosystems meet man's development. *Australian Journal of Emergency Management* Winter: 24–30.

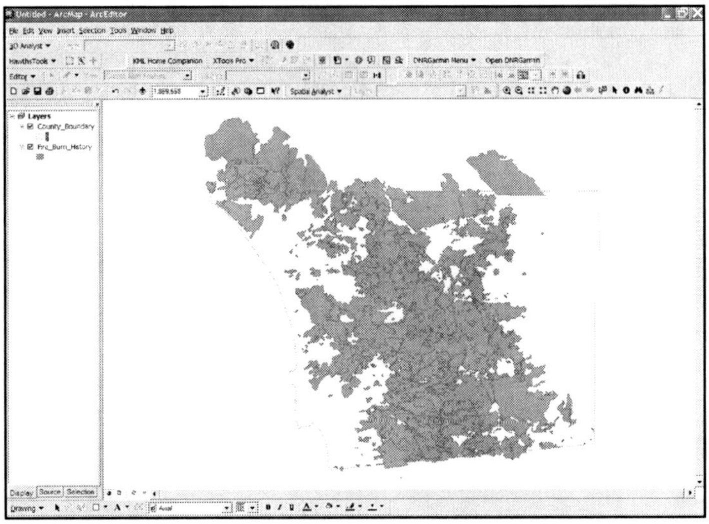

Figure 5:1. *Fire burn history from 1910–2007 in San Diego County.*

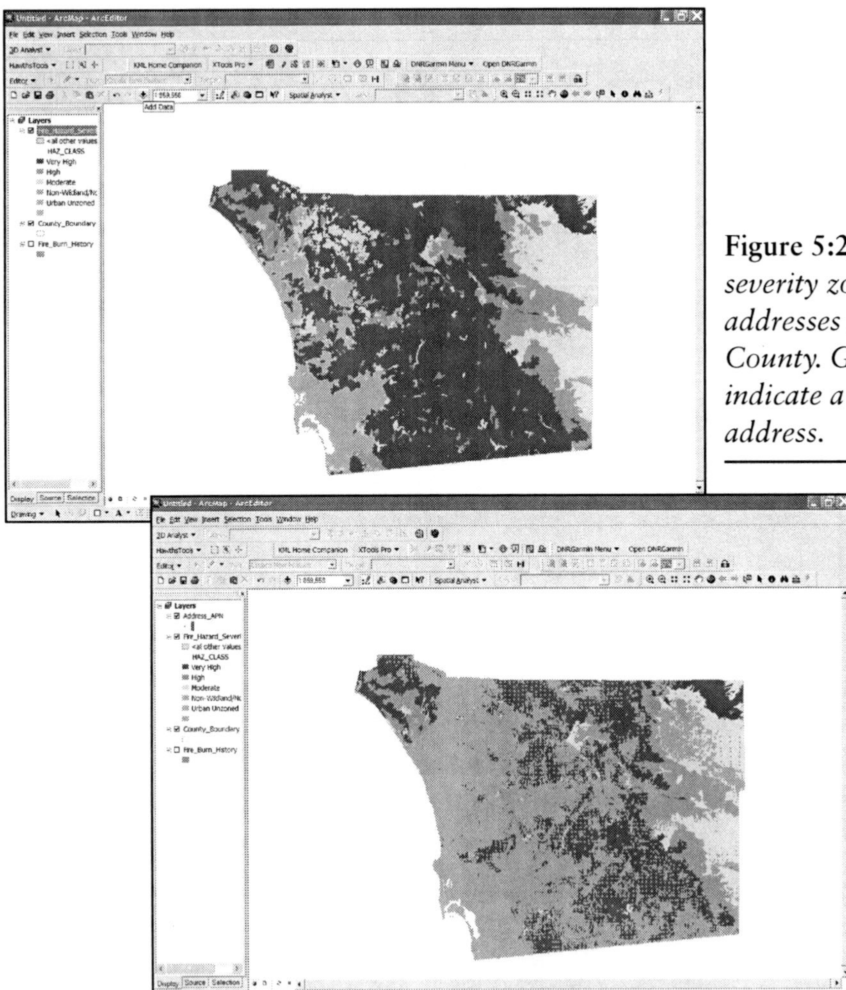

Figure 5:2. *Wildfire severity zones and addresses for San Diego County. Green points indicate a registered address.*

Geography at San Diego State University (SDSU) provides data online through an interactive map (http://map.sdsu.edu/FIRENET/) Also, SanGIS is one of the most comprehensive and user-friendly sources of free online GIS data (www.sangis.org). Finally, the San Diego Union-Tribune newspaper is an active user of Google Maps to display fire data (http://www.signonsandiego.com/firemap). In addition, they maintain an archive of news stories from the 2007 firestorm. In this chapter, we will draw from all of these sources to investigate aspects of human geography in wildfires.

BACKGROUND TO THE ISSUE

Though the focus of this chapter is human aspects of wildfire in the WUI, it is helpful to provide a baseline upon which we begin this discussion. This section will cover background on the basic issues of wildfire, specifically the fire triangle and policies of fire suppression. Then we will turn to the geography of Southern California, particularly, climate and vegetation, the Santa Ana winds, and then to the WUI.

Fire Triangle

Fire is a natural process and has been around well before humans began transforming the earth. It is a physical process that requires three elements: fuel (something to burn), heat, and oxygen (for combustion), as well as an ignition source. These elements are commonly represented in "the fire triangle." All three of these factors must be present for fire to exist; the absence of any one results in an inability for fire to occur. Fuel can be vegetation (plants, trees), or it can be human-created materials (such as houses, furniture, cars). Heat primes the fuel for ignition, while oxygen enables combustion. Finally, ignition sources can be either natural, such as lightning, or can be human-caused by either intentional (arson) or unintentional means.

Though humans do not affect the presence of oxygen, they are involved in the fuel load through what they plant, how they landscape, how wildlands are managed, how houses are built and with what materials. Also, through campfires, discarded cigarettes, and arson, humans contribute to ignition sources. These examples represent some of the ways in which people can impact the fire hazardscape[2].

Global warming has also been identified as potentially leading to more fires and fires with greater intensity. In particular, shifting precipitation patterns that

2 We will introduce this term more fully in the next chapter—but for here it can be defined as the combination of both biophysical and social vulnerability—in other words the combination of where someone lives with the social aspects that vary their exposure risk at that location.

have been predicted indicate changing patterns of drought. Drought in areas with a build-up of dead vegetation provides an ideal situation for conflagration. In Southern California this situation is disconcerting for the additional reason that the region constantly faces dwindling water supplies needed to support its growing population[3].

Policies of Fire Suppression

In addition to the physical components of fire, in order to understand the geography of this hazard in the United States, its political context must also be mentioned. Though fire is a natural process, the U.S. has a long history of fire suppression (quash all fires as soon as they start). This approach to management allows vegetation (fuel) to build up and as a result larger fires erupt rather than the smaller fires that would have acted as "cleaners" for the environment. Occasionally, prescribed burns (also known as controlled burns) have been used to reduce fuel load, but overall this approach has not been as widely practiced as fire suppression. However, fire suppression is known to have a negative impact on the Southern California environment[4]. Through our policies of managing fire, we have actually done more harm than good. As a result we face the potential for larger and more devastating fires. This situation is true across the United States, but due to the intersection of physical and human geography in Southern California, this region is particularly vulnerable to wildfire disasters.

Southern California Climate and Vegetation

Although wildfires are a problem in several regions throughout the United States, the combination of climate and vegetation make Southern California an amenable environment for this hazard. Based on the widely accepted Koppen Climate Classification System, the Mediterranean climate type (found in Southern California) is characterized by:

- Warmest month above 50°F; coldest month above 32° F but below 64.4° F
- Pronounced summer drought with 70% of precipitation in winter
- Hot summer with warmest month above 71.6°F

---

3 For readers interested in this topic we recommend the "classic" text Cadillac Desert: The American West and Its Disappearing Water by Marc Reisner.

4 As Goudie (2000) describes, "...at lower elevations the Mediterranean semi-arid shrubland called 'chaparral' has had its character changed. The vegetation has increased in density, the amount of combustible fuels has risen, fire-intolerant species have encroached and vegetation diversity has decreased, resulting in a monotony of old-age stands, instead of a mosaic of different successional stages." (45) Goudie, A. 2004. The Human Impact on the Natural Environment. Oxford: Blackwell.

What do these characteristics mean for the likelihood for wildfire occurrence? Let's think about this in terms of the fire triangle. With most precipitation in the winter, (which is also a mild winter), vegetation can grow lush from this point through the spring season. However, once summer arrives, warm temperatures combined with drought mean that this lush vegetation will dry resulting in an abundance of fuel. Furthermore, in humid subtropical climates, for example in the southeastern part of the country, moisture works with heat to facilitate decomposition of vegetation with the assistance of microbes, however the Mediterranean climate in Southern California does not provide such a rapid decomposition process. The result is that high fuel loads build and create a ready environment for wildfire potential.

However, not all vegetation poses equal potential for use as fuel. Southern California is composed of many types of flora, from grasses to shrubs to trees, some of which are natural and others that have been introduced by humans. Perhaps the most discussed native vegetation, especially in regard to wildfires, is chaparral. Chaparral vegetation is comprised of a variety of scrub-like trees and bushes that are adapted to the lack of water. In Southern California they have a long history (well before human occupation) of feeding wildfires. This is a natural process that will continue, even with humans living in the midst of this environment[5]. Due to this association with wildfire, chaparral carries a negative connotation in the minds of many. However, one must consider if it is the chaparral that is the problem or human management? Add the occurrence of Santa Ana winds to this cyclical pattern of precipitation, vegetation growth, and drought and a perfect storm for wildfires is created.

Santa Ana Winds

First, let's review some basic aspects of physical geography related to wind. Winds move from areas of high pressure to areas of low pressure. In the fall, coastal counties in Southern California tend to be warmer than their neighbor to the east, the Mojave Desert. This differential causes winds to blow from the desert towards the coast. They speed through the canyons in this process, sometimes reaching tropical storm or hurricane force, which can cause damage on their own without incidence of wildfires. Indeed, the San Diego Union Tribune article (from October 23, 2007) entitled "Santa Ana's Wreaking Havoc" provides the following description:

> Residents in many parts of the county talked of lying in bed late Sunday night and early yesterday morning, unable to sleep because of screen doors banging, windows shaking and tree limbs cracking.

5 Keeley and Fotheringham, 2006. *Wildfire—A Century of Failed Forest Policy*. Island Press: Washington, D.C.

However, in the context of this chapter, these winds 1) further dry out the vegetation (by warming the air) which creates a ready fuel load; 2) rapidly amplify any spark into flames; and 3) hinder response to the fires that develop because they create dangerous conditions for planes that are needed to drop water on the burn areas[6].

As we have discussed, there is certainly a physical geography to the occurrence of wildfires in Southern California, but there are also human aspects of interaction with the physical environment that lead to a specific geography of wildfire impact. The wind, vegetation, and topography in the WUI of San Diego County, for example, push fires through the canyons and up the hillsides. Homes built at the top of these hillsides offer spectacular views, but are also highly exposed to wildfire impact. In addition to being located at the edge of a hillside (top of a canyon), homes that are on south-facing slopes will face risk due to more sun exposure leading to more warmth and drier vegetation. Fires move up the sides of canyons due to local winds; because the flames pre-heat the vegetation above, they cause additional drying, in effect driving the fire uphill. As a result, some of the most expensive homes, with the most breathtaking views, are the ones at the highest risk of being impacted by wildfire.

Now that we have reviewed the basic physical components of wildfire and their physical geography in the region, let's turn to the human interaction with this environment.

Wildland-Urban Interface (WUI)

California is plagued with development at the wildland-urban interface (WUI)[7]. The WUI is defined in a variety of ways, but for the purpose of this chapter it will suffice to say that this is an environment which retains many of its natural characteristics with some human settlement intermixed or abutting this area. The creation of this area has not occurred blindly. Though some residents do not understand their risk to wildfire, others do acknowledge their choice to live in a naturally fire-prone place. The Union-Tribune published a story on October 31, 2007 entitled, "Weathering the storm" which provides the following excerpt:

> "This fire will help parts of the land up here," said Frank Mendenhall, part of the well-known Mendenhall family that homesteaded in the Palomar Mountain and Lake Henshaw area in 1869. "But it's hard to

6 For additional description of the Santa Ana winds and their role in wildfires listen to "Fabled Santa Ana Winds Fuel Wildfires in California" from National Public Radio's Morning Edition (http://www.npr.org/templates/story/story.php?storyID=15584420).

7 Indeed, Radeloff et al. (2005) count 5.1 million housing units in the WUI within California (799). Radeloff, V. C., R. B. Hammer, S. I. Stewart, J. S. Fried, S. S. Holcomb, and J. F. McKeefry. 2005. The wildland-urban interface in the United States. *Ecological Applications* 15(3): 799-805.

> say how many animals die, and how many are helped by new growth. That's the question. And then there's the new growth. A majority of it gets replaced by Ceanothus, which grows so high and thick it forms a barrier for everything, even animals."
>
> Mendenhall knows these fires are nothing new.
>
> "A hundred years ago they did the same thing, burned from the mountains to the beach," he said.

But 100 years ago human settlements were not in the fire's path. The backcountry has shrunk as the population has exploded out of the city of San Diego and other coastal towns in the county. As a result, what was once a fire process mostly in the domain of the natural environment has now come to be as much about human geography. This is particularly the case in San Diego County where the population has grown into what was once wildlands, and which still retains much of its wildland characteristics. Indeed, it is the "naturalness" that draws so many people to want to live in this beautiful, but dangerous environment.

In order to better understand the human geography of wildfire disasters, in this chapter we will use the San Diego Union Tribune newspaper to identify themes of importance to the public. Excerpts from stories published in October 2007 through November 1 (during the firestorm) will be used to bring human aspects into our understanding of wildfires.

THE 2007 FIRESTORM: ISSUES RAISED IN THE SAN DIEGO UNION-TRIBUNE

The San Diego Union-Tribune has published many stories on a variety of aspects regarding the 2007 firestorm. This section provides excerpts from these stories in order to gain an understanding of the issues surrounding this disaster. The following themes emerge: 1) spatial patterns of damage, 2) evacuation and animals, 3) health, 4) mitigation, and 5) recovery.

SPATIAL PATTERN OF DAMAGE...SPATIAL PATTERN OF RECOVERY

The 2007 wildfires affected a large portion of the population of San Diego County, not only those living in or proximate to the burn areas, but also those who volunteered, those who knew people at risk, and many who were (maybe unknowingly) breathing in particulate matter that could intensify existing health problems such as asthma. Figure 5:3 shows patients with asthma problems reporting to a small clinic close to one of the burn areas. Although there is not enough data here to perform any rigorous analysis, an increase in asthma cases coincides with the period of the Witch fire of 2007.

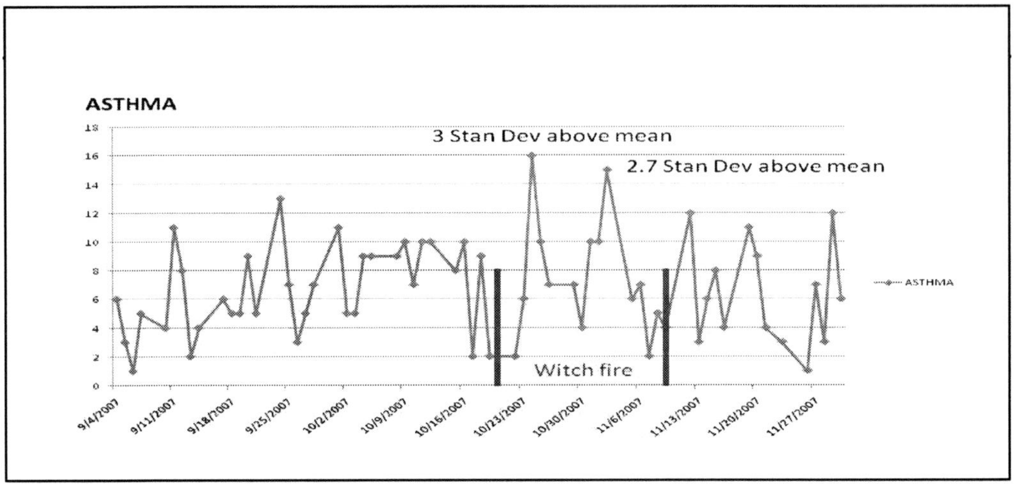

Figure 5:3. *Numbers of people suffering from Asthma attending a health clinic proximate to the Witch Creek Fire.*

Though this section focuses on direct impact from fires, be aware that other spatial patterns of impact also occur. The stories used provide a description of the seemingly random geography of wildfire damage. The following excerpts convey this fine scale pattern of impact; it is a geography that varies within a neighborhood and along streets. We will begin with October 23, 2007 and then select samples through to November 1, 2007.

Spatial Pattern of Damage

> The fire destroyed many homes on Aguamiel Road near West Rancho Bernardo Drive, but it also spared some—in certain cases scorching one house and leaving the one next door practically untouched. (October 23, 2007: Crews hope for break in Witch Creek blaze)

> "It's pretty smoky inside our house," said Michael Marlow, 62, a retired attorney.
>
> "It's bad," said Sharon Marlow, 67.
>
> It could have been worse. Flames leveled a neighbor's home and scorched the Marlows' yard, but did not touch the family's five-bedroom house. Their son David Marlow, 32, is studying to become a minister. The scene outside his parents' home brought him to a grateful and, given his training, obvious conclusion: "I'm inclined to say that's a miracle." (October 25, 2007: House is smoky, but still standing: It's 'a miracle.')

> In the Trails section of Ranch Bernardo, the randomness of flames from Monday could be seen in the large, estate-style homes that were destroyed.

One home on Bernardo Trails Drive had basically disintegrated, while a red Jeep with its top down in the driveway looked like it could be driven away.

Over at the neighboring subdivision of Montelena, jack-o'-lanterns made of paper and ghosts made of white sheets hung in the front yard of a home on Locksley Street.

Mark Huettinger was relieved when he called the fax machine at his Abra Place home in Rancho Bernardo on Tuesday and it answered. Then his brother brought him down to earth.

"You might come back to a chimney and a fax machine," his brother said. (October 25, 2007: Hope and heartbreak)

Like that, the wildfire was off and roaring. The Tullochs' house was spared, but two neighboring homes—including one that once housed the Witch Creek post office—were destroyed.

Hernandez got a hose and started watering down his house. Firefighters joined him, he said. His home was saved, but an uncle across the road wasn't so lucky. (October 26, 2007: Witch Creek fire fierce from start)

Bush's stops included a charred neighborhood and a crowded relief center in Rancho Bernardo and a command post for 2,600 firefighters in Escondido. In a series of brief comments during his four-hour visit, the president spoke of valor and comfort but declined to analyze the emergency response.

"I'm thinking about people whose lives are turned upside down," he said in a cul-de-sac where fire's fickle nature left some homes untouched but burned others to the ground.

He first visited a Rancho Bernardo neighborhood, where he surveyed the damage along Lancashire Way and met a couple, married for 39 years, whose home was destroyed by the Witch Creek fire. A ceramic tile displaying their street address survived, but beyond it a leaning spiral staircase led to nowhere. Roses bloomed nearby. Other houses on the street were husks, but some still stood. (October 26, 2007: Bush's visit cheers some, irks others)

Survivor guilt is common. Your house didn't burn, but your neighbor's did, prompting you to ask, "Why was I spared?" (October 26, 2007: Stress over crisis takes its toll; help is available)

> In the wee hours of Monday, a tornado of fire played a cruel game of hopscotch amid the hilly sprawl of Rancho Bernardo, taking one house while sparing the place next door.
>
> But nowhere was the firestorm's course more capricious than on side-by-side cul-de-sacs just east of Battle Mountain. (October 27, 2007: Fire erratic when choosing targets)

> Shortly before he lost his battle with cancer Oct. 15, Eric Noemer willed his house to his next-door neighbor, Velna Schneider. On Oct. 22, her house burned to the ground. His didn't.
>
> In Fallbrook, where a wildfire's random fury left some destitute and others untouched, Schneider, 80, says a miracle has occurred. (October 31, 2007: God's blessing seen in bequest from neighbor)

> Despite the heavy property losses in Rancho Bernardo, it could have been much worse, these firefighters say. On blocks where one or two houses were incinerated but the neighboring homes are still standing, "That's a clear indication that a heck of a fire fight went on there," Fennessy said. (November 1, 2007: Lessons of 2003 helped, but changes needed, officials say)

> Jeff Smith's house was about two miles from the nearest one in Escondido damaged by the Witch Creek fire last week.
>
> Yet to the astonishment of Smith and his neighbors, it caught fire and was gutted virtually without warning.
>
> "I'm fine; it's so far away," Smith recalled telling his daughter on the phone about the distant red glow in the hills. "Probably as my roof was burning above me."
>
> Firefighters and looky-loos are still driving by and talking about how flying embers must have settled on the home so far from the fire. Smith's shake-shingle roof may have been the reason he was so unlucky, neighbors said. (November 1, 2007: "It was so bizarre")

> "And then we flew over neighborhoods where, for the life of me, you'd see four houses clustered around one house that had burned. Why the others weren't touched shows you the erratic behavior and the threat that this creates." (November 1, 2007: Forecasters become first responders)

Several questions about the geography of damage are raised from these descriptions. For example, can any comparisons be made about locations that were damaged; do they contrast with locations that did not burn? What implications does the pattern of damage have for the way recovery will occur; specifically, is it easier to recover when those around you did not burn? What other questions does this geography of damage raise?

Spatial Pattern of Recovery

From January 2008 through May 2009, we have followed up on the recovery status of neighborhoods impacted by the 2007 firestorm in San Diego County, California. Using a Spatial Video approach which has already been described in this book, we mapped the pattern of damage transition to a pattern of recovery in most cases, and decline in others. Figures 5:4 and 5:5 demonstrate the geography of recovery over three time periods in the Rancho Bernardo neighborhood. Does the pattern of damage affect the pattern of recovery (e.g. do places with more damaged houses take longer to recover?)

Q: Can you describe the spatial pattern(s) of recovery presented in these maps.

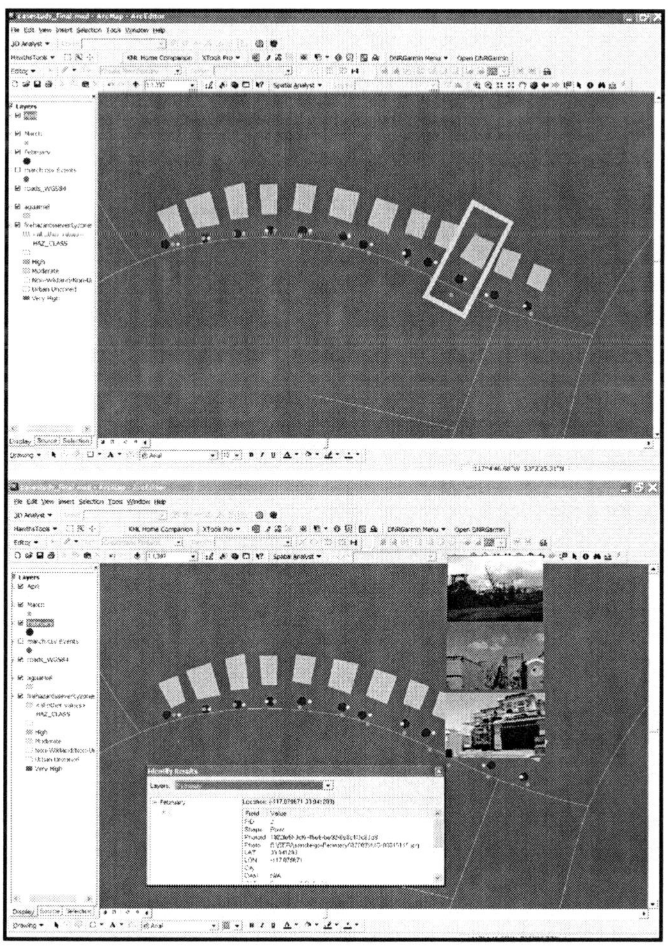

Figure 5:4. *Spatial patterns of recovery in the Rancho Bernardo neighborhood of San Diego County, California.*

EVACUATION AND ANIMALS

In addition to descriptions of spatial patterns of damage, another common theme that emerged was the importance of animals. Throughout the stories published during the week of the 2007 firestorm, twenty-two articles mention the importance of animals in a person's decision to evacuate and then their decision of where to evacuate.

a)

b)

c)

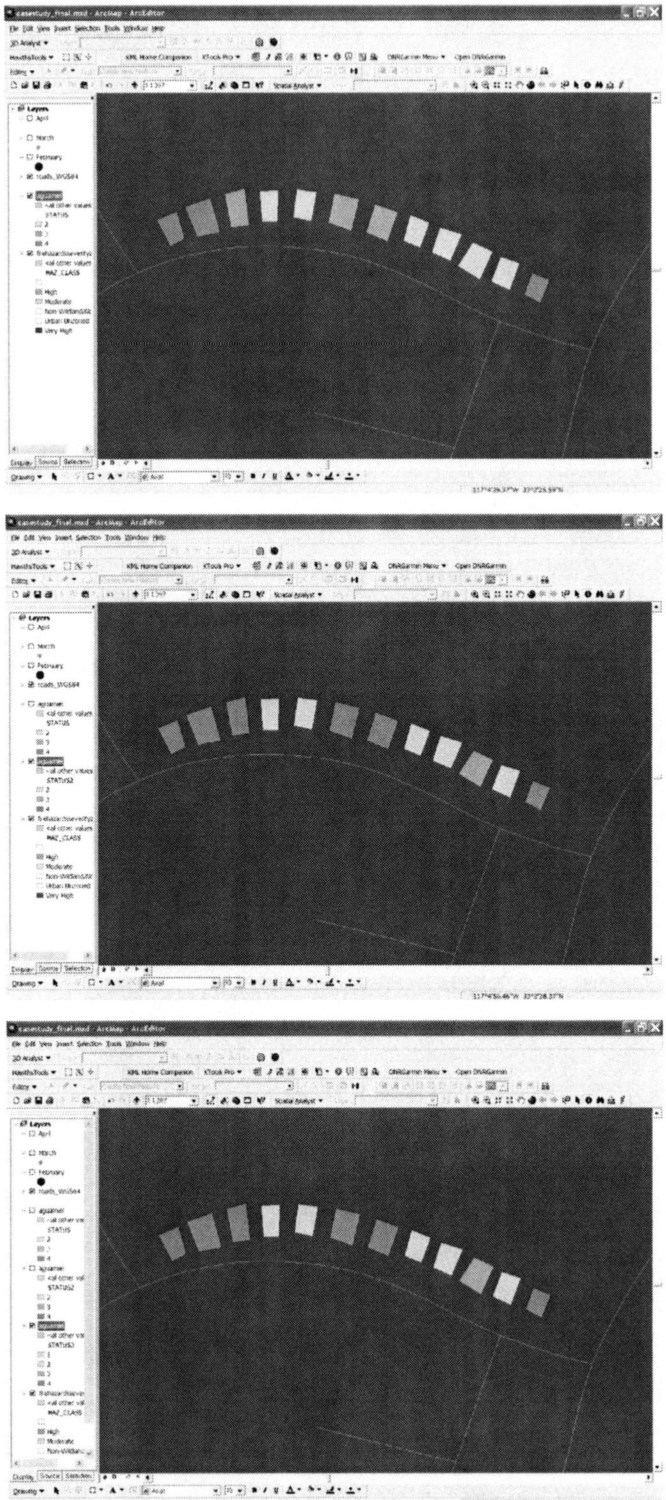

Figure 5:5.

George and Wendy Santamaria and their 17-year-old daughter received the call to evacuate from their Encinitas home about noon yesterday, and they packed up their possessions, including a collection of animals that would make Dr. Doolittle proud. Thirteen rabbits, five ducks, two cats and a dog were all piled into the family's vehicles as they headed for Solana Beach to stay with friends.

"We're going to have to check around to find a place to put everybody," he said.

Stan Smith and Ken Morris were two of the last people to leave the Del Dios area near Lake Hodges yesterday afternoon. They stayed behind to rescue a neighbor's two horses.

Smith and Morris coaxed the horses from a corral with handfuls of oats and reassuring pats on the neck. They stayed despite mandatory evacuation orders.

"I'd hear the loud speakers (of law enforcement officers), duck down and have another beer," Morris said. (October 23, 2007: Crews hope for break in Witch Creek blaze)

Much of San Diego County is horse country, home to thoroughbred breeding ranches and countless backyard stables housing family pets. As many as 300,000 horses live in the county, and when people flee from the fires, so do the horses. (October 23, 2007: Owners scramble to find safe haven for horses)

Animal Sheltering

Off in a far corner of the stadium lot, the Chargers' old practice field had been turned into an animal shelter for pets ranging from horses and dogs to rats and ducks. A sort of animal MASH unit, staffed by veterinarians, had been set up under a tent. (October 25, 2007: Prayers and Camaraderie)

Cats and dogs, as well as cows and other livestock, began showing up at shelters and in parking lots. Children wheeled shopping carts piled with hay through Parkway Plaza in El Cajon 7/8 food for the evacuated horses that were in pens made of rope. (October 28, 2007: Inferno)

Many guests have pets in their rooms, she noted. One guest set up a small tent on her patio for her dog. "There are tons of dogs all over the place, but they have been keeping them quiet." (November 1, 2007: Homes lost, they make the best of it at a hotel)

The allowance of animals sheltering with owners has been one improvement since the 2003 wildfires that occurred in San Diego County and also since Katrina. In both situations the owner-pet relationship caused problems with people not evacuating or evacuating and losing their animals in the process. For example the following quote refers to the situation faced in the 2003 San Diego fires: "A lot of people were just spray-painting their phone number on their horses and letting them go. I couldn't do that," Gutierrez said. (October 23, 2007: Firefighters make stand near Otay Lakes as Harris fire rages)

Furthermore, some stories mentioned the importance of animal health during the wildfires. For example, the October 25, 2007 story entitled, "Stress of fires affects animals, too" provides the following information:

> For the thousands of domesticated animals, large and small, that have been displaced by the fires, the main concerns are stress and dehydration.
>
> In horses, the trauma of leaving their homes, coupled with a change in diet, upsets their stomachs. In dogs and cats, the stress can take hold, too. They vomit and suffer from diarrhea.

Though health concerns with animals received some attention during the week of wildfires, by far the most common theme to emerge from Union-Tribune stories was human health.

HEALTH

From October 21 through November 1, 2007, the San Diego Union Tribune published thirty-seven articles that mentioned health concerns related to the wildfires. Of these, the majority focused on physical health implications (respiratory health, injuries, pregnancy), with only a handful of pieces dealing with mental health issues (primarily from stress). In addition, a theme of the geography of exposure emerged, in particular the shifting physical exposure to smoke due to wind patterns. Finally, the issue of health in vulnerable populations (the very young, the very old, illegal immigrants, migrant laborers, those with mental illness) is scattered through these stories. These themes will be expanded upon with the following excerpts.

Respiratory Health

> In San Diego County, normally a paradise for outdoor workout warriors, conditions are especially deceptive because the sun still shines through blue skies. The danger is what you can't see: noxious particulate matter

from the wildfires. (October 27, 2007: Fitness experts suggest a break from workouts)

More than 90 percent of the mass of wildfire smoke consists of carbon dioxide and water, but hundreds of other chemicals can be present, among them: carbon monoxide, mercury, aldehydes, nitrogen oxides, polynuclear aromatic hydrocarbons (PAHs), sulfur dioxide, volatile organic compounds, and ozone.

Many of these ingredients pose immediate and obvious health hazards. Smoke particles smaller than the diameter of a human hair can be inhaled deeply into the lungs, causing everything from lung irritation to an increased risk of cancer.

Carbon monoxide (CO) is an odorless, tasteless, invisible poisonous gas. Mercury is an extremely toxic element. Aldehydes are gases that irritate the eyes, nose and mouth. Some forms, like formaldehyde, are carcinogenic, as are PAHs.

Nitrogen oxides are associated with acid rain. In the presence of sunlight and hydrocarbons (organic compounds readily produced in fires), nitrogen oxides combine to create ozone, a gas that inflames and impairs the lungs and triggers asthma attacks.

These compounds can linger in seemingly clear air for days, even weeks. Perhaps the best cleanser is rain, followed by time and dilution. But the problem is that the increased frequency and severity of wildfires means there are more fires billowing more pollutants into the air more of the time. (November 1, 2007: Smoke gets in your skies)

Physical Injuries

Hospitals reported that a surge of evacuees began streaming in yesterday with respiratory problems, bruises, splinters, fractures, and injuries caused by items that fell on them.

Some had not realized they were wounded because they were racing to flee a fire, said Dr. Shawn Evans, an emergency room physician at Scripps Memorial Hospital in La Jolla. (October 25, 2007: Hope and heartbreak)

Countywide, hospitals have treated almost 100 people for burns, cuts, smoke inhalation and other health problems. That includes 27 firefighters, three of whom remain in critical condition at the UCSD Medical Center's burn unit. Many more people sought medical help at shelters. About a dozen clinics have closed countywide, along with an untold number of doctors' offices. (October 24, 2007: Help from above)

Pregnancy Outcomes

In Oceanside, Tri-City Medical Center was handling nearly double its typical number of maternity patients. Some of the deliveries were premature, perhaps because of stress from fire evacuations and the loss of homes, said Tri-City spokesman Jeff Segall.

Usually, the hospital's neonatal intensive care unit treats 11 newborn babies at a time. Yesterday, it had 21. (October 25, 2007: Hope and heartbreak)

Stress

"Some people will be emotionally numb. Others will be angry or sad or irritable. It's important to know that all of these emotions are OK. It's normal," said Jerry Gold, a psychologist who is director of Scripps Behavioral Health Services."

Other stress symptoms include anxiety and fear that the disaster will happen again; difficulty focusing and making decisions, hopelessness, spontaneous crying, increased or decreased eating, sleeplessness and nightmares.

There's also the physical side to stress, which if left untreated can lead to serious health problems.

"Things like increased heart rate, cardiac arrhythmias, headache, irritable bowels, muscle aches—these are all linked to increased stress hormones," said Mimi Guarneri, a cardiologist who is medical director of Scripps Center for Integrative Medicine.

Acute stress symptoms triggered by a catastrophe can continue for weeks after the trauma. But if symptoms persist much longer than a month, it may be time to see a doctor.

"If you continue to experience these problems much more than a month, then it crosses the boundaries of normal," Minassian said. "After that, it gets into post-traumatic stress disorder, and that's something we need to treat."

Don't be surprised if these feeling resurface at the holidays or at special occasions during the year. Hopefully, the depression or anxiety you may feel at these times will be "short-lived and resolve on its own when the (occasion) is over," Minassian said. (October 26, 2007: Stress over crisis takes its toll; help is available)

Geography of Exposure

By late tomorrow or Wednesday, the winds at the coast could shift to onshore. This could bring back smoke that had blown out to sea and redistribute it all along the coast. A slight wind shift late yesterday afternoon

> had a similar effect, bringing smoke to much of the city of San Diego. (October 22, 2007: Dry conditions, air quality expected to worsen today)

> Soon a hard, dry wind began blowing ominously from the east, and by yesterday, a haze had spread over every neighborhood, rich, poor or otherwise. The unavoidable smoke carried a message in this disparate region of 3 million: Even if our homes were spared, we are all in this together. (October 24, 2007: Regardless of ZIP code, we're all in this together)

> The worst air in recent days has been near the coast. That's because some smoke that blew out to sea early in the week returned when a weak westerly wind developed. (October 27, 2007: Air quality problems sticking around)

> Thick smoke enveloped the East County and pushed all the way to the Pacific Ocean. Ash rained down over hundreds of square miles. (October 22, 2007: Wildfires rage)

Exposure to the disaster, however, is not limited to smoke.

> "I haven't been sleeping," said Olivenhain resident Mary Mosse. "You wake up all night. You hear a fire engine go by and it's: 'What? What?'"
>
> Mosse was among the horde of evacuees. Her house is undamaged, but "it's going to take a long time before I feel like things are back to normal." The memories are too strong, too disturbing.
>
> Even those far from the flames felt a jumble of emotions last week. Fear. Alarm. Boredom even, from being cooped up because of the poor air quality. (October 29, 2007: After fiery week, return to normal won't be easy)

Timeframe of Exposure

> Lakeside resident Dennis Poe said he dealt with blowing ash nearly a year after the 2003 fires.
>
> "The ash that fell initially was one thing, but the ash that fell later was just as adverse," Poe said. "Every time a Santa Ana blew in, more ash would come." (November 1, 2007: Airborne ash expected to be lingering problem)

Health Issues With Cleanup

> Although everyone's eager to get the yard and house back in shape, health and environmental officials say it's important to make your cleanup safe.

"Remember, it's not just trees and homes that have burned. Synthetic materials are also in that ash, and that can be carcinogenic if inhaled and be potentially hazardous to your health," said Dr. Wilma Wooten, public health officer for San Diego County.

"Blowing it around just stirs up the dust, which gets into people's lungs. And the hose washes the ash into the storm drains and our water system," she said. "The best way to clean all outdoor hard surfaces is to lightly mist it with water and then gently sweep it up. Follow this up by wet mopping." (October 27, 2007: Officials stress cleanup safety)

Vulnerable Populations

Six illegal border crossers are hospitalized with burn injuries, and one of them is in critical condition, said Alberto Lozano, a spokesman for the Mexican Consulate in San Diego. He said the six men, all from central and southern Mexico, were rescued Monday by firefighters north of the shuttered Tecate port of entry and transported to UCSD Medical Center in Hillcrest. (October 24, 2007: Illegal border crossers brave flames, get burned)

As the homes above them have been evacuated in recent days, many of the agricultural and day laborers who live in the region's canyons have remained behind in their encampments, many of them afraid to leave.

Still others, cut off from television or radio, have smelled the thick smoke and wondered what's going on, but they have lacked the information to know what to do or where to go, if anywhere.

Others are afraid of contact with authorities because they are here illegally, said day laborer Jose Salgado, 32, a native of Acapulco, Mexico, who lives in Vista with roommates.

Some illegal immigrants evacuated from their homes have reported problems returning home through police checkpoints because they lack proper identification.

The fact that some agricultural employers in the area have remained open is a sore spot for the volunteers, who say it has been not only an incentive for these laborers to remain in a dangerous zone, but that it has kept them doing strenuous labor outside in unhealthy air. (October 25, 2007: Canyon dwellers not joining exodus)

In Ramona, flames were consuming houses just blocks away when the staff at Changing Options, a treatment center for 28 mentally ill adults, evacuated at 8 p.m. Sunday.

"We had one person hospitalized for anxiety, but everyone else has fared pretty well," said Angela Sturman, the center's clinical director. The clients and staff were evacuated twice, first to Poway High School and then to Mira Mesa High. (October 23, 2007: Thousands take refuge)

Older people and children are extremely susceptible to respiratory problems when air quality is bad, said Wilma Wooten, San Diego County public health officer. She said children should not play outside. (October 22, 2007: Dry conditions, air quality expected to worsen today)

As you can tell from this section on health, many concerns are raised. Often disaster related health outcomes are discussed in terms of physical injuries alone, but the issue is much broader. For example, given the border crossings that regularly occur in this area, and the illegal immigrants that live in Southern California, it is likely that these people's exposure goes unchecked. What issues/ concerns does this situation raise? What does this geography of air quality mean for those potentially exposed to health problems from the fires? Certainly, it is not limited to those who live in close proximity. Remember these issues when you read the upcoming chapter on health vulnerability. How might exposure and health vulnerability impact the recovery process in San Diego County?

REFLECTIONS ON MITIGATION

Now we will move from issues tied to the immediate event to look at policies put in place before and after a disaster. Mitigation is the phase where both structural and non-structural steps are taken to reduce the impact of subsequent events. Structural mitigation is mainly comprised of engineering activities, such as building levees, or using fire-resistant building materials, while non-structural mitigation strategies include education and the purchase of insurance, such as flood or earthquake policies, and the use of zoning policies to protect life and property.

Mitigation strategies for wildfires include land management (defensible space, neighborhood design), use of fire-resistant construction materials, and education about how to minimize the effects of a wildfire event, among other options.

Consider these statements:[8]

8 Taken from Winter, G. J. and J. S. Fried. 2000. Homeowner perspectives on fire hazard, responsibility, and management: Strategies at the wildland-urban interface. *Society and Natural Resources* 13: 33-49; Cook, Steve. 1997. Wildfire adapted ecosystems meet man's development. *Australian Journal of Emergency Management* Winter: 24-30.

- Winter and Fried (2000) find that homeowners in a Michigan Wildland-urban interface perceive damages from wildfire to be capricious, not related to mitigation strategies such as defensible space. However, they also believed that the benefits of the aesthetic quality of their residence were more important than creating a fire-resistant home environment. (40)
- "... potential wildfire hazards can be identified visually by trained observers, and this knowledge can be communicated to homeowners in the area involved." (Cook 1997:24)

Q: How can GIS be used to address these two mitigation issues? Keep these statements and this question in mind as you read the excerpts from the Union-Tribune on mitigation.

During the 2007 firestorm, sixteen articles from the San Diego Union Tribune mention mitigation. They focus on a mix of both structural and non-structure mitigation measures. Structurally, the focus was on building materials and planting. Non-structural topics focused mainly on policy for development in fire-prone areas. The other theme that is raised is whether or not mitigation matters.

> The owners of a huge home at the end of Valleyview Road seemed to have taken every precaution: They'd cleared most of the brush within several hundred yards of the structure. Their roof was made of fire-resistant material.
>
> Yet the house was completely ablaze, its multicar garage essentially melted to the ground and its owners nowhere to be found. (October 23, 2007: Flames create a surreal scene)

> In addition, it has been a struggle to find money for brush-management and code compliance officers whose job it is to keep city homes safe from fire before it breaks out. (October 24, 2007: Officials cite improvement since 2003 blazes)

> "A disputed land-use strategy designed to protect new developments from devastation in the county's exurban, fire-prone areas appears to have passed its first and most critical test.
>
> Officials with the Rancho Santa Fe Fire Protection District credit the lack of property damage to a defensive program called 'shelter in place.' It is a wildfire protection plan that imposes construction and landscaping standards intended to be so stringent—including mandatory interior fire sprinklers and broad swaths of protective

landscaping—that homeowners can remain sheltered in their houses in they cannot evacuate.

Crosby homeowner Dennis Samaritoni was relieved to learn that houses in his development were spared. But he said the fire-safety regulations didn't necessarily make him feel any safer. He and his wife evacuated from their 5,500-square-foot golf-course home Monday morning. (October 25, 2007: Exclusive homes emerge unscathed as fire-protection concept is tested)

Whatever the cause, remedies range from relatively simple to complex. Richard Halsey, director of the California Chaparral Institute in Escondido, said a good next step is to prevent homes from being built at the tops of canyons, where fires tend to get funneled. "We need to toughen up the rules," he said. "That is up to the politicians."

County Supervisor Ron Roberts said lessons can be learned from every fire, but that it's not time to criticize development decisions, some of which were made decades ago. "There will always be some kind of threat," he said. "All of San Diego is fire prone." (October 25, 2007: Building patterns are to blame, critics say)

On ABC's "This Week With George Stephanopoulos," Rep. Duncan Hunter, R-Alpine, said he predicts "a tug of war" between environmentalists and the firefighting community over laws that protect certain fire-prone vegetation from being cut down near homes. "We have brush that comes right up against housing subdivisions," Hunter said.

County regulations, however, say removing flammable vegetation is exempt from environmental restrictions if fire professionals recommend it in writing.

Sen. Dianne Feinstein, D-Calif., said cities and counties need to better use their zoning authorities as homes are increasingly built in fire-prone areas.

"I think you really have to understand what the fire pattern is, or what the flood pattern is," Feinstein said. "Local jurisdictions have to use zoning patterns more to prepare for fires." (October 29, 2007: Crews contain two major wildfires)

"You drove up the road. Did you see a single blade of grass left?" said Clay Hatch, who works at the nearby Mountain Valley Ranch. "These cows have nothing to eat. This is just enough hay to let them live."

Neal said that because there is nothing left in the valley for the cows to graze, he likely will have to close up his business or find another location.

> Sounding like the cattleman he is, Neal blamed politicians who bow to environmentalist pressures for the severity of the fire.
>
> "It's a mess," he said. "These fires are not preventable. You're going to have fires. But they can limit the damage and they aren't addressing limiting the damage."
>
> For instance, Neal said, in the past several years the federal government did nothing to remove brush in the surrounding mountains through controlled burns or other methods. (October 20, 2007: Did you see a single blade of grass left?)

RECOVERY

> San Diego County began its recovery yesterday as thousands of evacuees returned home and firefighters made momentous progress against this week's wildfires. But in an irrepressible image of the inferno's ferocity, officials found six more charred bodies. October 26, 2007: San Diego Union Tribune (http://www.signonsandiego.com/news/metro/wildfires/index.html)

The preponderance of studies that focus on recovery from wildfires, do so in terms of the natural ecosystem alone. Social science approaches to the study of fire are still in their infancy and there is still much to discover[9]. Furthermore, the studies that do exist do not usually address the societal process of recovery that takes place after a wildfire event. However, how this process unfolds has implications for the well-being of individuals and of the neighborhoods in which they reside.

> It's not over. But it's over for Bob Jacobs.
>
> Seeing his family's home upon returning from a one-night evacuation, he said he felt relief—not just because their possessions were safe, but because they would not have to move.
>
> "If it had burned down, I'm not sure we could afford to rebuild it," Jacobs said. "We might have had to pull up roots and leave the community.
>
> "People who live here tend to stay," Bob Jacobs said. "We're newcomers—we've only been here for 17 or 18 years." (October 25, 2007: Newcomers glad they won't have to move)

9 Cortner, H. and D. Field. 2004. Introduction to the special issue: Humans, fire, and forests: The reemergence of research on human dimensions. *Society and Natural Resources* 17: 473–475.

Four years and one day ago, the Cedar fire began its destructive march through San Diego County, charring hillsides and leaving blocks of smoldering rubble behind.

"The idea is to facilitate ideas that worked," said Bob Ilko, president of the Scripps Ranch Civic Association, which set up "Firestorm 2003—Project Phoenix" to oversee the rebuilding effort.

But from the ashes of the most destructive fire in state history, something else emerged: a possible blueprint for rebuilding houses and neighborhoods. The plan came from Scripps Ranch, one of the hardest-hit areas, where 349 houses were destroyed and damaged.

Karen Reimus, whose home was one of 46 destroyed on Pinecastle Street, said that one of the best steps they took was to establish an e-mail group of fire victims.

This was done through the civic association, a community group with a newsletter and broad presence in the community. The list, called "Firefolk," was invaluable, she said.

"It was one of the greatest resources," Reimus said. "It ended up being such a huge mechanism for communication. One thing I really learned after the Cedar fire is information is power. And the more ways you can get information to people who need it, the better."

"I know this is a harsh thing to say, but I would say you need to get set that you are two years out before you get in your home," he said. (October 26, 2007: Scripps ranch veterans of '03 fire ready to help)

While the government's response to Katrina was deplorable, there appeared to be a greater spirit of cooperation among local, state and federal governments in the California firestorms, said George Amedee, a political science professor at New Orleans' Southern University.

"There seemed to be a clear chain of command," Amedee said.

But he cautioned that the real test of governmental effectiveness will be the rebuilding process.

To speed up reconstruction in San Diego, the City Council will vote Tuesday on a package of exemptions to help city residents whose homes were hit by the fires. They include waivers on building permit fees and vouchers for recycling and disposal of debris. Building permit applications for fire victims would be fast-tracked. (October 28, 2007: Calm returns yet fires still rage)

By early next year, as money from insurance companies and government agencies pours into the region, there likely will be new work for hundreds of construction workers who have been laid off during the recent real estate slump. Home improvement and department stores will be doing more business as the fire victims stock their rebuilt homes.

"From an economic standpoint, the negative impact from disasters usually isn't long-lasting," said Kelly Cunningham, economist with the San Diego Institute for Policy Research. "Things look negative right now, but there will be a dip and then a rebound. Rebuilding spurs the economy."

Despite the disruptions, new money started to flow into the county. Insurers began paying for homes and businesses to be rebuilt. Federal and state agencies such as the Small Business Administration also pumped in money for reconstruction.

If 2003 is an indication, the influx of new money will not have a noticeable impact on the economy until early next year. After the Cedar and Paradise fires, it took four or five months before reconstruction in the burned areas began in earnest.

"When you've lost everything, it sometimes takes time to make up your mind about how to rebuild," said Candysse Miller, spokeswoman for the Insurance Information Network of California. "And even though there was an effort to streamline the permit approval process, rebuilding doesn't happen overnight. There are technical items involved in deciding on the house plan, having it approved, getting bids for contractors and making sure they are available."

Some homeowners and insurance analysts add that one reason the reconstruction effort took so long in getting off the ground is that the insurers dragged their heels on making payouts. By mid-April 2004, the insurers had passed out only $752 million of the $1 billion or so they ultimately paid out on claims from the fires.

Gary London, president of London Realty Group Advisors in San Diego, said the most positive impact of the fires may be that it provides the burn areas with a chance to rebuild more wisely.

"The upside is a modernization of homes, commercial properties and infrastructure," London said. "This is perhaps an opportunity for some better community planning and tighter communities." (October 28, 2007: Rebuilding expected to spur economy)

Yesterday morning, as a smoky haze painted the trees and hills of Ramona, the people of St. Mary's-in-the-Valley Episcopal Church arrived for breakfast greeted by a banner and the promise: "Recovery begins here."

Over breakfast, church members brainstormed about ways their church could help in the recovery. A deacon from the diocese was there

> to assure them that long-term support will come from the denomination. (October 29, 2007: Worshippers reflect on blessings, losses)

> The San Diego City Council and the county Board of Supervisors are trying to limit out-of-pocket expenses for fire victims.
>
> Supervisors voted last week to waive fees for fire victims planning to rebuild. And San Diego council members voted yesterday to waive building permit and dumping fees and expedite building permits for property owners whose homes burned. (October 31, 2007: Officials talk cleanup)

> Owners of houses that burned in the city in last week's firestorms may have one less headache to worry about as they begin to rebuild.
>
> Escondido is offering to demolish the remains of destroyed homes, clean up the debris, set up temporary fencing, and cut and haul away burned trees.
>
> It's all part of a plan to help fire-affected families as much as possible, said Steve Nelson, Escondido's assistant city attorney, and to speed up the process of rebuilding.
>
> Demolition permit fees are being waived, and building permit fees will be discounted, Nelson said. (November 1, 2007: City to cover home demolition, cleanup costs for fire-affected families)

Conclusions

What does the future hold for wildfires in the WUI in San Diego County? Population continues to grow in hazardous areas and climate change may lead to changes in heat and fuel components of the fire triangle through winds, precipitation and temperature impacting vegetation (perhaps more grass and scrub, different varieties?). Changing climate also raises the question of new insect introduction, such as the pine beetle which led to increased wildfires in the southeastern United States through killing pine trees, which in turn created a dry fuel source. Mitigation measures may be put in places which reduce the severity of the events and resources may be increased to improve response. However, as the final excerpts demonstrate, fires will continue and they will likely become worse.

> "Get used to it."
>
> That's what many climate experts are saying about catastrophic wildfires, including the dozen that have turned San Diego County into a disaster zone for the past week.
>
> They believe such blazes will become a regular part of life in Southern California because global warming is intensifying nature's cycles by

> lengthening fire seasons and prolonging droughts in parts of the West. The consequences would be more deaths, more houses consumed by flames and more budgets busted by firefighting costs.
>
> Not everyone is convinced that global warming explains the recent blazes.
>
> "I am a little skeptical," said Thomas Wordell, an analyst with the National Interagency Fire Center in Boise, Idaho. "The ecosystem (in Southern California)... has been very fire prone for a long time. We've had droughts...before anybody came up with the terminology of climate change."
>
> Wordell's view aside, international, national and local groups of scientists are increasingly alarmed about the potentially devastating effects of climate change. These include the United Nations' Intergovernmental Panel on Climate Change, which recently won the Nobel Peace Prize for its work, the National Research Council and numerous University of California researchers.
>
> They generally contend that the connection between global warming and more destructive wildfires is too clear to ignore." (October 30, 2007: Experts: Wildfires only going to worsen)
>
> "Given the ferocious nature of these Santa Ana-fueled blazes, these officials acknowledge, it is inevitable that people will lose their houses, no matter how well-equipped the fire department may be.
>
> The question, as usual, is one of money and priorities. How many homes will one additional fire station save during a wildfire? And for a city as broke as San Diego, can the money be better spent elsewhere?" (November 1, 2007: Lessons of 2003 helped, but changes needed, officials say)

This chapter has used a popular media source (a local newspaper) to describe the different spatial and social aspects of a disaster. It has also introduced two further disaster topics which are vital to understanding the human impact caused by disasters; social and health vulnerability. These will be discussed at length in the next chapters. This wildfire chapter will end with a series of exercises for the reader, one non-technical followed by a project requiring basic GIS skills.

WILDFIRE EXERCISE 1:

Pick another disaster which has occurred in the last 20 years and use only newspaper quotes to provide a comprehensive overview. This overview should include a description of the actual event, the response and clean-up, recovery and rebuilding, as well as the social and political story. In addition, focus on any health outcomes described, and pick out references to the different types of social vulnerability described in the next chapter. Finally, could all these accounts be used to create a GIS/map?

WILDFIRE EXERCISE 2:

For the following project some basic GIS skills are required.

1. Access the following datasets for San Diego County, California:
    - Population by blockgroup from the U.S. Census
    - Wildfire Severity Zones from SanGIS
    - Digital Elevation Models (DEM) from the U.S. Geological Survey (USGS)
    - Streets from SanGIS
    - Address points from SanGIS

2. Bring these data into a GIS

3. Answer the following questions:
    a. In San Diego County, how many people live in each of the following areas: moderate, high, and very high wildfire severity zones?
        - Now, zoom into one neighborhood, Rancho Bernardo. This area is delimited by the following roads: Aguamiel Road and AlboradaDrive, West Bernardo Drive, Oculto Road and Capilla Road.
        - Look at the topography for this area. Add the address points that are in this area.

    b. Which houses are most at risk for being burned?
        - Export the address points for Aguamiel Road into Google Earth.

    c. How much defensible space surrounds each house?
    d. Based on this intersection of physical and human geography, what mitigation measures do you recommend for this area, in particular regarding defensible space and where or where not to build neighborhoods in San Diego County?

SIX

The Geography of Vulnerability

A QUESTION SOMETIMES ASKED by people who don't live on the Gulf Coast is why didn't everybody just evacuate New Orleans if they knew a major hurricane was heading their way? Unfortunately, for many people this decision is never simple. Confounding this process of evacuation is a person's vulnerability which includes physical impediments, such as disabilities, along with social and financial factors, and finally "choice" itself. Though this term is used in slightly different ways by different academic disciplines, in this book vulnerability will be used to describe the complex sequence of living conditions that will increase the chance of experiencing a disaster (what we call exposure) and the ability to limit the effects of the disaster after the event. Vulnerability involves both physical characteristics, such as where a family lives or even the structural characteristics of their home, as well as various social aspects[1], such as does the family own their home? Are they well represented politically? Do they have the financial ability to limit exposure and then recover? In the previously mentioned Ohio River flood of 1997, families owned homes on the flood plain because that was all they could afford. They didn't have the opportunity to buy elsewhere. They lacked a political voice to bring attention to their predicament (remember the levee wall on the other side worsened their flood), and they would know that when the flood came, they would lose everything because no insurance company would offer policies on such a risky proposition. This was a disaster scenario poverty trap.

The topic of social vulnerability has long been a topic of interest to academics because it is a predictor to what will happen in a disaster. By knowing the mix of socio-economic characteristics of a neighborhood[2], in combination with the

---

1 Morrow, B., "Identifying and Mapping Community Vulnerability." *Disasters*, 1999. 23(1): p. 1-18; Jaspers, S., "Targeting the Vulnerable: A review of the Necessity and Feasibility of targeting Vulnerable Households." *Disasters*, 1999. 4: p. 359-373.

2 Although there are many papers discussing various aspects of disaster related vulnerabaility, these will provide a good starting point;

Blaikie, P , T Cannon, I Davis, and B Wisner. 1994. *At Risk: Natural Hazards, People's Vulnerability, and Disasters*. London: Routledge.

Blaikie, P. and H. Brookfield. 1987. "Land Degradation and Society." New York: Metheun & Company, Ltd.

Bolin, Robert and Lois Stanford. 1991. "Shelter, housing and recovery: a comparison of US Disasters." Disasters 15.

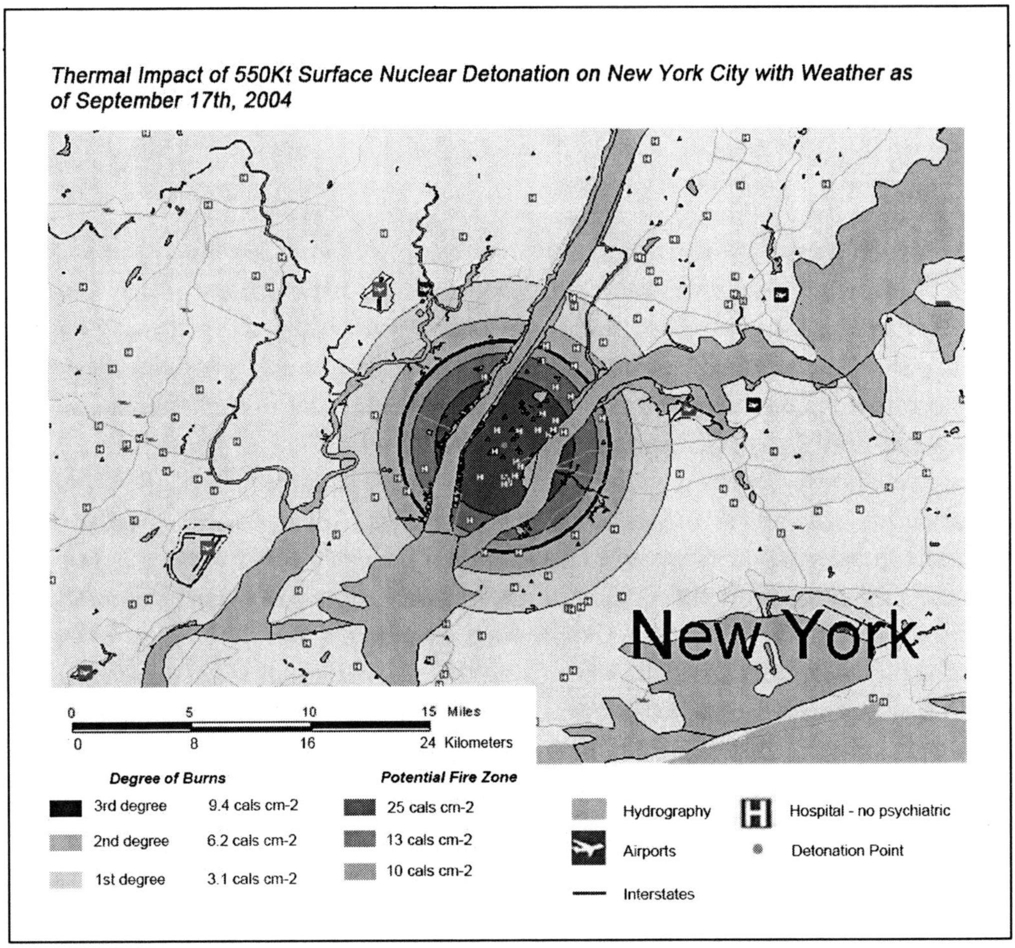

Figure 6:1. *Vulnerability of populations and the urban health care systems to nuclear weapon attack—examples from four American cities, International Journal of Health Geographics.* Source: Bell, William, A Dallas, Cham (2007)

historical, cultural, political and economic processes that shape an individual's or social group's capacity to cope within the neighborhood, and then on top of these layer physical proximity to a hazard, it should be possible to develop a "future geography" of what will happen when a disaster strikes. This future geography can

Cutter, S. 1995. "The Forgotten Casualties: Women, Children and Environmental Change." *Human and Policy Dimensions* 5:181-94.

Cutter, Susan L., J.T. Mitchell, and M.S. Scott. 1997. *Handbook for Conducting A GIS-Based Hazards Assessment At The County Level.* Columbia, SC.: University of South Carolina.

Cutter, Susan L., B. Boruff, and W.L. Shirley. 2003. "Social Vulnerability to Environmental Hazards." *Social Science Quarterly* 84:242-261.

Hewitt, K. 1997. *Regions of risk: a geographical introduction to hazards.* Harlow, Essex: Addison Wesley, Longman.

Mileti, D. 1999. *Disasters by Design: A reassessment of Natural Hazards in the United States.* Washington DC: Joseph Henry Press.

be foreseen not only in terms of predicting situations during the disaster itself, but also about how the city and its neighborhoods will recover afterwards. Although much of the focus of this book has been Katrina, and understandably so given its magnitude, the same precursors and post-effects could be imagined for any other city, based on its social characteristics and the type of hazard(s) it faces. Of course disaster scenarios do not have to be be natural. In keeping with an all hazard approach we should also consider the spatial patterns of vulnerability with regards potential accidents (such as a train derailment carrying a toxic load), or even a terrorist attack.

Figure 6:1 shows the physical impact of a 550Kt surface nuclear detonation on New York City. The purpose of the academic paper from which this map was taken was to assess causality estimates based on geographic proximity to the blast. The authors note that many of the area hospitals in the four city study (the others being Atlanta, Chicago and Washington DC) would be rendered inoperable by the detonation. This study does not include the additional component of social vulnerability—what would be the social/demographic breakdown of those losses? Addressing this question would be a natural next step in a GIS analysis.

The combination of these social and physical elements is often referred to as a hazardscape,[3] and this probably will change between different neighborhoods of a city, as well as between cities and regions. There are, however, certain characteristics which are widely acknowledged as being common to most places. Those living around the poverty level (though any family without reasonable savings may soon enter hardship if the effect of a disaster is prolonged and involves relocation), the young and elderly, the disabled, female-headed households, families with young children in the home, and ethnic minorities have all been identified as cohorts of concern.[4] What makes these groups vulnerable?[5] It is not just one reason, but usually a series of direct and indirect factors that increase exposure. For example the very young and elderly may have less ability to fend for themselves and are therefore reliant on help from others. The elderly may not choose or not be able

3 Cutter, Susan L. 1996. "Vulnerability to environmental hazards." *Progress in Human Geography* 20:529-539; Cutter, S L , J T Mitchell, and M S Scott. 2000. "Revealing the Vulnerability of People and Places: A case study of Georgetown, County, South Carolina." *Annals of the Association of American Geographers* 90:713-37.

4 There is some evidence that a high density of a minority population actually acts as a buffer in a disaster, through a common bonding through adversity—therefore more minority = less social vulnerability, but this varies by location.

5 There is a large literature on the disproportionate impact of disasters across society. Readers are pointed to the following "classics": Mileti, D., *Disasters by Design: A reassessment of Natural Hazards in the United States*. 1999, Washington DC: Joseph Henry Press; Blaikie, P. and H. Brookfield, eds. *Land Degradation and Society*. 1987, Metheun & Company, Ltd.,: New York; Cutter, S.L., B. Boruff, and W.L. Shirley, "Social Vulnerability to Environmental Hazards." *Social Science Quarterly*, 2003. 84(2): p. 242-261.

to evacuate because of physical immobility.[6] Both the young and elderly are also more susceptible to infections that might result from the disaster. People living with a disability might need assistance in evacuating a hazard for similar reasons as the elderly. For many cities a plan is supposed to be in place to facilitate their evacuation. Unfortunately, anecdotal evidence suggests this system did not work effectively for Katrina. During our walk through New Orleans we saw at least one home where search and rescue had marked a death inside, and the vehicle outside displayed a disabled sticker. Similar risks are faced by people who suffer other physical and mental challenges, such as being blind. If we again return to the 911 calls after Katrina, these "vulnerabilities" are all too evident:

- Trapped in house, unable to get out, elderly need oxygen, oxygen running low.
- T*** 89 has diabetes, Phone of victim stopped working. Victim will be on ground
- Has special needs. She has multiple sclerosis and is a quadriplegic
- Victim in wheelchair and oxygen is running out. OUT OF OXYGEN EMERGENCY NEED HELP—large woman

Figure 6:2 shows two homes in St. Bernard Parish on October 26th, 2005. In the top picture a message has been written asking if the elderly person who lived inside was alright, and to contact the letter writer if possible. The large black "Ken" sign on the wall tells the tale that a body had been found inside and had been removed (there were variations to the meaning of this sign, but when found in isolation in a neighborhood this was the usual interpretation). "Ken" stands for Kenyon International, which was responsible for recovery of bodies in the aftermath of Katrina. Had the note been placed before "Ken" had been sprayed, or was the letter writer unaware of its implication? In the bottom image a single "1" on the wall indicates a body was found inside. The car (which has been enlarged and the license plate covered) shows that the occupant was disabled.

6 A disproportionate number of the elderly who died during Katrina were elderly. Of the 800 deaths attributed to Katrina by the CDC, 60% were 65 years or older. The average age for all men who died was 64, and 71 for women. Only 12% of the population of New Orleans was over 65 (Scripps Howard News Service). A horrific example of this elderly vulnerability was in a St. Francisville, Louisiana nursing home where 35 elderly people were left by the proprietors to drown. The victims were either in wheelchairs or beds. Unfortunately we heard many stories from New Orleans about elderly people discovered during September and October after Katrina. Most were discovered by a neighbor or clergy who had returned and simply wondered what had happened to a neighbor. The most heart wrenching of these experiences was a faded note left on a door (see Figure 5:2). Researchers had previously described before Katrina that the elderly were less likely to evacuate along the Gulf Coast when facing a hurricane Howell, Susan E. 2005. "Citizen Hurricane Evacuation Behavior in Southeastern Louisiana: A Twelve Parish Survey." Survey Research Center, University of New Orleans, New Orleans, LA.

Figure 6:2.
Photographs by Curtis

In keeping with the theme of the "hazardscape" this chapter will present a series of cartographic overlays representing both physical and social vulnerability in New Orleans. Figure 6:3 and 6:4 display two expressions of physical exposure (average flood depth in meters), and the average amount of building damage per census tract. Several things should be mentioned about these maps. Firstly, the three "neighborhoods" included later in this book as exercises are added to allow for future social vulnerability context (these are identified by boxes). Secondly, the color scheme is reversed from normal, with the highest/worst category being the lightest. This is in order to help with map interpretation—the white of the highest category will help identify any coinciding spatial layers. Thirdly, a measure of actual exposure during the disaster, a contour surface of the total number of 911 calls generated for up to five days after Katrina are displayed. Fourthly, the map itself may seem a little strange but this is because the political boundaries cover all space, including the Mississippi River and Lake Pontchartrain. We have not discussed the area extending south of the Mississippi River in this book, though it is presented here. The area to the north of the map, which appears as black in most images, actually marks the (approximate) edge of the Lake.

The way to interpret a map overlay is to see where different layers intersect. For these maps we are looking at the surface of the underlying map (the physical or social vulnerability), the disaster exposure, and our three study neighborhoods.

Figure 6:3.

In Figure 6:3 we should first be aware that flood depth varies within each of these census tracts. Other flood maps in this book (Figures 2:2 and 2:3) do a better job of depicting the spatial heterogeneity of flood depth. However, for this exercise this flooding layer still shows that the worst flooding was apparently around the edge of Lake Pontchartrain, though the Lower 9th Ward also contains one section of considerable depth (over five meters). The highest concentration of 911 calls (the way to interpret these is, the closer the lines are packed together, the greater the number of calls) does not necessarily coincide with flood depth. If we compare the three study neighborhoods by flood depth, we see that the Lower 9th Ward neighborhood suffered the worst, with the 7th Ward[7] (the top box) and Holly Grove (the left side box) both mixing between the second and third highest categories—meaning their flood depths were still at least 2 meters (average) depth.

In Figure 6:4 we should again remember the loss in detail from a census tract aggregation. If we refer back to the map in Chapter 4 (Figure 4:12), there were areas of considerable damage even in the sections of the city that escaped flooding. Another point to be aware of with this map is the highest category of damage is actually light grey as white is reserved for sections of no data. In correct cartographic tradition this should not be part of the same grey scale—but introducing another

7 This neighborhood also has other names, including the Bernard area.

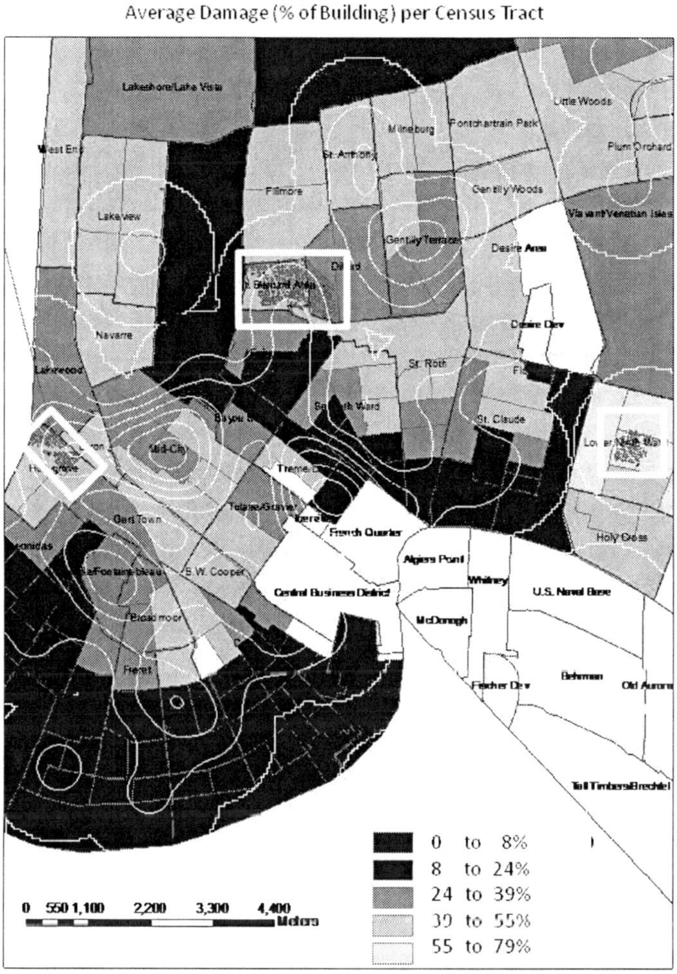

Figure 6:4.

color scheme will confuse the map further.

In this map, the contour distribution actually follows the building damage layer closer than for flooding, though again, it is not a perfect match. All three of our neighborhoods fall in one of the top two categories for building damage, meaning that structures in these neighborhoods have at least an average of 38% damage.

It is widely acknowledged that women suffer disproportionately in a disaster for multiple reasons. For example, women living close to or below poverty tend to be over reliant on the social network support system of their neighborhood[8]. As a consequence, they will suffer its loss more so than men. In addition, women may be more attached to the home (especially items with family or memory significance). As a result the loss of the home may affect the sexes in different ways, leading to depression in women, and possibly even physical risk as male negative coping strategies manifest as spousal abuse[9]. A further disproportionate exposure occurs with women who spend more

8 It has been shown that for a woman maintaining this support structure can be more important than preserving the actual home itself Fordham, M. H. (1998). "Making Women Visible in Disasters: Problematising the Private Domain." *Disasters*, 22(2), 126-143.

9 Relocation and / or the loss of items associated with the memory of the relationship can lead to arguments, separation, abuse and depression. See Cutter, S. (1995). "The Forgotten Casualties: Women, Children and Environmental Change." *Human and Policy Dimensions*, 5(3), 181-194; Enarson, E. (1998). "Through Women's Eyes: A Gendered Research Agenda for Disaster Social Science." *Disasters*, 22(2), 157-173; Enarson, E., & Morrow, B. (1997). "A Gendered Perspective: The Voices of Women."

time in the home, meaning she is perpetually reminded of the situation and its losses (especially so if the family is living in a FEMA trailer). In addition, there may also be the anxiety of not knowing if the home environment was safe—the most obvious example being from mold toxicity.[10]

There are several examples of environmental exposure with disproportionate impacts to socially vulnerable populations[11], however one situation in particular can help illustrate this gendered exposure risk. Over a ten year period in Southern Mississippi, Methyl Parathion (MP)[12], a pesticide used to control the boll weevil in agricultural fields was used in poor communities to control pest infestation in their own homes. MP was used in 1,800 buildings, mostly homes but also in social gathering points such as day care centers and churches. The chemical was not supposed to be used in domestic situations, but this did not stop unscrupulous pesticide sprayers. Once it was determined that exposure to the pesticide was harmful to health, people left their homes. Unfortunately, families that were not adjudged to have been exposed enough, in other words not reaching a critical threshold level, received no financial assistance to relocate. This situation resulted in residents making a choice of paying for the clean-up of their homes, or living with the consequences. It is not surprising that the authors of this study found that those living in poverty, African Americans and women, suffered these consequences disproportionately. Simply put, families that could not afford to move, or clean-up, were left to either be further exposed, or believe they might be exposed. If a woman spent more time in the home, then she would disproportionately face this burden, with a reasonable chance that anxiety could manifest to a mental health problem. Furthermore, another correlated factor with families living in poverty is that only 16% of those diagnosed with depression went on to receive treatment.

It is not hard to imagine that a similar situation faces women who have returned to New Orleans, because as they move through the clean-up stage, a reasonable question they may have is, is my home safe?

Disproportionate gendered exposure to risk increases if the woman is also the head of the household, and especially if she has small children under her care. In this case she is more likely to suffer a higher chronic stress load simply due to a lack of support in performing normal day-to-day activities, for example earning a living,

In W. G. Peacock, B. Morrow & H. Gladwin (Eds.), *Hurricane Andrew: Ethnicity, Gender, and the Sociology of Disasters*, Routledge.

10 In one post-Katrina survey, 93% of all surveyed believed that mold in the home could make them sick (2006).

11 Indeed this opens up a connected subfield of Geography focused on environmental (in)justice.

12 Rehner, T., Kolbo, J., Trump, R., Smith, C., & Reid, D. (2000). "Depression Among Victims of South Mississippi's Methyl Parathion Disaster." *National Association of Social Workers*, 25(1), 33-40.

Figure 6:5.
Source: October 19, 2005 by Curtis

attending medical appointments, shopping, or transporting children. Of course, this situation is worsened if she has little money (a likely situation) not only because of the normal anxieties of living within a tight budget, but also because the parent has less ability to pay for childcare[13]. Therefore the normal decision making process, such as attending regular family clinic visits, shopping for nutritious food, even facilitating her child's outdoor activity come with an opportunity cost—often these must be sacrificed for more work hours.

Figure 6:5 shows a daycare center in the St. Roch neighborhood. There are two points of interest on this photograph. The first is the flood mark at the top of the sign—an indication of the physical exposure. The second are the hours of operation, with pick-up times of 1am—an indication of a one-parent family area and a service that would allow late work hours if necessary.

Figure 6:6 continues our series of vulnerability overlays with "Percentage of Families with Female Householder, No Husband Present, With Related Children under 18 Years, Below the Poverty Level in 1999." In other words, three layers of vulnerability are captured here, female as head of household, with children, and living in poverty. Obviously this surface would cause concern with regards to potential disaster exposure. In terms of our three study neighborhoods—all three sit in the second to third highest categories.

If a single parent decided to stay through the storm, the additional stresses of being stranded, waiting for rescue and then staying in the evacuation centers, while also caring for a child, would have been immense. There have been several reports suggesting that the mischievous action of children in the shelters, a common occurrence because of the lack of child-friendly activities, only increased the stress burden on the parent. Once again, the poverty effect comes into play because the situation resulting in this single parent sitting in the shelter is also likely to have left him/her with one of the many chronic diseases common to the urban poor of New Orleans, such as diabetes,

13 Which helps explain why women are more attached to the social network structure of their neighborhood. To "evacuate" from this is not just an exercise of moving location, but removing a safety net.

Figure 6:6.

Percent of Families With Female Householder, No Husband Present, With Related Children Under 18 Years, Below the Poverty Level in 1999

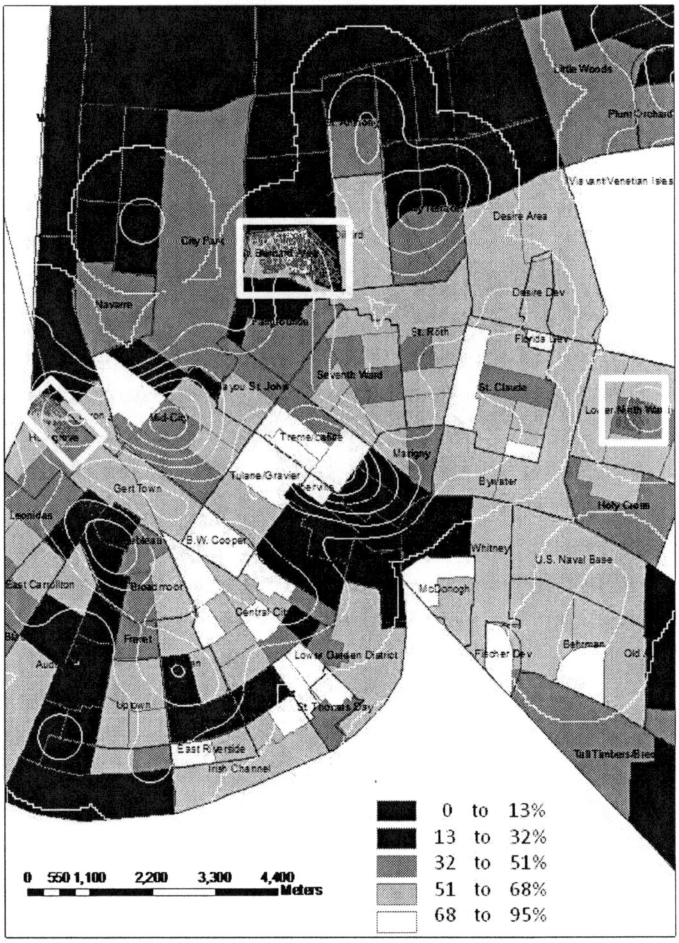

hypertension or obesity. It should also be remembered that the children also suffer these same situations[14]—they come from the same neighborhoods and domestic environments as their parent, meaning they share some of that stress burden. They have just gone through the same disaster related exposure, and now they face shelter hardships, relocation problems, and "bouncing around" new education environments. As a result it would not be surprising if the impact of Katrina is felt by an entire generation as they carry these effects on into later life.

POVERTY

Being a female head of household is often tied with two other measures of vulnerability—living in poverty and being African American, both of which will be considered here. People living in poverty are more susceptible to the impacts of a disaster for a variety of reasons. Their exposure can be greater simply because they don't have the financial resources to evacuate. A family must have a car to act on media warnings that a hurricane is approaching. If it doesn't, then there must be accessible public transport. Before Katrina a lack of available transportation left many communities apparently stranded. However, work we conducted in the Lower 9th Ward using the spatial video showed that many cars had been left

14 Madrid, P. A., R. Grant, M. J. Reilly, and N. B. Redlener. 2006. "Challenges in meeting immediate emotional needs: short-term impact of a major disaster on children's mental health: building resiliency in the aftermath of Hurricane Katrina." *Pediatrics* 117:S448-53.

behind in that neighborhood. Although an abandoned car may be an indicator of a two-car family, it also suggests another poverty connection—actually having a vehicle doesn't mean a family can afford the gas, or a place to stay on the other end (such as an extended stay in a motel).

Figure 6:7 continues our series of overlays with "Percentage of Occupied Housing Units with No Vehicles Available." The central section of the city again has the highest percentage of this variable, which coincides to some degree with the 911 hotspots. In terms of the three neighborhoods, all again fall in the two highest categories.

A lack of money will also hinder the recovery process, limiting the number of journeys a family can make back into the city as well as the financial ability to actually rebuild. Even if the family had flood insurance—which is doubtful if they were living close to the poverty level, then the gap between what they were offered in an insurance settlement and what it takes to rebuild will also hinder recovery. Finally, although a post-Katrina assistance plan was created, the Louisiana Road Home, the education needed to navigate through this was probably beyond many indigent families (again education and poverty tend to be correlated). In addition the system itself was not effective.

Figure 6:8 shows data collected using the spatial video that PolicyLink used for the Churches Supporting Churches group to map the "rebuilding gap" found in the Lower 9th Ward. Simply put, this average value of $40,000 for the neighborhood is the total difference in funds received from all

Figure 6:7.

Figure 6:8.

sources, and how much it would cost to fully recover a property. The spatial video included returnees, blight, and important social buildings such as churches (highlighted in Figure 6:8).

We can also discuss the vulnerability of poverty as an indicator of exposure. For example, during Katrina not having the ability to leave New Orleans not only resulted in suffering through the storm, but then having to survive in the city for several days before rescue. The exposure of the hurricane was compounded by a flood that left many on rooftops, without adequate food or medicine, and helplessly waiting for rescue in the sweltering Louisiana heat and humidity. This exposure continued through an evacuation to a shelter that held its own horrors, and then on a long uncomfortable bus ride to (for some at least) another state. For many this was the start of a string of other stresses—a second hurricane-related evacuation from Houston as Rita approached, living in FEMA camps for extended periods, and being separated from the social network of their old community which is especially strong in New Orleans. Although the experiences of Katrina were not limited to the poor of New Orleans—they were disproportionately felt by the poor. In other words, the events described in this paragraph were more likely to have happened to those living in the poorest neighborhoods. A further series of health vulnerabilities associated with poverty is also linked to these exposures, both in terms of increasing the likelihood of not evacuating (obesity being one example), and exaggerating every experience that followed by exacerbating the symptoms. We will return to these issues shortly.

RACE

A further correlate with poverty is race, especially in the Deep South of the United States. Academics have frequently drawn the connection between disaster related vulnerability and race. Generally speaking, African Americans are more likely to live in neighborhoods that have greater risks associated with them, while at the same time suffering from a higher chronic health burden and a higher background chronic stress load (such as living with high crime and generally fewer opportunities)—all of which will again increase disaster exposure[15]. In addition, mitigation approaches and especially response strategies have long been criticized as being culturally and ethnically insensitive[16]. This situation again means that the degree of exposure[17], even if physically experienced in the same way by two families, one white and one black, will still leave a different lasting impact. It is easy to understand the anger caused by the scenes of African Americans waving for help from rooftops as helicopters buzzed by[18]. Partly as a result of these images a common perception held both at the time, and still today, is that the abandonment of the citizens in New Orleans was mainly due to the color of skin. Whether this is true or not, the perception is real for many residents of the city who often ask—would a poor white population have suffered in the same way? From the perspective of another area of the United States with a looming hazard and a considerable political and academic dialogue regarding race—how would Latino populations fare if a catastrophic earthquake were to hit Los Angeles?

---

15 Before Katrina there were several papers writing about this race-related vulnerability, most notably Pulido, L. 2000. "Rethinking environmental racism: white privilege and urban development in Southern California." *Annals of the Association of American Geographers* 90:12-40; Peacock, W G, B H Morrow, and H Gladwin. 2000. *Hurricane Andrew and the Reshaping of Miami: Ethnicity, Gender, and the Socio-Political Ecology of Disasters*, Florida International University, International Hurricane Center. Miami, FL Bolin, R.C. and P. Bolton. 1986. "Race, Religion, and Ethnicity in Disaster Recovery." Boulder: University of Colorado though the events of Katrina have focused far more attention on this topic, resulting in books such as:

16 Phillips, B.D. 1993. "Cultural Diversity in Disasters: Sheltering, Housing, and Long-term Recovery." *International Journal of Mass Emergencies and Disasters*, 11:99-110.

17 This has been likened by some academics as the modern day equivalent of a lack of lifeboats on the Titanic, Atkins, D. and E. M. Moy. 2005. "Left behind: the legacy of hurricane Katrina." *BMJ* 331:916-8.

18 68% of respondents in one post-Katrina survey believed this abandonment as reported by Atkins, D. and E. M. Moy. 2005. "Left behind: the legacy of hurricane Katrina." *BMJ* 331:916-8 We have also witnessed many occasions when our African American community collaborators have been overcome with emotion when discussing the plight of their community – the simple question they ask is, how can this be allowed to continue? Not surprisingly the answer soon becomes one of race.

Figure 6:9 displays the proportion of African Americans across New Orleans. What should immediately be obvious is how the entire middle swath of the city, which coincided with many of the other vulnerability measures, as well as the 911 call hotspots, is heavily African American, many census tracts being above 85%. All three of our neighborhoods fall into this highest category.

IMPEDIMENTS TO EVACUATION

Many of the reasons why African Americans didn't evacuate from New Orleans can be linked to the previously mentioned vulnerabilities[19]. Other reasons why people don't evacuate when facing an impending disaster can be socially non-specific including having "evacuation fatigue" though of course carrying a chronic illness may also influence this burn-out, to a concern about the safety of pets in the home[20]. Previous evacuations from hurricanes along the Eastern Seaboard of the United States had resulted in 20 hour journeys—even the trip from New Orleans to Baton Rouge tended to take 8 rather than 1.5 hours—so it is hardly surprising when people look for reasons not to evacuate. It should also be remembered that evacuating can bring its own risks—for example, 20 died on a bus from Galveston, TX, full of elderly escaping from Hurricane Rita.

In one Texas shelter survey, 61% said they hadn't evacuated before Katrina's landfall, and of these 36% claimed the single biggest reason why they didn't leave was because they had no means to evacuate[21]. Figure 5:7 presented an overlay of where census tracts with few vehicles were found—in other words, showing the geography of this vulnerability. However, for many families in New Orleans they simply had nowhere to go, even if they had a car. The geographic extent of friends and family for many of the highest risk census tracts is relatively small meaning there were few options in terms of evacuation destinations. Given just these reasons alone, it is not hard to understand why people chose to "weather out" the storm.

19 Research before Katrina had already noted that African Americans were less likely to evacuate along the Gulf Coast when facing down a hurricane—see Howell, Susan E. 2005. "Citizen Hurricane Evacuation Behavior in Southeastern Louisiana: A Twelve Parish Survey." Survey Research Center, University of New Orleans, New Orleans, LA

20 Dow, K. and CL Cutter. 1998. "Crying wolf: repeat evacuation issues: repeat response to hurricane evacuation orders." *Coastal Management* 26:238-52.

— 2000. "Public orders and personal opinions: Household strategies for hurricane risk assessment." *Environmental Hazards* 2:143-55.

— 2002. "Emerging hurricane evacuation issues: Hurricane Floyd and Souht Carolina." *Natural Hazards Review* 1:12-18.

21 According to the 2000 census, approximately 250,000 people in New Orleans did not have access to a private vehicle, and of these, only 10% could be evacuated by city buses (though this was not attempted D Atkins, EM Moy: "Left behind: the legacy of hurricane Katrina." *BMJ* 2005, 331:916-8.

Figure 6:9.

Percent of African Americans per Census Tract

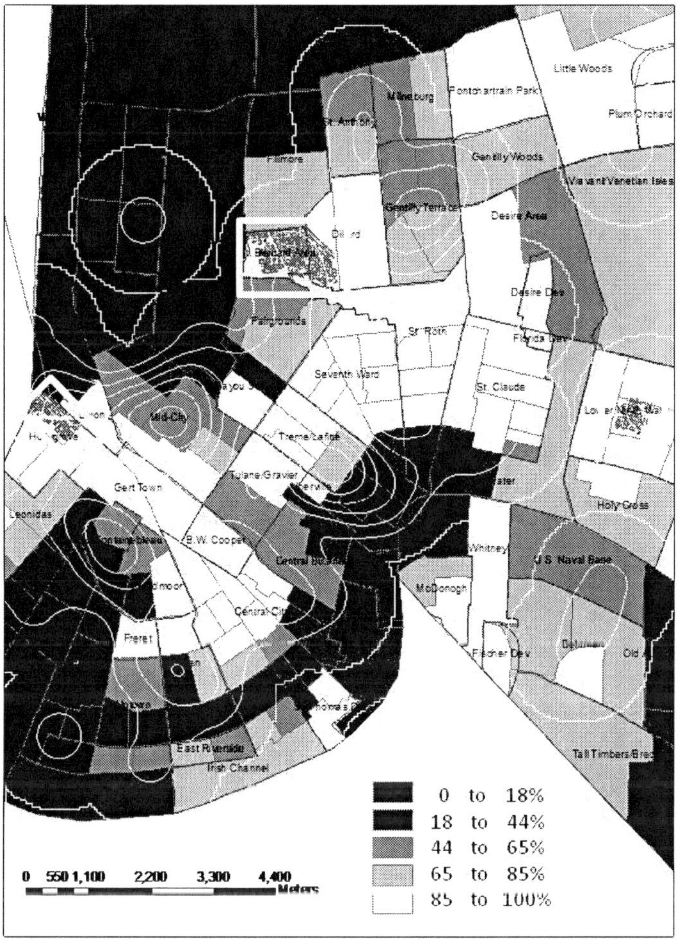

The final combinatory reason for not evacuating is a mix of economic hardship, not completely understanding the severity of the situation and simply believing the government had measures in place to protect them. Many families couldn't afford flood insurance (or any insurance), so they felt a need to stay and protect their belongings. They also did not believe that the situation would be as bad as it got (29% in that previous survey said they had underestimated the storm), and also believed that the government had suitable flood protection measures in place (mainly the flood control and levee system).

Figure 6:10 shows the areas with the least educational attainment (Percentage of persons 25 years and over with less than 9 years of school). Although this could be a factor in increasing exposure, in terms of not understanding the potential risk, it is probably a variable of concern more for the recovery process as families try to navigate through the various federal and local programs offering assistance. Of our three neighborhoods, the 7th Ward (also called the St Bernard Area) sits within the highest category.

ADDING GEOGRAPHY

Throughout this chapter we have been using cartographic overlays to move our discussion of the hazardscape into the geographic realm—so that we can see where

Figure 6:10. Percent of Persons 25 Years and Over With Less Than 9 Years of School

problems exist, and how these spatially intersect. It is also possible to describe these surfaces in terms of a classic geographic concept: site and situation. Site is the physical location of a neighborhood and situation is the context of that location. The site would be elevation¬—how far below sea level, how close to a levee, or if we change hazards, does it sit on a landslide zone, or in the urban wildlife interface zone (for wildfire risk). For Katrina, the area that escaped most of the flooding was also the oldest, built on the highest land available (see Figure 6:3). The situation for each of these neighborhoods involved the degree to which that physical vulnerability coincided with a set of social conditions that would accentuate the risk. For example, the area around Lakeshore in New Orleans was proximate to both the lake and levees, and as a consequence received considerable flooding. The homes here were also large and expensive. It is likely that more residents from this neighborhood had the social ability to evacuate, thus lessening their physical exposure, and had more resources to recover afterwards, thus reducing their exposure length in comparison to similar flooded neighborhoods in the Lower 9th Ward.

It is an interesting exercise to extend this "site" and "situation" discussion to the other maps of the chapter, and then to other cities facing a different set of hazards.

DID IT REALLY HAPPEN? ANALYZING VULNERABILITY

We can detail the various social components of vulnerability as we have done in this chapter—we can also spatially present these results as maps. However, as academics we often want to go beyond this and model the spatial interrelationships among these layers. For example, can we explicitly state that the general impression held about Katrina, that the disaster disproportionately affected African American neighborhoods, is true? And if so,to what degree?

The social geography of New Orleans before Katrina was a "perfect storm" of vulnerability. The city had excessive poverty, crime, drug use, poor health and a social service structure that placed severe impediments on the ability for many citizens to succeed, or even live normal productive lives. At the risk of oversimplifying the social, economic, and political conditions that contribute to this urban situation, we can map these geographies using cartographic overlay. Simply put, neighborhoods in poverty, which also sit in areas of high crime, high drug use, poor health and failing schools, are more likely to hold families with the least ability to cope when faced with the externalities of a catastrophe. However, instead of just "visualizing" the problem, we can apply analytical techniques to model these interactions. Possibly the most common analytical technique in social science is a regression analysis. There are many types of regression approaches, ranging from models that don't directly consider geography, to those that include the subtleties of space in their calibration[22]. This book will not describe how these techniques work instead it will interpret the results of one such regression approach. The important point to remember here as we interpret these results is that we are looking for relationships among variables—not just where flood and social vulnerability measures coincide, but to what degree does that association occur.

Figure 6:11 shows a combination of a regression approach used to associate flood height with different social vulnerability measures[23] extracted from the United States Census Bureau 2000 Summary File 3 (SF3) for 481 block groups in Orleans Parish. In keeping with our previous cartographic overlays, these indicators include: percentage of households with incomes less than $25,000, the percentage of nonwhite population, percentage of households without a vehicle, percentage of single parent households with children under the age of 18, and the percentage of households

22 Two excellent and easily readable papers on the issue of local statistics and regressions in geography are: Fotheringham, A. S. "Trends in quantitative methods I: stressing the local." *Progress in Human Geography* 21 (1997): 88-96; Fotheringham, A. S. "Trends in quantitative methods II: stressing the computational." *Progress in Human Geography* 22,2 (1998): 283-292

23 The work presented in this chapter is a precursor to a paper published in the journal Curtis, A., Pine, J., Marx, B., Li, B. and J. Mills forthcoming "A Multiple Additive Regression Tree Analysis of Three Exposure Measures during Hurricane Katrina: Implications for Recovery in Orleans Parish," *Disasters: The Journal of Disaster Studies, Policy and Management.*

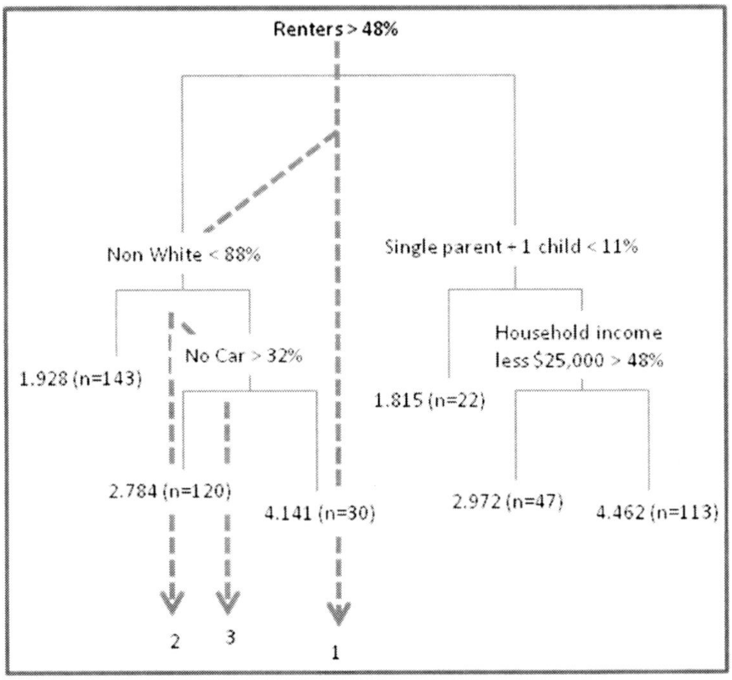

Figure 6:11. *Source: Modified from an original regression output produced by Brian Marx and Bin Li Department of Experimental Statistics at Louisiana State University.*

who rent. For this analysis, census block groups with no population (City Park and New Orleans East) were not included.

The way that this regression tree can be interpreted is as follows—the "branch" on the left side of the diagram covers 143 census block groups—and it shows that the average water depth for predominantly non white census block groups (over 88% being non white) and more than 48% of families being renters was 1.928 meters. Following another sub-branch—30 census block groups with a high non-white population, high proportion of renters, and where over 73% of the families had no vehicle, had an average flood depth of 4.141 meters in 30 census block groups, and 2.784 meters in 120 census block groups. These values give us something concrete to begin a discussion with, such as what was the flood height for populations that were vulnerable and could not evacuate. The diagram also shows a simple additive process of vulnerabilities which mirrors our overlays. In this simplest of additions, where no one vulnerability is given a higher weight than another[24], any place on a map can have a vulnerability score created based on reaching threshold points (for example greater than 75% living in poverty, 75% being non white, 75% not having a car, etc.—in this case the highest score would be "3"). The regression approach is more pleasing from a statistical perspective, especially as it reduces our need to choose these arbitrary thresholds.

24 This is unrealistic as different vulnerabilities are likely to have a greater impact on exposure if flood depth is held constant – for example being elderly is likely to be of more concern than being a renter.

Even more sophisticated regression analyses can be performed on these same data, and a second round of output can be seen in Figure 6:12. The way to interpret these diagrams is to read the flood depth on the "Y" axis, and the percentage of the vulnerability variable along the "X" axis. For example, in census block groups with more than 60% of families renting there was actually minimal flooding.

As the neighborhood proportion of renters increased, so lower flood heights were found. Although this is counter-intuitive to the traditional finding in the hazards literature, it makes sense from what we know of the post-Katrina landscape, as many neighborhoods with the deepest flooding, for example Lakeview and New Orleans East, also had the highest proportions of home owners. What is more

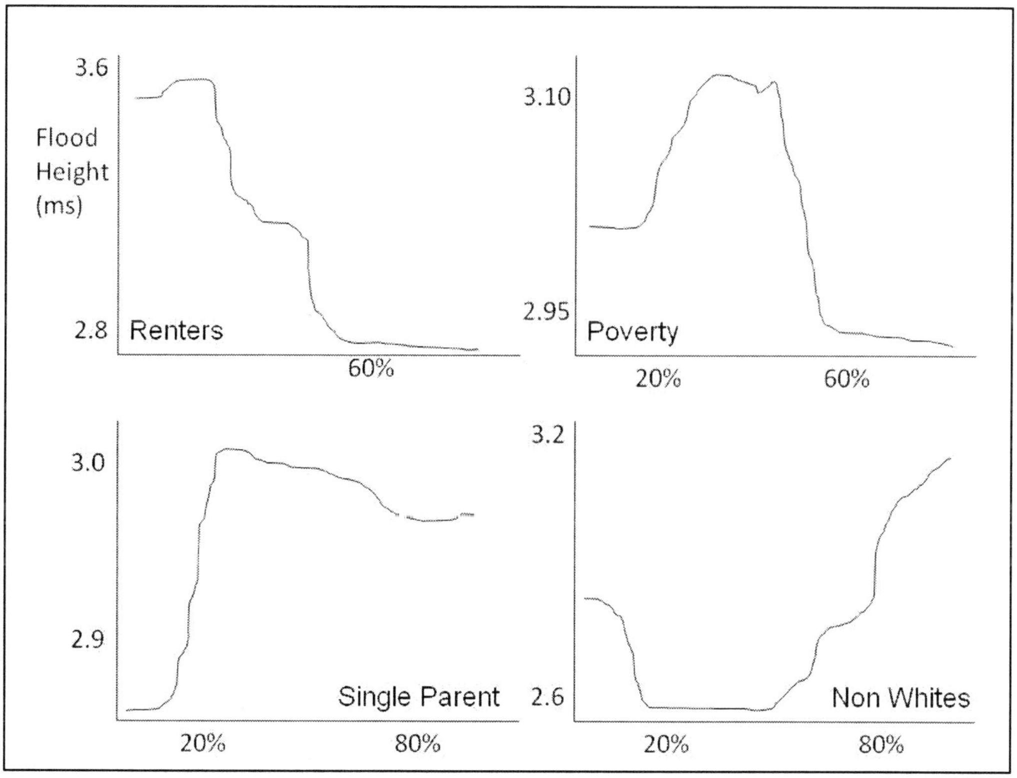

Figure 6:12. *Partial dependence plots from a Multiple Additive Regression Tree Analysis Original regression output produced by Brian Marx and Bin Li Department of Experimental Statistics at Louisiana State University. These are precursor results for a subsequent published paper.*[25]

25 Similar model output for spatial aggregations at the census tract level can be found in Curtis, A., Pine, J., Marx, B., Li, B. and J. Mills forthcoming A Multiple Additive Regression Tree Analysis of Three Exposure Measures during Hurricane Katrina: Implications for Recovery in Orleans Parish, Disasters: The Journal of Disaster Studies, Policy and Management

interesting is that the Lower 9th Ward, which has been characterized in the media as being comprised of a poor minority population, and by extension, one would assume a high rental occupancy, also contains many home owners.

Figure 6:13 displays not only the renters but also those living close to the poverty level (as captured by the measure Percent of Specified Renter-Occupied Housing Units in Which Monthly Gross Rent Is 35 Percent or More of Household Income in 1999). The majority of Holly Grove and our Lower 9th Ward neighborhood fall within the highest two categories of poor renters.

Several interesting renter-related questions can be asked, such as, is there a greater post disaster stress load for a poor homeowner rather than a renter? The poor homeowner is left with the chronic stress of having to return home and rebuild, which, in combination with other vulnerabilities such as a poor education, may result in a prolonged fight for financial help. The renter has less tying him/her to one location, and although appearing to be forced to relocate, this may actually reduce the stress load, and the associated mental health problems as he/she is forced to move on. According to our regression output, there was still a high association between flooding and neighborhoods where 1 in 5 homes were rented. This loss of rental stock, in combination with a shift by some homeowners into temporary rental accommodations, resulted in the demand exceeding supply for units, and an increase in rental prices.

The regression output also reveals that as the percentage of non-whites (starting at around 50%) in

Percent of Specified Renter-Occupied Housing Units in Which Monthly Gross Rent Is 35 Percent or More of Household Income in 1999

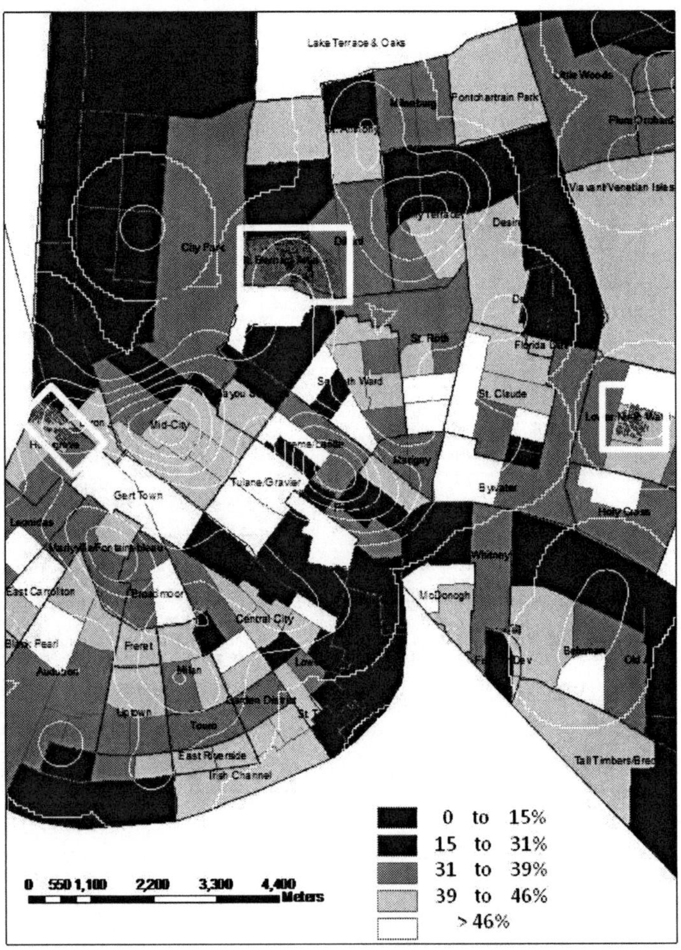

Figure 6:13.

a census block group increases, so the flood height also increased. However, census block groups with the highest percentage of whites (the beginning section of that plot) also experienced high levels of flooding. This is understandable given the heavy damage experienced in the Lakeview neighborhood by the break of the 17th Street Canal. However, flood height in the highest percentage non-white block groups exceeds those in the highest percentage white block groups. Of course this is where our previous discussion as to the complexity of the hazardscape comes into play, because even though these two neighborhoods suffer from flooding, one is considerably more financially adapt at recovering from the situation.

The other plots can be interpreted in the following way: flood height deepens in block groups with increasing poverty until a threshold of 60% (in homes with less than $25,000 annual income) is reached, at which point the flood depth falls off dramatically. It would be wrong to interpret this finding as the poorest neighborhoods did not flood as dramatically as portrayed in the media. The highest flood levels were associated with census block groups with between 20 and 50% of households in poverty. These would still represent neighborhoods with extremely high poverty rates.

Arguably the most interesting of the plots is for single head of household families with children in the home. For this variable the highest amount of flooding is found in block groups where greater than 20% of families are headed by a single parent. Although the flooding level does decrease slightly from this high, there was still considerable flooding for all block groups with a high percentage of single parent families.

The main deficiency of the regression analysis is that the results are not explicitly spatial. We can interpret results "generally" in terms of the average flood height for different vulnerability matches—but we do not see these spatial associations displayed on a map. There are several ways to incorporate geography into a regression[26]—and although these techniques have not yet been applied to Katrina—or other disasters, it is only a matter of time. Until then, however, normal statistical or regression analysis, in combination with the type of cartographic overlays presented here, still provide powerful insights into the patterns of physical and social vulnerability, and the exposure suffered during the disaster.

FINAL THOUGHTS

The geography of vulnerability is complex, and works at different geographic scales—individual, neighborhood, and city. It is important to remember that

26 For those of you interested, Geographic Weighted Regression is arguably the most common "spatial" regression approach.

although different neighborhoods may be thought of as vulnerable because of the population characteristics of the people who reside in these places, there is further disproportionate risk or exposure for sub-groups within these communities. For example, of particular importance to poor African American women in New Orleans is the loss of a neighborhood support structure, which traditionally extends to other females in the household (mother and grandmother). The church also plays an important role in this community support network. In both our Lower 9th Ward and 7th Ward neighborhood areas there are sometimes two or three churches per road. I asked one of our collaborators, how could the population support so many houses of worship? He told me, people would return to services at the church of their childhood, irrespective how far away they lived. Where you live now is not as important as the neighborhood of your childhood. This also makes it more understandable why people feel loathe to leave this geographic community, either for an approaching hurricane or later as rebuilding becomes so challenging. The neighborhood support mechanisms are expansive and multi generational. As stated earlier in this chapter, the neighborhood social structure is often viewed as being more important than personal possessions.

What we must learn from this disaster is that a single coarse-scale response (or recovery) plan is not enough. Time must be spent understanding the cultural and social needs of neighborhoods and the special needs of sub-groups so that all aspects of disaster planning and response are fair (and therefore effective).

The next chapter will consider a set of variables arguably more important than social vulnerability in terms of predicting both exposures to the actual event and to a long term successful recovery. It is therefore surprising that health vulnerability is often assumed away as an unfortunate side-effect of the other measures described here.

SEVEN

Health Vulnerability

IN THE PREVIOUS CHAPTER we discussed the link between social vulnerability and disproportionate exposure to a disaster. What was not considered, at least not in detail, was the associated condition of health vulnerability. Subsequent chapters will describe issues of public health associated with a disaster; from the perspective of pre- and post-event vulnerability, disaster-caused morbidities and mortalities, and the mental health issues that may arise in the impacted population.

Many of the traditional measures associated with disaster related social vulnerability are also linked to health disparity—especially in regard to race and living in poverty. For example, an estimated 84,000 African Americans die per year because of health care deficiencies. Another way to describe this disproportionate health burden is by geography, with many of the chronic diseases afflicting the United States also being located within poor neighborhoods of large cities. A further geographic way to consider this health problem is by region—with the southern states, especially Louisiana and Mississippi consistently listed in unenviable positions on any health table. It is hardly surprising then that a survey conducted seven weeks after Katrina found that over 55% of households had at least one family member with a chronic health problem. Similarly, over 50% of evacuees in one shelter also suffered from at least one chronic disease. For New Orleans, examples of chronic diseases include Asthma, Diabetes, Hypertension, and Obesity, as well as a range of

Miami, FL	53.8
New York, NY	41.9
Baton Rouge, LA	35
Washington DC	35
Baltimore, MD	32.8
New Orleans, LA	31.9
Poughkeepsie, NY	31.3
Orlando, FL	31.2
Jackson, MS	30
Jacksonville, FL	29.9

Table 7:1. *AIDS case rates (per 100,000) for Metropolitan Statistical Areas. Source: CDC, HIV/AIDS Surveillance Report, Vol. 16, 2005*

Washington DC	11.42
Mississippi	10.33
Louisiana	9.95
Tennessee	9.05
South Carolina	8.98
Delaware	8.88
Alabama	8.82
Georgia	8.65
Arkansas	8.47
North Carolina	8.35

Table 7:2. *Infant mortality rates (per 1,000) for the worst ten states*
Source: CDC, National Vital Statistics Report, May 2, 2007.

sexually transmitted diseases, especially HIV/AIDS (See Figure 7:1 and Table 7:1—Louisiana has two Metropolitan Statistical Areas (MSAs) falling in the top ten list, while Mississippi has one, and Florida has three). In addition, Louisiana has one of the highest infant mortality rates, especially within the African-American population.

There are many reasons why social vulnerability is correlated with poor health outcomes including a lack of access to health care, poor nutrition (partly because of limited access to appropriate foods and not having the money to afford fresh produce), inadequate exercise opportunities (many neighborhoods in poverty also have high crime rates tending to keep people indoors), and a multitude of chronic living stresses. As a result chronic illnesses such as diabetes and obesity are reaching epidemic proportions for many U.S. cities. Suffering from any of these conditions also means a family is at a higher risk of an exposure to a disaster. A hurricane provides one of the clearest examples to demonstrate this point. People suffering from illness are less likely to want to undertake an arduous evacuation. As a result, this means that an already health-burdened population is more likely to suffer additional disaster related morbidities and mortalities. In terms of Katrina, people died or were injured mainly because of the flood waters, building collapse, hurricane-wind driven projectiles, and later heat exhaustion.

Starting at the top of Figure 7:2, we see, as was just explained, that disaster exposure can be linked to both social and health vulnerability. If a person survived without any immediate injuries, then he/she would be evacuated to a shelter. Unfortunately, the conditions in these shelters were often inadequate. Three possible medical outcomes of the "shelter stage" are the worsening of existing health conditions (such as diabetes) because of mental duress compounded by limited access to appropriate medication, and exposure to new infections especially skin infections. Of course the link with social vulnerability is that living in these shelters for a prolonged time is not a matter of choice but rather an outcome of a lack of financial ability and a reliance on "the system." Therefore, moving down the left side of the diagram, we see that the combination of initial health vulnerability, compounded with disaster exposure and shelter experience, may lead to a further widening of the health gap. Existing or even new conditions may be caused leaving a "Katrina

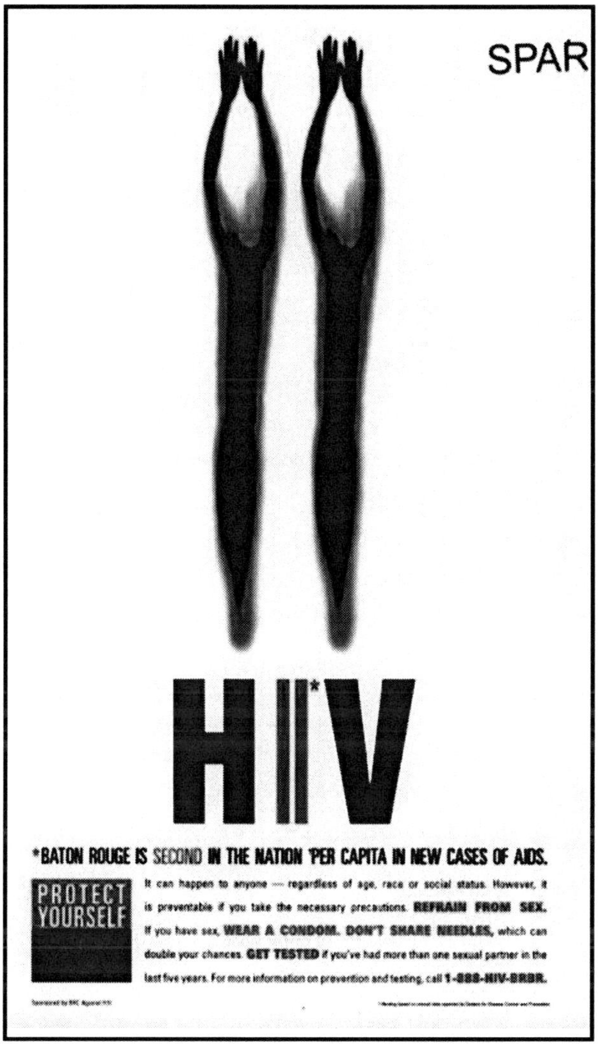

Figure 7:1. *A warning poster distributed to health clinics in Baton Rouge*

health legacy" such as a greater likelihood of future poor birth outcomes in the impacted population.

After leaving these shelters, and especially if the next move is into a FEMA village, the evacuee may now have an exacerbated health problem, may be suffering from a range of mental health problems, and could also be using inappropriate coping strategies, such as alcohol or drug use, which are also detrimental to his/her health. In order to "recover," the person needs to return to normalcy as quickly as possible, which means returning to his/her home, and returning to the neighborhood and social networks that helped him/her cope with the chronic stresses of life before Katrina. As long as this return is prolonged, so the health problems, especially stress-related conditions, continue. Linking back to the beginning of Figure 6:3, many of the measures used to define social vulnerability will also prolong the ability to return and recover. Not only will the family have less money to rebuild, but pre-existing health vulnerabilities may make it physically difficult to undertake manual labor. At the same time, these vulnerable populations have fewer resources to cope with the stresses which increases the likelihood for mental health problems.

As demonstrated in previous chapters, newspapers can provide an excellent source of disaster related material—especially with regard to the "human aspect" of the event. For example the following story nicely captures the elements of post-disaster health vulnerability:

In the two years after Hurricane Katrina pummeled the New Orleans area, Tulane University doctors found a threefold rise in heart attacks among their patients, and they put the blame on continuing storm-related stress, according to a study to be presented Sunday. Calling this sustained rise in heart problems part of Katrina's "lingering legacy," researchers found these patients showed post-storm increase in factors that are evidence of stress, including homelessness, unemployment, loss of insurance, smoking, poor eating habits, failure to take prescriptions and living in temporary housing such as FEMA trailers.

Living under such conditions "creates a high stress level," said Dr. Anand Irimpen, a Tulane cardiologist and one of the investigators. Norman Morales can vouch for that. The Chalmette man, whose home was drowned by floodwaters, suffered a heart attack in January 2008 after enduring a two-year odyssey that included time in shelters and a FEMA trailer. "It was pure misery. I would say the stress caused a lot of it," said Morales, whose case is part of the Tulane study. His treatment included quintuple bypass surgery. Although there has been a consensus that post-Katrina stress is responsible for an array of health problems, "the only way to prove it was to collect data and conduct a study," he said.

While the Tulane doctors were treating different patients after the storm, their characteristics were "very similar," Irimpen said, explaining that each group of men and women had similar rates of high blood pressure and diabetes, which are two risk factors for heart attacks. They were "well

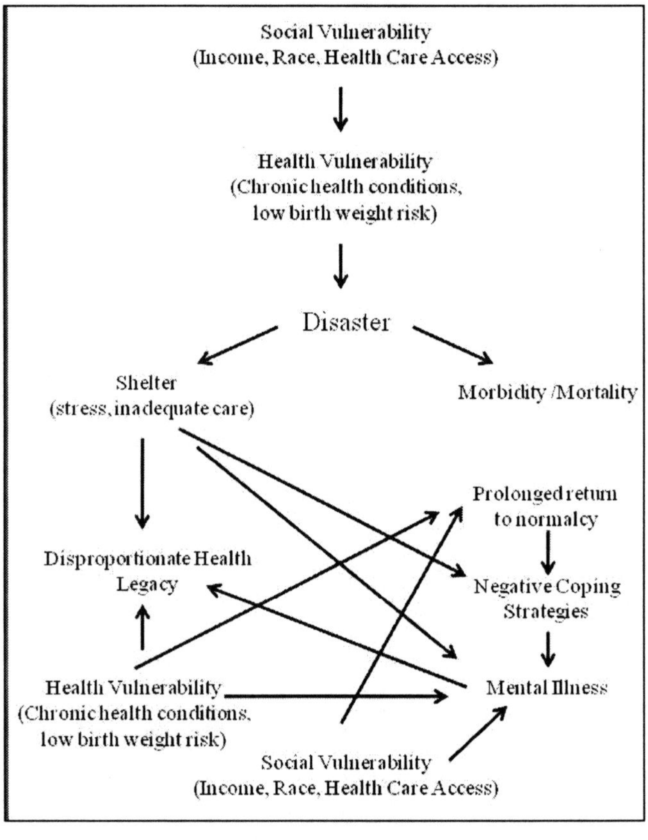

Figure 7:2. *Pathways connecting health and social vulnerability with disaster exposure*

matched" for those two conditions, he said, as well as for factors such as age and gender. The risk factors that rose after the storm were cholesterol levels, which Irimpen attributed to poor eating habits, and smoking. The men and women the team studied probably were stricken because they were too busy trying to put their lives back in order to pay attention to their health, Irimpen said.

"For those of us in New Orleans, we know what the effects are," he said. "You're constantly trying to get things done, people are still rebuilding, and they have to get through the stress of getting contractors to come or the workers to come on time. They're living in those homes and trying to get to work and their regular activities. When you're caught in a situation like that, how helpless you feel," Morales, 64, said. Researchers studied two groups of patients: 21,229 men and women who entered Tulane's hospital from Aug. 29, 2003, to Aug. 28, 2005, the day before the storm hit; and 11,282 people who were admitted from Feb. 14, 2006, the day Tulane's hospital reopened, through Feb. 13, 2008. Among the pre-Katrina patients, heart attacks struck 150 people—0.71 percent—compared with 246—2.18 percent among the people who were hospitalized after the storm. "We expected the increase," Irimpen said, "but I was a bit surprised by the threefold increase. A threefold increase is very significant." In most cases, "Disasters are over, and people move on," said Charles Figley, a Tulane professor of social work who is a renowned scholar of catastrophe and trauma. He was not part of the study. "Here, there are so many people who were affected and for so long because there was such widespread destruction," Figley said. "The Katrina effect will be with us for a long time."

Studies the team consulted followed patients for only a short time, Irimpen said, and most of them showed that stress-related problems abated relatively quickly. "We were impressed to see that even more than two years after Katrina, there was a significant increase in heart attacks," Irimpen said. "Katrina keeps rewriting the textbooks," Figley said, "and this is another chapter because there's nothing like it to compare with."

Although researchers found increases in such risk factors as smoking and substance abuse, Irimpen said they couldn't find one trend they had expected to see: an increase in alcohol consumption. The investigators found that stress afflicted more people than those who, for instance, lost their homes or their jobs. "You may not have been affected by Katrina," Irimpen said, "but every day, you're dealing with people who have been affected by Katrina." For instance, he said, "We had a contractor who came here to do work, and he presented with a heart attack. He probably was going through a lot of stress. We all think contractors have it easy, but some of those who take their work seriously are under a lot of stress."

"Tulane study finds threefold jump in post-Katrina heart attacks; stress blamed" by John Pope, The Times-Picayune, Thursday March 26, 2009.

From a geographer's perspective, it would be interesting to link the data in the above article to an actual location, and then to identify the spatial distribution of these stressors. Just as in the previous chapter where a series of cartographic overlays was used to create a holistic surface of disaster exposure and social vulnerability, so too could health variables be overlaid. Unfortunately, post-disaster health data are even harder to get than other demographic layers, so we often have to reach back before the event to see where health vulnerabilities initially clustered and then project these patterns forward.

Figure 7:3 shows the distribution of gonorrhea cases from 2004 by Census tract. In order to preserve patient confidentiality data are classified into ten categories, which basically means that each category is equivalent to 10%. Therefore the lightest category in the map shows areas falling into the highest 30% for this disease. Source: Louisiana Office of Public Health

Q: By comparing this map to those in the previous chapter on social vulnerability, what spatial patterns can you identify?

THE LINK BETWEEN DISASTERS AND POOR BIRTH OUTCOMES

One future health outcome that can be linked to previous disaster exposure (albeit the research is still in its infancy) is poor birth outcomes. Why is this topic worthy of investigation? Simply put, the disparities revealed during a disaster may leave future echoes that further widen the gulf between those who have, and those who have not. Infant mortality is often used as a societal indicator because the stresses faced by mothers during pregnancy, and in the first year of the infants life, in addition to the medical care system and other basic qualities of life, correlate with the infant mortality rate. In other words, it provides a life snapshot of the conditions in which the mother lives. It is not hard to imagine that a disaster, including the unsettled path to recovery, may have an impact on this rate. In addition, other problematic birth outcomes, such as short gestational and low birth weight deliveries are also susceptible to these same environmental "insults." This leaves the possibility of future generations also suffering because of the disaster. Many of these other poor delivery outcomes are linked to life-long developmental, behavioral, and even medical problems which will lay impediments along the path to a productive adult life.

Sexually Transmitted Diseases Rate Classified by 10% Categories

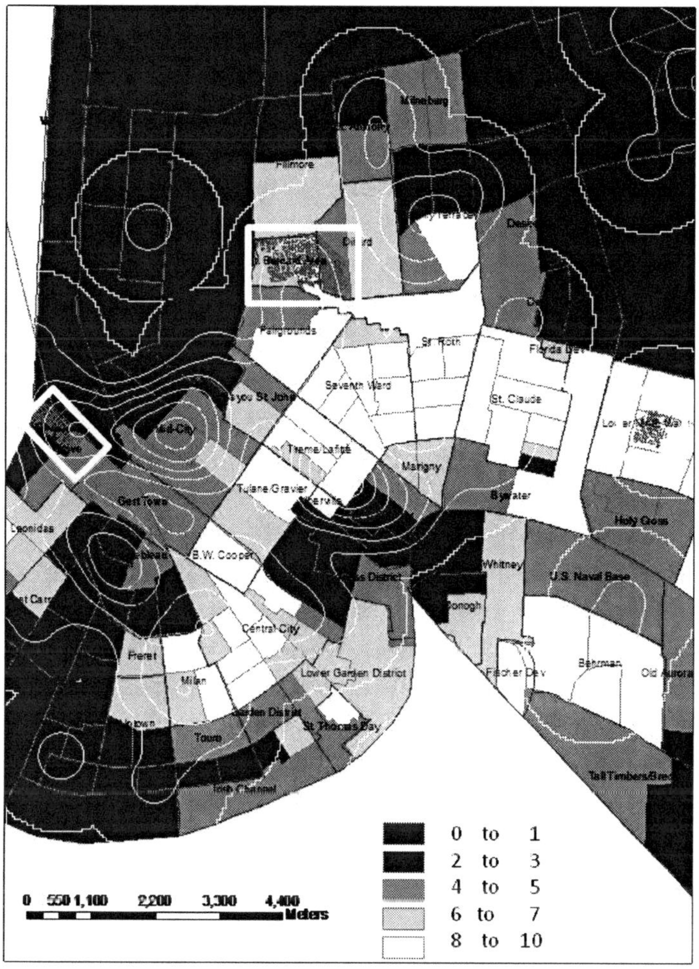

Figure 7:3.

In order to better understand how the post-disaster environment might affect birth outcomes, it is useful to outline traditional risks associated with pregnancies. Pregnancy risks can be broken into three general areas: behavioral, environmental, and medical. Behavioral risks can include smoking or drinking alcohol during a pregnancy, or choosing not to attend prenatal visits. These categories are often linked together by common neighborhood factors such as high crime and therefore a subsequent impact may be that a pregnant woman does not want to risk the trip to a clinic. In addition, the clinic may also operate on limited opening hours because of neighborhood crime. Therefore, what may appear as a behavioral decision is in reality an outcome of her environment. Similarly, another behavioral "choice" such as smoking during pregnancy may be a coping mechanism in response to neighborhood stress. This probably explains why many single message programs (such as just smoking cessation) are not successful because of all the associated problems which are not addressed.

Environmental risks, which we define here as originating outside of personal choice or biology, or in other words usually coming from a source in the home or the neighborhood, are the easiest to "map" without original medical data. Many of the cartographic overlays presented in the previous chapter could be used to create a comprehensive map of birth risk, as these are also traditional social risk factors associated with unsuccessful pregnancies. However, although these maps may provide useful insights (and they are often the only available data), understanding exactly how a neighborhood, or even a street, negatively impacts a pregnant woman is a geographic scale impossible to extract by only using census data.

Typical medical risks of concern to pregnant women include bacterial vaginosis, or chronic health conditions such as obesity or hypertension. Also, women who have previously experienced a poor birth outcome are more likely to suffer a similar outcome. What this last point means is that a woman who has previously had an infant death, a miscarriage, a low birth weight delivery (less than 2,500 grams), or an early birth (less than 37 weeks of gestation), is more likely to experience the same problem in subsequent births than someone who has a "clean" pregnancy history. From a geographic standpoint, although these factors are obviously individual, the root causes are often shared in the neighborhood¬—meaning that "hot spot" maps can still be made. For example, obesity may partly be caused by similar cultural practice but also by neighborhood stresses resulting in the individual having a more sedentary lifestyle. This in combination with poor local access to healthy food options, a poor educational system explaining risks, and even a poor local health support system could in effect be used to produce neighborhood risk maps. In other words, we may be able to rank neighborhoods for poor birth outcomes. Any woman suffering from an illness known to be a pregnancy risk, and who was also pregnant during the disaster, would be of particular concern. In the case of Katrina, patient records including medical histories were lost. Vaccination status of children, the correct medication, and with regards to a pregnant population, previous pregnancy problems and risk behavior, all had to be generated from the memory of the evacuee.

Irrespective of the disaster type or location, the following scenarios are likely to exist. Some women will be pregnant when the disaster occurs, some of whom will be in later stage pregnancies requiring special attention, while a further subset may have already suffered a previous poor birth outcome, making them more susceptible to a similar subsequent delivery.

Some of these women will also have been receiving counseling, medication, or other regular medical supervision. In a worst case scenario these pregnancy risks will be precipitated by the disaster, especially if the woman is close to her delivery date, possibly resulting in a preterm birth. Although there is research that suggests a delivery in a shelter may actually benefit the mother and child due to the critical mass of medical expertise associated with a disaster, this would not have been the case for births occurring in the squalor of the evacuation shelters located in New Orleans. It is therefore somewhat surprising that relatively little attention has been paid to the link with future birth outcomes. Although patterns have been established between poor birth outcomes and the terrorist attacks of September 11th , little pregnancy related research has been conducted on geographically extensive events such as Katrina, especially in the United States. However, if members of these vulnerable populations were also pregnant, or became pregnant soon after the hurricane's landfall, their vulnerability was likely to have exaggerated their disaster exposure. Given this situation it is possible the effect of Katrina could extend beyond the immediate response and recovery phase to include a subsequent generation.

Here are a few other sobering facts:

The African-American population of New Orleans in particular suffered high rates of poor pregnancy outcomes. This racial disparity is largely a result of secondary factors, for example a sizeable portion of African-American women also live in poverty, are single and raise children, have less access to suitable medical resources and healthy living (including food) options, and live in more highly stress-generating neighborhoods.

Although African-Americans only comprise 32% of the total population in Louisiana, they represent 73% of all new HIV cases, 63% of all people living with HIV (83% in Baton Rouge), In addition 89% of all HIV positive women who are pregnant are African-American.

In Health Region 2 which contains Baton Rouge, 94% of HIV positive women were African American, 11% received no prenatal care, and 22% received inadequate prenatal care. A common reason given for this inadequate prenatal care (especially in rural areas) was "pregnancy is not an illness"—therefore why the need to see a doctor.

The following sets of experiences were recorded by case workers at Family Road of Greater Baton Rouge, and the Healthy Start program located there. These groups traditionally provide pregnancy-specific services to pregnant African-Americans living in Baton Rouge. Prior to Katrina, Healthy Start would provide outreach, counseling, education and social worker/nurse help on a one-to-one basis for women originating from an eight zip code region identified as suffering from consistent poor birth outcomes. With Katrina's landfall, Family Road, Healthy Start, and numerous other non-profit organizations worked together to serve the unprecedented influx of people from New Orleans to Baton Rouge, while still trying to support their existing clientele. For several weeks the population of Baton Rouge grew dramatically, stretching existing service provision beyond expected capabilities. Several of the temporary shelters housing those relocated from New Orleans fell inside the Healthy Start program area, meaning that the size of the service clientele grew substantially.

Many of the evacuated women had no prenatal care, which increases the risk of a poor birth outcome. While working with this population, caseworkers found that Katrina related depression was partly to blame, one symptom of which is a lack of energy to seek a medical provider or attend prenatal appointments. Other depression related symptoms included arguments and a generally "short fuse" with family members or anyone else. These symptoms often had a genesis from the storm and the rescue/evacuation period, though subsequent conversations found that they were compounded by grief and anger resulting from the feeling of abandonment and the loss of material possessions. Expressions of anger witnessed by case workers included aggressiveness, including violence and rape, and self-destructive behaviors involving increased drug use. Using these "coping" mechanisms to deal with the stress of the situation also has a direct impact on pregnancies, with drugs, tobacco, and alcohol all being traditional pregnancy risks. Also, the desperate nature of the situation generally lowered inhibitions, increasing promiscuity as a comforting behavior. These actions, in combination with a lack of appropriate prophylactics, increased the risk of both sexually transmitted diseases and unwanted pregnancies. From the limited data collected in the Healthy Start surveys a mini "baby boom" developed where conception occurred after September 2005.

Case workers were concerned at the time that a lack of suitable sleeping conditions could even result in overlay deaths. This concern extended beyond the shelters to the trailers in the FEMA villages as the evacuated population lacked the financial resources to purchase cribs or bassinets. There were also unsuitable sanitary conditions for pregnant and postpartum mothers in the shelters, partly due to the overcrowding, with mothers in some instances having to use lavatories to wash the families clothing; take sponge baths; and clean baby bottles (the lack of feed formula in the shelters being another problem). In some instances parents were physically separated from their infants in the Neonatal Intensive Care Unit (NICU), usually due to a lack of available housing close to the hospital (remember almost the entire population of New Orleans had evacuated out into the surrounding areas). In some instances this separation was considerable (the distance from Lafayette to Baton Rouge, being over 60 miles). Many of the subsequent trailer villages were also located away from services, which led to a feeling of isolation, compounding residents' depression, and making it hard for pregnant mothers to keep a suitable prenatal care sequence.

Attempts were made to assess the general needs of the pregnant and maternal population as demonstrated by the PtoP survey (Figure 7:4). Much of the information contained on the survey was relevant to pregnancy needs, such as the date of delivery, number of children in the home, father's name, various traditional pregnancy related risk factors and former and current address. These data were collected based on several existing forms utilized by Family Road of Greater Baton Rouge, and the Healthy Start program. It was well known that this evacuated population who had just experienced considerable trauma, also originated from areas of New Orleans known for their serious social and birth problems. Indeed,

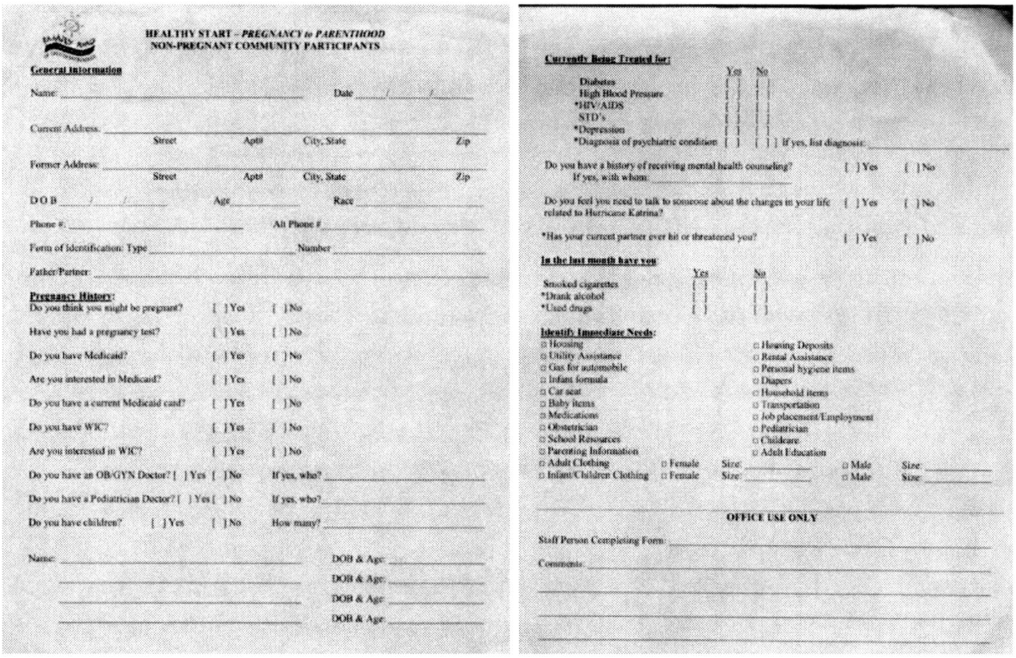

HEALTHY START – *PREGNANCY to PARENTHOOD*
NON-PREGNANT COMMUNITY PARTICIPANTS

General Information

Name: ______ Date ___/___/___

Current Address: ______ Street Apt# City, State Zip

Former Address: ______ Street Apt# City, State Zip

DOB ___/___/___ Age ______ Race ______

Phone #: ______ Alt Phone # ______

Form of Identification: Type ______ Number ______

Father/Partner: ______

Pregnancy History:

Do you think you might be pregnant? [ ] Yes [ ] No

Have you had a pregnancy test? [ ] Yes [ ] No

Do you have Medicaid? [ ] Yes [ ] No

Are you interested in Medicaid? [ ] Yes [ ] No

Do you have a current Medicaid card? [ ] Yes [ ] No

Do you have WIC? [ ] Yes [ ] No

Are you interested in WIC? [ ] Yes [ ] No

Do you have an OB/GYN Doctor? [ ] Yes [ ] No If yes, who? ______

Do you have a Pediatrician Doctor? [ ] Yes [ ] No If yes, who? ______

Do you have children? [ ] Yes [ ] No How many? ______

Name: ______ DOB & Age: ______
______ DOB & Age: ______
______ DOB & Age: ______
______ DOB & Age: ______

Currently Being Treated for:

	Yes	No
Diabetes	[ ]	[ ]
High Blood Pressure	[ ]	[ ]
*HIV/AIDS	[ ]	[ ]
STD's	[ ]	[ ]
*Depression	[ ]	[ ]
*Diagnosis of psychiatric condition	[ ]	[ ] If yes, list diagnosis: ______

Do you have a history of receiving mental health counseling? [ ] Yes [ ] No
If yes, with whom: ______

Do you feel you need to talk to someone about the changes in your life related to Hurricane Katrina? [ ] Yes [ ] No

*Has your current partner ever hit or threatened you? [ ] Yes [ ] No

In the last month have you

	Yes	No
Smoked cigarettes	[ ]	[ ]
*Drank alcohol	[ ]	[ ]
*Used drugs	[ ]	[ ]

Identify Immediate Needs:

□ Housing	□ Housing Deposits
□ Utility Assistance	□ Rental Assistance
□ Gas for automobile	□ Personal hygiene items
□ Infant formula	□ Diapers
□ Car seat	□ Household items
□ Baby items	□ Transportation
□ Medications	□ Job placement/Employment
□ Obstetrician	□ Pediatrician
□ School Resources	□ Childcare
□ Parenting Information	□ Adult Education

□ Adult Clothing □ Female Size: ______ □ Male Size: ______
□ Infant/Children Clothing □ Female Size: ______ □ Male Size: ______

OFFICE USE ONLY

Staff Person Completing Form: ______

Comments: ______

Figure 7:4. *The Pregnancy-to-Parenthood (PtoP) survey form.*

the sister program of Healthy Start, Great Expectations in New Orleans, had scattered in the evacuation, with case workers and clients having no means to contact one another.

Family Road remained at the Health Unit to provide community resource information in the months after landfall, however the majority of the survey participants became more self selecting in terms of either expressing interest in the program or through community resource referrals and attendance at Family Road itself. Of the women interviewed, approximately one quarter was pregnant when Katrina hit, with a sizeable number becoming pregnant after landfall. The interviews ranged from September 2005 to January 2006. The PtoP survey revealed that some of the pregnant women in the shelter had previous low birth weight deliveries, (including women who had also miscarried), and previous preterm deliveries (for some multiple times). Some of these highest pregnancy risk women were also in the last trimester when Katrina made landfall. These women fall into a high-risk category irrespective of Katrina. The events of the catastrophe are only likely to have accentuated these risks.

Although described in the last chapter, it is worth reemphasizing the additional risk of being a female head of household. Approximately 50% to 60% of those in the PtoP survey and who had children, didn't give the father's name. Single parent families are acknowledged as having a greater risk of a poor birth outcome, especially if combined with being in poverty and having other young children in the home. In the PtoP survey, over two thirds of women had one child, a quarter had two children, and almost one tenth had five or more. Unfortunately, the risks keep multiplying for this one particular cohort.

Q: By referring back to what you have read, what would the disaster-related stresses be for a female, head of household with young children?

In the data-poor post-disaster environment geographers can help in the prediction of where the greatest future pregnancy risks may be found by adopting a similar cartographic overlay of vulnerabilities adopted in the previous chapter. Although limitations exist in terms of these data being old (pre-disaster census information), they still provide the best guess at what the chronic stress levels were like for people (in this case mothers) living before the storm. These surfaces can be further developed by adding other known stressors such as crime, or even access to health care. Finally, as we have also discussed in this chapter, previous birth outcomes are predictors of future successful pregnancies. Maps showing the spatial distributions of past infant mortality rates, or low birth weight percentages, give some indication of which neighborhoods will already be starting with a health disadvantage even before adding in the Katrina experience.

This is what is shown in Figures 7:5 and 7:6.

For both mapped surfaces we are again looking at the lightest areas having the worst outcomes. Our three neighborhoods are

Figures 7:5. *Infant Mortality Rate for Orleans Parish*

also included allowing us to link these maps back to the vulnerability surfaces of the previous chapter.

Q: How well do our three neighborhoods perform in terms of previous birth outcomes?

Q: What spatial patterns do you see with the previous cartographic overlays?

Q: What are the limitations of using Zip Code level data?

One reason to perform this type of geographical analysis is because the risk of a poor birth outcome is not limited to the immediate hours or days after the disaster. There were significant increases in pre-term deliveries extending beyond a nine-month period after Hurricane Andrew. In other words the post-disaster environment had a negative impact on mothers who were not yet pregnant at time of landfall. Reasons for such a prolonged impact might include stresses associated with the loss of neighborhood and possessions, the inability to return home, and even once home, the anxieties associated with a lack of safety in the residence. As we have already mentioned, poor minority populations often rely greatly on the social networks in their neighborhoods—the disruption of these can both be stressful (not a good situation for a pregnant woman), and literally problematic as logistical help such as transport and child minding are absent,

Figures 7:6. *Low birth weight deliveries for Orleans Parish*

both of which may result in a lack of visits to a prenatal clinic (refer back to the newspaper story from earlier in this chapter). This situation is further worsened because of the additional stresses faced in a "rebuilding" New Orleans, especially high crime rates and an inadequate health care system. In addition, there must be real concern about the use of negative coping approaches used by a mother, especially as alcohol, drug and tobacco use have all been linked to negative birth outcomes, stillbirths and birth malformations.

HEALTH VULNERABILITY EXERCISE

Earlier in this chapter the non-profit support organized by Family Road of Greater Baton Rouge was described. In the weeks following Katrina multiple non-profit organizations, community groups, health care providers, religious centers and a host of other groups rising to the challenge of a huge influx of vulnerable women and families from New Orleans, would meet to discuss the condition of shelters, what was needed, and who was there to help. The last section of this chapter provides an insight into the complexity of the array of services needed to serve an already vulnerable and displaced population still suffering from disaster exposure.

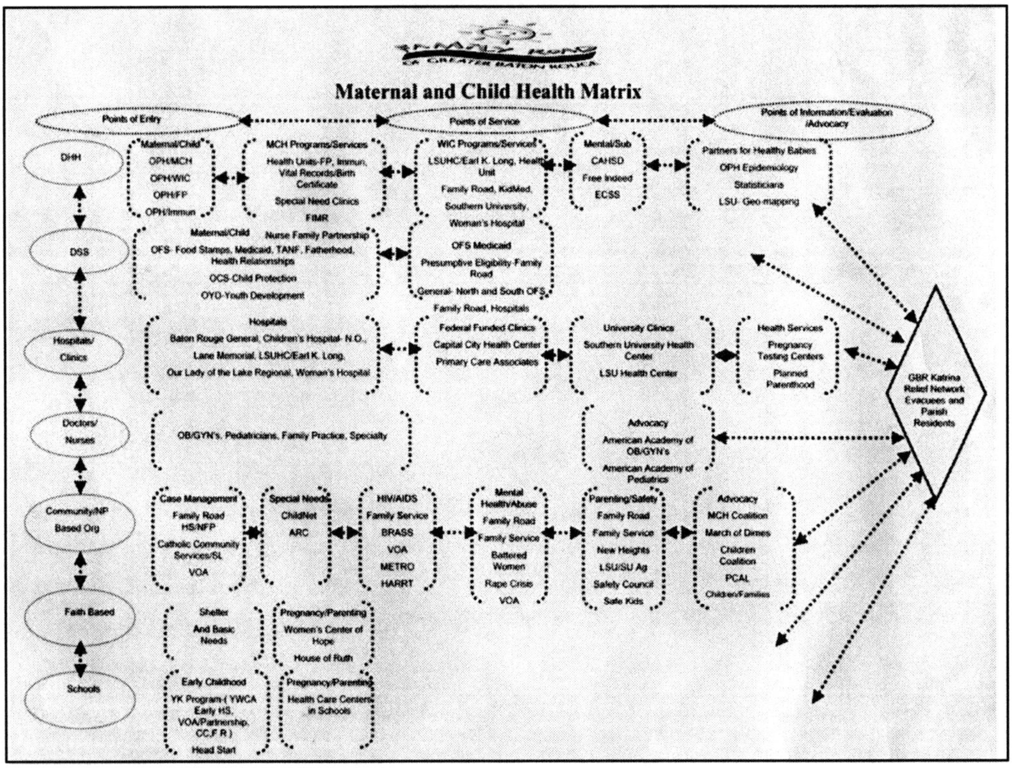

Figures 7:7

Figure 7:7 shows a hastily drawn service provider matrix in the weeks following Katrina—the attempt here being to provide a cheat-sheet for any volunteer/counselor to know who to go to, and how each provider fell within the general scheme of evacuee support. This was complimented by a data table on September 21, 2005 which provided more detailed help (found at the end of this chapter).

Using both these resources, answer the following questions:

Q: Name five vulnerable cohorts described in the last two chapters that are mentioned in the following table. Also name the specific provider support.

Q: Are there any other vulnerable cohorts here we have not mentioned? Why have they been missed?

Q: Which of the following sections in the table might benefit from a GIS? How?

Organization name	Types of services	Population serving / targeted
Physical Health Needs and Medication		
Ochsner Clinic	Medical Clinic- Extended hours of operation- Monday–Friday, 7 a.m. to 7 p.m. Saturday, 8 a.m. to 5 p.m. - seeing walk in patients, accept most insurance.	all
Lake After Hours	Medical Clinic- Normal business hours 3:00 p.m. to 11:00 p.m. Monday–Friday and 9:00 a.m. to 6:30 p.m. on Saturday-Sunday, except the Drusilla location is open from 9:00 a.m. to 11:00 p.m. on Saturday and Sunday. Most insurance accepted.	all
Baton Rouge Primary Care	Family Practice- pediatrics to adults. All insurance accepted. Working and no insurance-fee on sliding scale, bring check stub to appointment	all
Capital City Family Health Center	- Pediatrics and Adult Medicine - Dentistry - Accept most insurance - Fees on sliding scale, if no insurance`	all
Med Aid	Medical Clinic-Business hours- 8:00 a.m. to 5:00 p.m. - Most insurance accepted	all
American Cancer Society	- For an evacuee who is a cancer patient and needs help.	Cancer patients - evacuees

Organization name	Types of services	Population serving / targeted
St. Vincent DePaul	- Medications	all
Wal-Mart, CVS, Rite Aid, Walgreen's or Kroger's pharmacy in Louisiana or around the country	- Emergency prescriptions filled at no cost (depending upon patient need). - Give the pharmacy 24 hours to fill prescription, only for those who do not have insurance.	all

Mental Health Needs and Medication

Capital Area Human Services District	- Sending mobile response teams to the shelters. - Concentrating on individuals with severe mental health problems.	All- severe/ chronic mental illnesses
Family Service	Counseling Services	all

Prenatal/Infant

Woman's Hospital	- Prenatal Services - Better Beginnings: This program offers access to the services of an obstetrician (doctor) for prenatal care, delivery of your baby at Woman's Hospital and a follow-up check-up 6 weeks after delivery. Must have a valid Medicaid card.	Evacuees- pregnant women, and women looking for a Pediatrician
ARC/ Early Intervention Program/ ARC Day School	Infant and Preschool classes	All- including evacu-ees in the community / Children and special needs children
Early Head Start	- Multi-services to be provided to get children out of shelters, into schools and increase the capacity for Early Childhood Services - Will be the hub for EBR	
Family Road of Greater Baton Rouge	- Prenatal classes - Case management - Resource/referral center	all

School

West Feliciana Parish School Board	- Early Head Start, Head Start, Even Start, School Based Health Center, Adult Education, Salvation Army	•All, •evacuees
EBRP School System.	- 4,500 displaced students have registered, 1,100 students have been processed.	• Children • Evacuees

Organization name	Types of services	Population serving / targeted
St. Vincent DePaul	- Uniforms for all children that are registering for school as a result of the hurricane. - Guidance counselor will get the sizes needed and contact St. Vincent DePaul. - St. Vincent DePaul will fill the order and return the uniforms to the guidance counselor for distribution to the student. - Uniforms will be provided to children in the 12 civil parish areas.	• Children • Evacuees
School-based health clinics		
Health Care Centers in Schools (HCCS)	- Vaccinating health responders and looking at the general public - LINKS is working and can be assessed by providers	All Special needs children
Early Intervention Program of Baton Rouge (EIP)	-Center and community based early intervention for developmentally delayed 0–3 year olds -Child care for special needs- 6 weeks to age 4	All Special needs children
McMains Children's Developmental Center	Evaluations and therapy -Counseling for children with special needs (AD/HD, dev. Delays, speech, learning disabilities and physical disabilities)	Evacuees in the community / Special needs children
Children		
Youth Oasis	Provide emergency shelter to homeless and runaway youth- ages 10–17 -Transitional living shelter for youth ages 16–21	All / Children
EBR Cooperative Extension Services (LSU & Southern University Ag Center)	Youth development services -Information on stress management, disaster relief/ recovery, food safety and family strengthening	Evacuees in the shelters / Families and children
Louisiana Partnership for Children and Families	Working on having trained security for shelters in regards to protecting children - Working with PCAL on mandatory reporting	Children

Organization name	Types of services	Population serving / targeted
Prevent Child Abuse Louisiana (PCAL)	Children's activities -Trauma/Aftermath information -Security (shelters) -KIDLINE 1-800-CHILDREN	All / Families and children
Boys and Girls Club	Working as Teacher's Aids at Scotlandville Middle - Educational/recreational programs in shelters - Provide daily enrichment classes - Scotlandville Middle/ Mayfair in need of the following: school supplies, teachers' supplies, classroom supplies, book bags, carpet for floors, uniforms. Please call (225), if able to assist with supplies - Will be providing educational development programs—after school programs - for children after October	Children in schools/ shelters
Big Buddy	Big Buddy and Boys and Girls Club are currently staffing the River Center - Need volunteers with Early Childhood experience/ love for children	Children in River center shelter

HIV/AIDS

Organization name	Types of services	Population serving / targeted
Baton Rouge AIDS Society	HIV/AIDS counseling - Will provide transportation for clients to get them the services that they need - Assistance with housing at St. Anthony's - Prevention Services	
Family Service of Greater Baton Rouge	Case management - Direct services, including medications, transportation, housing assistance and mental health/substance abuse counseling. - Prevention Services	All
Volunteers of America	Case management - Direct services, including medications, transportation, housing assistance and mental health/substance abuse counseling. - Prevention Services	All

Organization name	Types of services	Population serving / targeted
Girl Scouts - Audubon Council	Large pool of volunteers and members focusing on community service in 10 parish area. Councils across the world sending goods to be dispersed through the community. Creating programs fro children in shelters and other areas.	All / Families and children
Rape		
Rape Crisis Center	Notify agencies working with evacuees about Rape Crisis Center services -Counseling / Forensic exams/ Law enforcement / Follow up -24 hour crisis hotline	Survivors of sexual assault
Housing		
Housing Authority	- Professionals who are looking for homes to move into	Evacuees
HUD	Located 5600 units in the area (600 in Baton Rouge) - List of HUD units that are available - 100% financing, programs being offered to people who need homes- call	Evacuees
Youth Oasis	Homeless shelter for children 10–17 years old - Openings are available	Children
Housing for Homeless/ Mentally Challenged	- Housing for homeless and individuals that are mentally challenged	
Families		
Sacred Heart of Jesus Church	- Referral to St. Vincent de Paul for clothes and clothes - Voucher for mattresses when available	Evacuees
United Methodist HOPE Ministries	Client choice food pantry- evacuees can shop twice a month -Family mentoring program -Community education initiatives	Evacuees and/ or families hosting evacuees Zip Code 70805
New Heights Family Health Center (BRASS)	Parent GED prep, Parent Technology Certification, Career training, Becoming a Responsible Teen, Future Fathers of America program, Mentoring, Parents as Teachers, Tutoring, Book distribution, Referrals, Information library, Technology lab, Parenting classes, and Bi-lingual services	All

Organization name	Types of services	Population serving / targeted
State Basic Needs Services		
Emergency Food Stamp Benefit Information	- Persons applying should bring: Proof of identification, Current address and proof of parish from where they have been displaced. - Once the two-page application is processed, you will receive a food stamp card and will be able to purchase food, including hot meals. - Households determined eligible for Disaster Food Stamp benefits will receive one month's benefits for September 2005 at the maximum allotment for the appropriate household size for families in effected parishes. - Disaster Food Stamps Applications will be taken for the following parishes only: Jefferson, Lafourche, Livingston, Orleans, Plaquemines, St. Bernard, St. Charles, St. Helena, St. John, St. Tammany, Tangipahoa, Terrebonne, Washington and West Feliciana. - If you are currently receiving Food Stamp benefits, you will receive your regular benefits automatically and on Sept. 15, a full disaster allotment for your family size will be automatically added to your electronic benefits transfer card. **There is no need for you to come to the office.**	Evacuees
Medicaid Benefits	Medicaid benefits still apply as long as you are treated by an out-of-state provider who is willing to accept Louisiana Medicaid as payment. For more. - For Medicaid Pharmacy Program benefits, go to a pharmacy that accepts Medicaid benefits with your Medicaid card; call Medicaid for temporary enrollment. For Medicaid providers who need to get recipients' prescriptions filled in other areas, call (out-of-state) or (in state).	Evacuees

Organization name	Types of services	Population serving / targeted
WIC		
WIC Clinics- Call for times that WIC services will be provided. Kid Med South- not accepting new WIC clients at this time. LSU Mid City- accepting WIC clients. Capitol City Family Health Center- accepting WIC clients. Southern University - accepting WIC clients.	The East Baton Rouge Parish Health Unit will be issuing vouchers for the New Orleans area people on a walk in clinic basis. Replacement vouchers can also be issued if they were lost during the storm. Prenatal (ONLY Prenatal) patients may also make WIC appointments with their proof of pregnancy, check stubs, picture id and proof of residency. No other appointments are available through the Health Units until further notice.	Evacuees
Birth Certificates	The Department of Health and Hospitals is offering no-cost birth cards (small versions of birth certificates) to evacuees and their children—Birth cards can be obtained from Parish Health Units in the following regions: East Baton Rouge, Lafayette, Lake Charles, Alexandria, Shreveport, and Monroe.	Evacuees
Vital Records	Issuing copies of birth records at no charge to those individuals devastated by Hurricane Katrina. - Will work with each person to ensure that they have what is needed to obtain a driver's license and/or identification card when our department cannot determine identity thru photo retrieve.	Evacuees
Disaster Unemployment Assistance:	People who lose their jobs due to the disaster may apply for Disaster Unemployment Assistance (DUA) which provides weekly benefits to individuals who are unemployed and not eligible for regular Unemployment Insurance compensation.	Evacuees
Federal Assistance		
FEMA Disaster Assistance		

Organization name	Types of services	Population serving / targeted
	Provides disaster housing assistance to those whose homes are damaged or destroyed. - Be prepared to give your Social Security number, describe your losses, provide financial information, and give directions to the damaged property	Evacuees

Information

United Way	Information for assistance, volunteer opportunities and donations. - Command Center takes donation calls and directs distribution to appropriate agencies for distribution.	All

Domestic Violence

Battered Women	Crisis counseling and advocacy - Posting poster in shelters - 24 hour hotline - Shelter Safety Planning - Education and training of Domestic Violence Men in battered relationships	Battered women and their children

Financial Assistance

American Red Cross	Red Cross is providing Hurricane Katrina survivors with emergency financial assistance to meet urgent needs until state and federal aid is available. - 125 Shelters open in the Capital Area - Training and sending volunteers to shelters on daily basis with resource information	Evacuees

EIGHT

Art Therapy and Hurricane Katrina

IN THE LAST TWO chapters we outlined the challenges faced by a large vulnerable population having to evacuate from their homes for an undetermined amount of time. We have also concentrated on certain aspects of vulnerability, especially women as head of household with young children, and we have also stressed the importance that families such as these place on social support networks. In this chapter we turn our attention to a support approach used to great effect in one large FEMA trailer park. This chapter is written by a guest contributor, Karla Leopold, who was part of a team using Art Therapy to help evacuee children in Renaissance Village, a FEMA trailer park located just outside of Baton Rouge, Louisiana. What has always fascinated me about this art, from a geographer's perspective, is the importance of place in these pictures. See how many actually look like maps.

ART THERAPY AND RENAISSANCE VILLAGE

Figure 8:1. *"Picture 4"*

Look closely at the drawing. Underneath the violent swirling of red, pink and brown colors, are a house, a store, and a dead woman. Behind the anxious swirls drawn by this six-year-old girl are the images stuck in her head for over six months before she was able to draw her story to an art therapist. This experience will have a defining impact on her childhood and the rest of her life.

The six year old girl explains: "Brown in the center is the hurricane: the rest of the brown is the water. The buildings are houses and stores. The big one is my house. In the corner is my dead auntie who died. She was eighty nine. We went to a shelter to drop stuff off and when we got back she was already floating. My uncle and grandmother were waiting too but they were in a safe place—in the closet. We tried to take my auntie with us but she wanted to stay with my grandma, then she drowned. "

For three years after Hurricane Katrina, more than 1,500 people, including 600 or more school age children, lived in a large trailer park called Renaissance Village. The trailer park was set up by the Federal Emergency Management Agency (FEMA) near Baton Rouge, Louisiana two months after the storm. Despite its hopeful name, the trailer park was far from anyone's idea of home sweet home. Schools were far away, safety was a constant concern even as security guards with guns patrolled the fenced grounds and conveniences most of us take for granted, such as restaurants, parks, and transportation, were not to be found. This is where this young girl and her family of five lived in a small, crowded travel trailer. The grounds of the facility had been a cow pasture two weeks prior. Many of these trailers were later to be determined toxic by the Center for Disease Control and Prevention.

Before moving into the trailer park, the evacuees had been housed for two months at large shelters, such as the Baton Rouge River Center. They were transferred by

Figure 8:2. *Renaissance Village Trailer Park*

buses with their meager belongings in garbage bags to their new homes, twenty five to twenty seven foot trailers. The art therapy team greeted them as they arrived. The therapist's tables filled with art supplies were set up in the dirt to be immediately surrounded by mobs of displaced children ready to tell their stories.

Katrina affected the lives of many, but especially the children. They lacked the skill to recover without guidance from adults. The physical and mental affects on this traumatized population were overwhelming. These families lost everything. They were in shock. The focus became the children. We used what we knew as the best treatment possible for trauma, art therapy.

I am an Art Therapist and also a Marriage and Family Therapist with over thirty years of experience working with families and children as the leader and organizer of teams. My practice in California was focused on treating families experiencing untimely death and loss. In addition, I had a background in natural disasters having worked with the families after the California earthquakes and fires. I was recruited by a friend who had recently returned from Louisiana delivering food to the survivors. The friend was concerned about the lack of mental health treatment for the children who were obviously suffering from their traumatic experiences. We discussed the idea of organizing a team of art therapists who could address the problems of the young survivors. When it was agreed to bring together a group of professionals, the funding was secured and the first trip was finalized.

The success of the first trip led to three more years of visits to Louisiana. The art therapy teams were composed of therapists from around the country. During the three years, art therapists, professors and students from universities with art therapy programs worked together in teams of eight to twelve members. Funding provided on-site work for bi-monthly visits, lasting seven to ten days for the first year. In the following two years, visits were in three or four months intervals.

It was fortunate that funding was found for this project. Currently, there is no funding by the federal government to provide mental health services for natural disasters. The reason given by the Center for Disease Control is that no large, evidence-based studies exist proving the effectiveness of mental health treatment after a natural disaster. The Red Cross provides only immediate crisis response. Funding for the art therapy teams were provided by multiple, private sources. Team members were recruited outside the state of Louisiana because local therapists were unavailable. They were dealing with their own traumas after the hurricane.

The success of the project depended on finding a community liaison with contacts and clout. The Baton Rouge children's advocate, Sister Judith Brun, was able to open many doors that otherwise would not have been available to us. The art therapists were outsiders in this community. Sister's Judith's knowledge, dedication and relationships with the local families, politicians, universities, personnel and press made our work possible. She worked with the families continually between

Figure 8:3.

the art therapists' visits. After Renaissance Village closed, she has continued to work with the Katrina families providing food, shelters, support and counseling.

Children are at a greater disadvantage during a natural disaster. They depend on adults for survival. They also need advocates for healing and recovery. The art therapy teams became the advocates for the children at Renaissance Village. Trust became a big issue while dealing with this population. The chaotic and unsuccessful evacuation process taught them not to trust what they had been told or promised. The trust had to be built before healing could begin. It took several return trips before a firm relationship was established between the evacuees and the therapists. The art therapists were one of the few consistent entities working at Renaissance Village. The children and adults looked forward to and depended on regular visits.

Three years of work was done in a tent that often had moldy sides or no sides at all (Figure 8:3). The tent was donated by another private organization to provide a place for the families to eat and work together. This tent was our home regardless of massive rain storms, thunder, lighting, cold, dust, or soaring temperatures over 100 degrees. When the art therapists arrived, the tent filled with adults and children wanting to work with the art materials. Often more than sixty people crowded the tent seeking attention and an outlet for their anxieties.

Art is a natural form of communication for children because it is easier for them to express themselves visually than verbally. For many of these children there were no words to describe what they had lived through. There were only images reoccurring in their heads. The art provided the tool to externalize and contain their

The original cardboard city and graveyard was constructed by the children of Renaissance Village in December 2005.

Using piles of cardboard boxes and many rolls of duct tape, the children created a city that spanned almost one hundred yards. Their city included a governor's office, a mayor's office, and a church including a graveyard for lost pets, friends and relatives.

Their cardboard city was destroyed by a storm on the afternoon it was completed. What they had built was left in a soggy, formless mess like the city they left behind.

Under the supervision of the art therapists, the children were able to process their feelings of loss by planning and rebuilding a tabletop city.

Figure 8:4. *Photographs of the Graveyard*

trauma stories. The survivors were empowered by their ability to create and detail their experiences on paper. It allowed them to communicate difficult issues, reduce stress and reconcile feelings. The healing process began as the images appeared in physical forms. The traumas and feelings no longer just lived in their heads. They could begin to develop a positive outlook for the future.

After a twelve-year-old boy created a square sculpture covered with plastic flowers, crosses and candles, he shared his thoughts;

"The dead bodies don't have to live in my head and dreams anymore. I buried them in this grave I built. I buried the people that died in Katrina here."

The next day he returned and said: "I forgot to bury the dead animals so I added them in the grave last night."

It took six months before the girl with the dead auntie was able to draw her story. She had to build trust with the therapists. It was important to find a safe place to tell her story. Once she began the process, the dam had broken. She drew her story over and over. She couldn't stop explaining her drawings to anyone who would listen.

It was important to provide her tools beyond telling the story on paper. For the remainder of the week she was able to create images that provided safety and reassurance. Before the therapists left she created an angel from clay. She took it home to put it above her bed. She said; "This is my auntie angel gone to heaven. She will help me not to have bad dreams."

THE TRIANGLE HOUSES

Probably the most profound images drawn by the children of Katrina were the triangle houses. Art therapy provided a vehicle for the traumatized children to express themselves, especially for those whose verbal skills were compromised. They could not verbalize what they were feeling or what they needed. But they could draw about it. Through these images we were able to see the depths of the traumas suffered. It gave us a deeper understanding of what these children experienced. The drawings also provided a tool to measure their recovery.

The art therapists were puzzled by one aspect of the drawings we received—many children drew large triangles instead of houses. The therapists first thought the triangle houses were coming from the youngest children, but then they started to notice the older children were drawing the triangles as well. These children no longer drew a house like most children do with the rectangle bottoms, pointy tops with windows and doors. These children just drew the triangles with a window or door in the middle. The roofs had become their symbol for safety and security. For these children the roof was the only safe place not under water. They had been so traumatized by their experiences that their internal schemas had been changed.

Figure 8:5.
The Triangle House

They no longer drew houses as normal children would but they drew roofs.

It was many trips later before the therapists saw fewer and fewer triangle houses. Before that began, however we constructed, painted and drew houses repeatedly. The houses would include food, strong roofs, walls surrounding the houses, boats in the garage, antennas on the roofs to detect hurricanes, and other safety devices. Together the children and therapists built and rebuilt the city of New Orleans. The city created by the children had many bridges, airports, multiple police stations, fire stations and graveyards, graveyards for pets, relatives, friends and their overwhelming losses.

When working with the children, the therapists found it important to supply healthy food and water. For many, the week the art therapists arrived was the only week that they would eat healthy food and have complete meals. For some this was the only week that they would leave their trailers. Isolation and safety were problems. The food became a metaphor for the feeding they were doing. The therapists provided for both the physical and emotional needs.

The food would also attract the teenagers and adults into the tent to participate in art therapy. There was never enough food or art supplies. The needs were so great. They had so little. The holes created by their losses were so deep that there was never enough to fill them.

THE LEGACY OF KATRINA

It was reported by a group of psychiatrists at Louisiana State University that one year after the storm of the three thousand children affected by Katrina that were tested, over half met the threshold for a clinical referral for treatment of trauma and loss. Post traumatic stress, depression, anxiety and suicide ideation were cause for clinical referrals. People feel crisis after a traumatic event because it disturbs their safety, security and survival. No one can adjust to trauma until they have

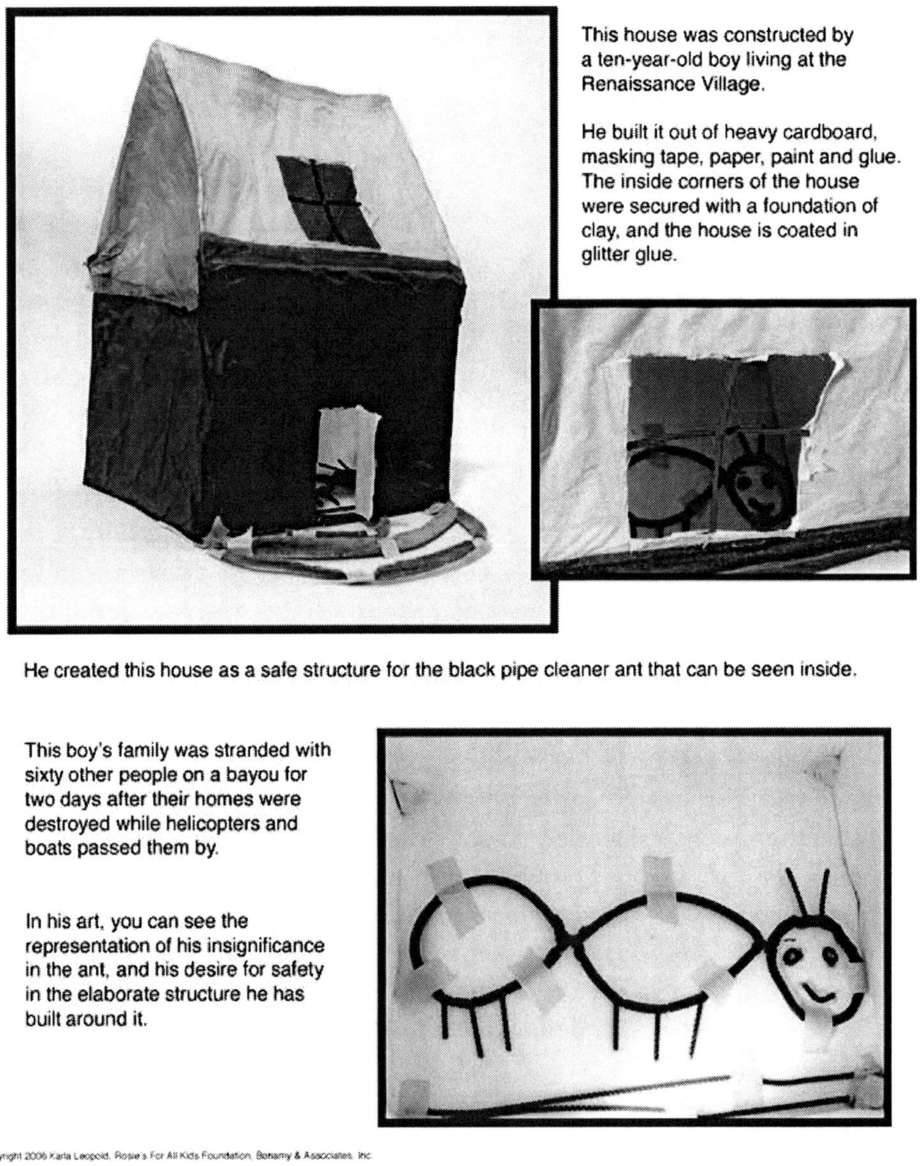

Figure 8:6. *Blue House and Ant*

those back. This is a difficult process for those who lost everything including their homes, churches, jobs, contacts, community and resources.

Two years after Katrina, another group of researchers from Harvard University reported that the mental health problems had risen to an alarming 8% of the population. It had affected the rich and poor, black and white, male and female. It had crossed all lines. Generally two years after a trauma the mental health issues are on a decline. Not so for those affected by the storm; post traumatic stress, depression, anxiety, and suicide had continued to rise. They attributed some of this to the slow recovery rate and lack of trained mental health workers.

Tens of thousands of children who are traumatized by the devastation and loss caused by Katrina and similar disasters lack treatments to deal with the traumas. If these children are left untreated, the consequences become a larger generational problem for America. Many of these children will grow to be adults unable to hold jobs, they may be mentally ill or fall foul of the law.

Prior to the closing of Renaissance Village the number of children not going to school was alarming. At the last count of the four hundred and fifty children remaining, on a good day, only one hundred and twenty got on the bus to attend school. There is a direct correlation between the number of school transfers and the dropout rate especially when dealing with teens. There is also a direct correlation between children suffering depression and/or post traumatic stress disorder and the drop out rate.

The following is a story to illustrate the depth of the traumatic experience of a young boy after Katrina:

During our first visit the children were asked by the art therapists to create an animal that would represent themselves. The therapist noticed a ten-year-old boy very quietly creating an ant out of blue pipe cleaners. The therapist had seen a lot of animals created by children but never an ant. This boy had obviously been traumatized but could not find words to talk about his experience. She sat down next to the boy and asked how his ant was feeling. His reply took her breath away.

He said: "The ant was lonely and scared. The ant felt like he didn't count for anything. The ant thought he was going to be stepped on and killed."

The therapist told the boy she was very concerned about this ant and she wondered if he could think of anyway to make the ant feel safe. With the help of the therapist, he began the process of creating a safe house for the ant to live in. He spent several days using layers upon layers of glue, paper, paint and glitter. He used clay to make the foundation strong. He cut a door in the roof for an escape exit.

When he was finished, he was very proud and stated: "This ant is safe now."

Later in the week the father came to talk to the team because he was elated to see a smile on his son's face. His son had also begun to talk again. He told us of the family's heroic and dangerous evacuation. Through the muddy waters, by hanging

Figure 8:7. *Picture: This is the Sahara Desert with cactus and a road that goes up into the mountains. It's hot.*

onto a wooden door the family reached a levee, a dirt mound, with over sixty other evacuees waiting to be rescued. For two days as helicopters and boats passed them by, they were stranded on the levee. They were among the very last to leave the dirt mound. This is why the young boy lacked words to tell his story. However, he could illustrate the experience by creating his ant.

Over the course of the next two years this young boy continued to create numerous houses, often with multiple toilets relating to the difficulty of lacking a bathroom while being stranded on the levee. The house would also have food and beds on the roof level.

It should be noted that this young boy suffered multiple traumas even before Katrina. He is representative of the desperately poor children of America. He, like so many, is several grades below his age group and has difficulty with school. For many of these children, this natural disaster will be their first memory.

One young teen, at age fifteen, after sharing her losses in her art work stated: "I changed very quickly from a child to an adult after Hurricane Katrina."

Among the families at Renaissance Village, some were able to recover. They regained their safety, security, and survival skills. These families were generally the first ones to move from the trailers. Many had families or support systems to help with the relocations.

The desert image painting displayed in Figure 8:7 was created by a boy as the family was planning on moving from Renaissance Village to another city. It portrays a sense of hope when looking at his road up the mountain. It also shows a sense of loneliness and the difficulty of moving to a new home, a new school and making new friends. There is no reassurance his move will be better but his picture, with the use of such beautiful colors, tells us he thinks his future will be brighter. It is good that he has hope for the future. After living through such difficult traumas, for many of the children hope will be a difficult concept to achieve. With the resiliency of childhood, some will achieve it. Unfortunately for many of these children, fear becomes the feeling they are most familiar with.

Approximately twelve hundred people remained at Renaissance Village in May 2008 when the notice was given that they needed to vacate the premises. Even though the Federal Emergency Management Agency (FEMA) provided assistance and funding to expedite the relocations, many of the evacuees lacked the resources and skills to move forward with their lives. For some of the remaining population, their paper work was lost or never found. Therefore it was deemed that they did not qualify for assistance. It was declared that they had lived at Renaissance Village illegally. There were no funds available to assist with their relocations. Many of them were the sick, the elderly and single moms with multiple children. These children were transferred to another school, for some this was their forth or fifth transfer in three years. Fortunately, Sister Judith Brun and her teams were there to assist.

In the aftermath of Katrina, there is much to learn about the effect on the lives of the families that live through a natural disaster. However, it seems that there are more questions than answers:

Why did these people who lost so much, receive so little help and support?

The initial response to Katrina was positive but long term efforts were lacking.

Where were other mental health teams? The art therapists were the most consistent mental health workers on site. For the first year we were there for six visits. Mental health treatment was and still is, so desperately needed.

Is there help available later when these children need it? Our original goal was to provide immediate treatment for the traumatized children by supplying art materials and our expertise. After three years, we left and eventually the families relocated. The experience of surviving Hurricane Katrina will shape their lives forever. They should continue receiving help.

It took more than a year before the therapists began to see a positive change in the children's art. Signs of healing and progress began to appear in the drawings. Still, these children were stranded in Renaissance Village. The teams could not fix where they lived but they could help instill a sense of hope.

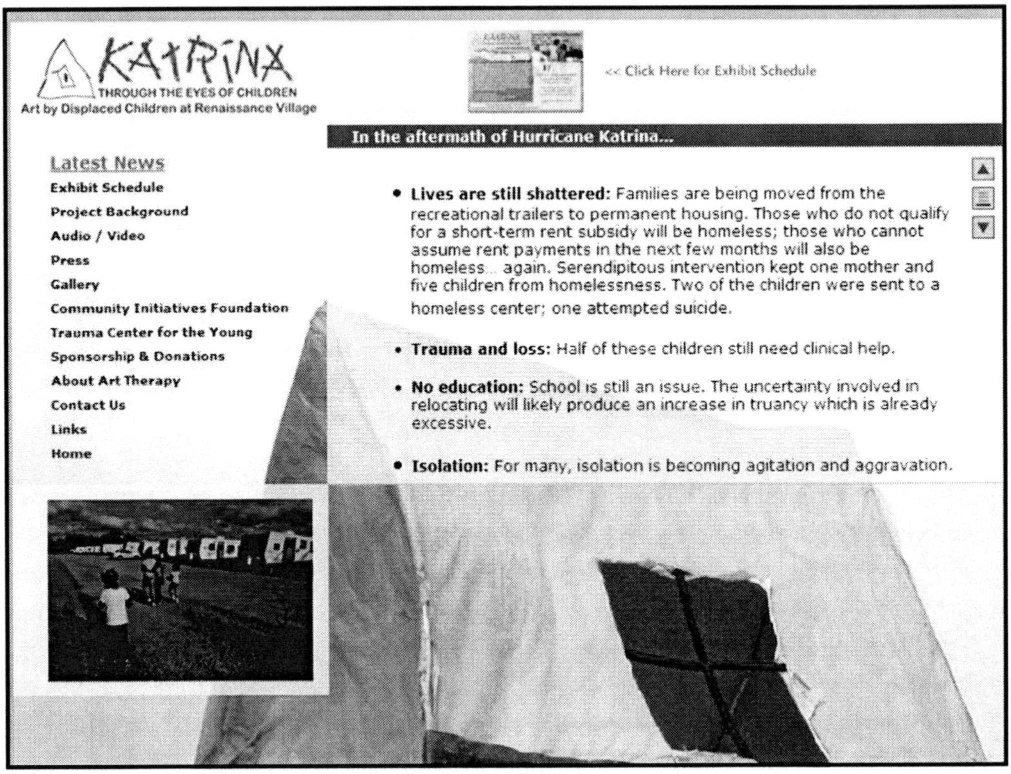

Figure 8:8. *www.katrinaexhibit.org has more information, updates, and images created by the children.*

The six year old girl who drew her auntie floating in the brown waters after surviving Hurricane Katrina will have these memories for the rest of her life. Her violent swirling allowed the therapists to see how anxious and frustrated she felt about the experience. This child lost everything including the pink purse she would draw. Living at Renaissance Village for three years surrounded by bleakness and worry added to the negative impact on her life. It is yet to be seen as she grows older, if the sense of fear and frustration she drew about the trauma will be replaced with the feelings of hope and happiness she felt when she created her auntie as the angel.

Throughout the three year project, the team members often felt depressed and frustrated with the surrounding conditions and the hopelessness they felt about the future of these children. The team leader would share the message that she would repeat to herself: "Their job was not to change the journey for the children but to hold their hands, listen to their stories and ease their pain-even if it was just for a moment." The art therapy teams can proudly say that this was a goal they did achieve.

NINE

Life Interrupted: The Stress Process in the Face of Natural Disasters

THE MAJORITY OF this chapter has also been written by a guest contributor, Louisa Holmes, who has an extensive background in the realm of urban stress and public health. At the time of writing this chapter she is working on her doctorate at USC—focusing on the geography of stressors and vulnerable populations. In several chapters we have mentioned the mental health of a disaster. The following two chapters extend this connection further, with the first of these providing a detailed frame of the "mental health consequences" of a disaster. There is a lot of information to digest here, and it is not a chapter to skim quickly. However, we believe it is important to expose the reader to the complexity that underlies a simple statement such as mental health problems follow disasters. Finally, as one criticism of post-disaster mental health research is that it does not extend far enough in time beyond the actual event, with most studies being conducted within one year of the disaster, we have supplemented the text with photographs taken approximately one year after Katrina showing that normalcy has not been achieved.

The health implications of a disaster extend well beyond the immediate aftermath. In addition to the more acute and physically apparent maladies discussed in the previous chapter, the experience of a natural disaster may have varied and long-term health impacts. In particular, natural disasters have psychological implications and may exacerbate existing stress or contribute to acute stress, a condition that can lead to chronic illness and mortality if not properly addressed. The purpose of this chapter will be to: 1) provide an overview of the stress process with attention to the potential for increased vulnerability among certain populations; 2) briefly review the findings regarding the associations between natural disasters and stress and psychological health; and 3) discuss gaps in the existing literature with an eye toward identifying where geographers might best contribute.

THE HUMAN STRESS RESPONSE

The General Adaptation Syndrome and Allostatic Load

Walter Cannon's research in the early 1900s first introduced the word "stress" into the vernacular of medical research. However, stress first began to gain real traction in the academic literature when in 1936 Hans Selye[1] discussed it as a syndrome produced in the body as a nonspecific response to "noxious stimuli," (or what we would presently call stressors) stamping it with a decidedly more physiological than psychological interpretation. As Selye continued to work on stress, this general definition evolved into what he termed the "general adaptation syndrome," in which the body 1) reacts to stressors by activating the hypothalamic pituitary adrenal (HPA) axis, and 2) adapts to the stressors. The third stage, exhaustion, occurs when the stress response persists and takes an adverse physical toll on the body[2]. The first two stages naturally occur in response to traumatic events such as a disaster and these are the stages that receive the most attention in the literature with respect to examining people's psychological health in the immediate, and occasionally intermediate, aftermath of a disaster. It is the third stage however that creates the potential for long-term psychological and physical illness.

More recent studies of stress in neurology and endocrinology focus on allostatic load, a comprehensive concept that speaks to the manner in which the body's various neuroendocrine systems process and adapt to the stressful situation impacting it (the previously mentioned noxious stimuli); allostasis is "the ability to achieve stability through change"[3] while allostatic load is "the wear and tear that results from chronic overactivity or underactivity of allostatic systems."[4] In other words, allostatic load is akin to a physiological index indicating the body's ability to properly respond to a stressor and adapt—the first two stages of the general adaptation syndrome. The extent, to which the body cannot or does not adapt, most often because an individual is consistently experiencing some degree of stress, is the extent of its allostatic load. The study of allostatic load is therefore important for understanding the implications of the stress process, including psychological distress, for the body; people who consistently experience stress, either from vulnerability to regular new exposures, or the ongoing impact of a highly stressful experience, such as a disaster, or experiencing a prolonged recovery period, may be at much greater risk for developing chronic illness. Recent investigations in this area indicate that much like psychological distress and disorder, certain populations are more at risk for suffering the adverse affects of high allostatic load—ethno-racial minority populations and those of lower socioeconomic status (SES) in particular

1 Selye, H. 1936. A Syndrome Produced by Diverse Nocuous Agents. *Nature* 138:32.

2 Selye, H. 1946. The general adaptation syndrome and diseases of adaptation. *Journal of Clinical Endocrinology* 6:117–173.

3 Sterling, P., and J. Eyer. 1988. Allostasis: a new paradigm to explain arousal pathology. In *Handbook of Life Stress*, eds. S. Fisher and J. Reason, 629–49. New York: Wiley.

4 McEwen, B.S. 1998. Protective and Damaging Effects of Stress Mediators. *New England Journal of Medicine* 338:171–179.

tend to have higher allostatic load[5]. Unfortunately this also represents the majority of people affected by Katrina in New Orleans.

Psychosocial Stress

While researchers in the tradition of Selye have made great strides in understanding the stress impact on physiological systems, the study of stress as a social phenomenon has been an equally important development. This field of research, which investigates the psychological implications of social processes and relationships, emerged in the early 1900s through the work of psychiatrists like Adolf Meyer and Harold Wolff. The study of psychosocial stress is multivariate, but there are a few primary topics on which the literature tends to concentrate that are also especially valuable for understanding the impact of disasters on human health: 1) life events stress; 2) coping resources; 3) social support; and 4) exposure to stress and social vulnerability.

Stressful Life Events

In the 1960s, Holmes and Rahe developed the Social Readjustment Rating Scale (SRRS), which is basically a checklist enumerating a series of changes or developments that may occur over the course of a person's life (e.g. changing jobs, death of relative, getting married). They used this scale to examine the relationship between these "life events" and the types of stress study subjects experienced as a result. The relatively straightforward approach employed with the SRRS spawned an entire field of research devoted to measuring psychosocial stress and life events with attention to gaining more precision and a more sophisticated theoretical understanding of the correlates between life events and psychological illness in particular.

In terms of understanding the link between natural disasters and stress, some of the more important issues raised in this literature are 1) whether traumatic life events may actually induce mental disorder (as opposed to simply being correlated with it) (Dohrenwend 2000); 2) how the accumulation of stressful life events relates to the level of exposure to stress mediated by socioeconomic and demographic characteristics (Turner and Turner 2005); 3) how personal resources, such as coping, interact with stressful life events to moderate the stress reaction (Billings and Moos 1981); and 4) whether the cumulative effects of stressful life events or the intensity of certain singular events better predicts mental disorder (Lorimor, McBee, and Weinman 1979)[6].

5 Taylor, S.E., R.L. Repetti, and T. Seeman. 1997. Health Psychology: What is an Unhealthy Environment and How Does It Get Under the Skin? *Annual Review of Psychology 48 (1):411–447*

6 Dohrenwend, B.P. 2000. The Role of Adversity and Stress in Psychopathology: Some Evidence and Its Implications for Theory and Research. *Journal of Health & Social Behavior* 41 (1):1–19; Turner, H.A., and R.J. Turner. 2005. Understanding Variations in Exposure to Social Stress. *Health: An*

Psychological Distress and Daily Hassles

In contrast to work stemming from Cannon and Selye, psychosocial stress research has focused almost exclusively on the implications of stress for psychological health, and clinically measured psychiatric disorder in particular. Furthermore, the emphasis has more often been on cumulative or chronic stress (although the trigger may be an episode of acute stress) than on the daily insults that may lend themselves to mental perturbance[7]. These are important tendencies to note in the case of natural disasters as the incidence of post-traumatic stress disorder (PTSD) may increase significantly among the affected population in the weeks immediately following the event, and this is where the post-disaster research has been drawn[8]. However, PTSD symptoms tend to subside for most of the population within a few weeks after the disaster occurrence. It may therefore be especially useful to track less visible signs of mental distress that may not rise to clinical significance but nevertheless take a serious toll on health. Additionally, comparatively minor stressors, such as anxiety about household finances, may not be severe enough to contribute to clinically-diagnosed mental disorder in the absence of trauma, but in the face of disaster such considerations may become paramount.

Hence, there is a need to measure nonspecific psychological distress as well as psychiatric symptoms and clinically-diagnosable cases. Approximately six percent of the U.S. population suffers from serious mental illness, and while this is certainly problematic, it accounts for only a small segment of society. However, 26% of the U.S. population experiences a diagnosable mental illness in any given year and an additional three percent endures ongoing psychological distress[9]. Furthermore, African Americans and Hispanics have a higher prevalence rate of psychological distress than non-Hispanic whites, and lower SES also predicts higher rates of psychological distress . Following a disaster, therefore, it seems essential to measure not

Interdisciplinary Journal for the Social Study of Health, Illness and Medicine 9 (2):209-240. Billings, A.G., and R.H. Moos. 1981. The Role of Coping Responses and Social Resources in Attenuating the Stress of Life Events. *Journal of Behavioral Medicine* 4 (2):139-157. Lorimor, R.J., G.W. McBee, and M. Weinman. 1979. Weighting Events in Life-Events Research. *Journal of Health & Social Behavior* 20 (3):306–308.

7 Daily insults may be geographic in terms of living in a high crime area, or living close to constant loud noise (such as a busy road),.Therefore, it is possible to collect chronic stress data in such a way that it can be incorporated into a GIS and mapped.

8 Kessler, R.C., S. Galea, M.J. Gruber, N.A. Sampson, R.J. Ursano, and S. Wessely. 2008. Trends in Mental Illness and Suicidality after Hurricane Katrina. *Molecular Psychiatry* 13 (4):374–384.

9 Kessler, R., W. Chiu, O. Demler, and E. Walters. 2005. Prevalence, Severity, and Comorbidity of Twelve-month DSM-IV Disorders in the National Comorbidity Survey Replication (NCS-R). *Archives of General Psychiatry* 62 (6):617–627. National Center for Health Statistics. 2007. Serious Pyschological Distress. In Early release of selected estimates based on data from the 2008 National Health Interview Survey. Atlanta, GA: Centers for Disease Control and Prevention.

Figure 9:1.
A selection of Baton Rouge news headlines highlighting stress in the weeks after Katrina, including "Support group offers assistance on sharing space," "Coping after Katrina," and most worryingly, "Gun sales surge still going strong."

only the most severe and prominent signs of mental illness but to pay more attention to subtler measures of stress and disturbance over the long term. As Berkman and Kawachi point out, distress is the biological link between sociogeographic factors and long-term health[10].

Additionally, some researchers have recently pointed out the necessity of distinguishing between "daily hassles,"[11] such as the financial anxiety mentioned above, and chronic stress for understanding the mechanisms by which social circumstances impact individual health. In the context of disasters, such daily hassles may grow from nuisances to obstacles threatening survival or recovery, thereby combining with the experience of the traumatic event itself to exacerbate health problems. As we have read elsewhere in this book, one survey of Katrina survivors found that 34% of respondents failed to evacuate from New Orleans because they did not have access to a car—while the inability to afford a car in most circumstances may constitute a frustration, in the case of this event it became a life-threatening issue for a substantial proportion of the affected population.

10 Berkman, L.F., and I. Kawachi. 2000. A Historical Framework for Social Epidemiology. In *Social Epidemiology*, eds. L.F. Berkman and I. Kawachi. New York: Oxford University Press.
11 Serido, J., D.M. Almeida, and E. Wethington. 2004. Chronic Stressors and Daily Hassles: Unique and Interactive Relationships with Psychological Distress. *Journal of Health and Social Behavior* 45:17–33.

Coping Resources

In the social stress literature, coping resources play an essential role[12], and while these are more often conceived of as emotional or intellectual resources, natural disasters are perfect examples of when both emotional and physical resources become crucial to cultivating health. Personal resources, or coping mechanisms, operating at both individual and population levels may intervene to interrupt the stress process. At the individual level for example, subjective well-being researchers pay special attention to factors such as self-esteem, positive and negative affect, and aspects of personality such as extroversion or perfectionism to explain why some individuals and populations manage to stave off chronic stress and illness while others do not.

At a more aggregate level, those who study the psychosocial correlates of stress also take individual attributes and cumulative experience into account but situate these in social-environmental contexts. Specifically social stress research examines how the social arrangements and structural foundations of populations interact with individual characteristics to create opportunities for coping. Thoits breaks the literature on coping into three theoretical areas: coping strategies, coping resources, and social support[13]. Coping resources are simply the arsenal of tools and supports that people can draw on in stressful situations. Coping strategies, or styles, are behavioral or cognitive techniques that individuals employ in an attempt to alleviate stress, and therefore fall under the umbrella of coping resources. The strategy may be an attempt to address a stressful situation directly, or it may be an attempt to effectively process the emotional response to a stressful event. Self-esteem and perceived control are intertwined with the idea of coping strategies in that people with internal loci of control are expected to approach stressful situations proactively and in problem-solving fashion. In this way coping strategies are highly predicated on social circumstance as research has demonstrated that women and groups of lower SES exhibit lower self-esteem and lower mastery than their counterparts.

The issue of control, or mastery, deserves additional mention as a disaster creates a situation that initially seems beyond the immediate control of everyone affected. In 1966 Julian Rotter published an influential text introducing social learning theory, which included an explanation of the locus of control spectrum. Locus of control describes the extent to which an individual believes she determines her own circumstances and bears responsibility for her own outcomes in life. An individual

12 A resource may be an intangible property, such as feeling in control of one's own future, or a tangible material resource, such as a car, the lack of which threatened the immediate well-being of people in New Orleans during Hurricane Katrina thereby creating the circumstances for longer-term health effects as well.

13 Thoits, P.A. 1995. Stress, Coping, and Social Support Processes: Where are We? What Next? *Journal of Health & Social Behavior Extra Issue*:53–79

who falls toward the internal locus of control side of the spectrum believes she is largely in control while an individual falling toward the external locus end of the spectrum believes that she is largely subject to extrinsic forces. Applied at an epidemiologic level, populations sharing an external locus of control may view their social circumstances as largely predetermined, e.g. low-income minority populations living in seismic hazard areas may view their fates as largely predetermined by structural and social barriers and therefore neglect to prepare for a disaster in the same way high-income residents on a hillside who feel in control of their fates would. Locus of control has weighty implications for psychological health in a variety of situations, but the above example demonstrates how crucial it may be for understanding disaster preparation as well as recovery. For example, the extent to which people feel in control of their future may determine the types of coping strategies they employ, which may in turn influence whether they physically return home and attempt to rebuild after a disaster and whether they continue to experience distress.

Social Support and Social Networks

Social support, or the idea that some aspect of social relationships has a protective effect for individual and population health, has received a great deal of attention in the stress literature as well as across a variety of disciplines in recent years.

In particular, the social stress literature has found the existence of social contacts on which individuals can rely for emotional support, as well as the perception of support, to be crucial for protecting against mental illness. While these findings are important, they are relatively egocentric, and it is important to note that there are also a variety of theoretical constructs applied to broader social interaction that may have implications for health. For example, civic group participation is often used to measure people's engagement with their community, while collective efficacy is used to measure a community's propensity to work together to achieve collective goals.

Furthermore, investigating the structure of social networks may be especially important for gaining an appreciation of the longer-term health impacts of a disaster. It is possible that individuals gain psychological benefit simply from participating in community activities regardless of the types of contacts they make or relationships they develop. However it is likely that the instrumental and emotional resources they accrue as a result of participation and engagement with others in the community have a great deal to do with any consequent health impact. For example, people who attend church services on a weekly basis and participate in church functions may not maintain intimate friendships with most of the other participants, but in a period of crisis they have the comfort of knowing that they belong to a community on which they can rely for material as well as emotional

Figure 9:2. *A message of encouragement outside an abandoned church in the Lower 9th Ward.* Photographs by Curtis

support. Ultimately, it may be that in the face of a catastrophic event such as a disaster, each of the constructs for describing social interaction may provide insight into health consequences and provide clues about how to facilitate recovery.

It is important to investigate the existence, quality and structure of social networks and their capacity for support in order to gauge their effects on health. In a disaster situation social networks and support likely play important roles in influencing preparedness, response, and prospects for long-term recovery, just as the disruption of these networks may exacerbate the trauma experience.

Differential Exposure

The common backdrop to all of the stress research discussed above, including physiological as well as psychological health outcomes, is the social environment.

There exists a debate in the psychosocial stress literature as to whether psychological illness is the result of social selection or social causation. As Johnson et al. succinctly describe, "Social causation theory hypothesizes that environmental adversity, disadvantage, and stress associated with low SES (interestingly all components of the "hazardscape") contribute to the onset of psychiatric disorders. Social selection theory hypothesizes that constitutional and environmental factors contribute to the onset of psychiatric disorders, which in turn cause individuals to experience downward drift in SES or to fail to rise out of low SES."[14] The difference of opinion regards the place of SES in the equation, i.e. does low SES precede psychiatric disorder or does it come about as a result of psychiatric disorder (which in this case is partially induced by the very conditions contributing to a segmented economic structure to begin with); but the important point here is that low SES is taken as given—regardless of which comes first, inhabiting the lower end of the socioeconomic structure is bad for your health. The continually revealing findings of the Whitehall studies in Great Britain[15] further punctuate this point; Marmot

April 2007 Volume 1, Issue 1

HCNA

Holy Cross Neighborhood Association Newsletter

Established in 1981

1st Issue Since Hurricane Katrina

Holy Cross Neighborhood Association – P.O. BOX 3417 New Orleans, LA 70177

Grand Opening, Lower 9th Ward Center

The Holy Cross Neighborhood Association will celebrate the grand opening of the Lower Ninth Ward Center for Sustainable Engagement and Development (CSED) on Monday, April 2, 2007. The event will take place from 10:30 – 11:30am in the back of Greater Little Zion Baptist Church, located at 5130 Chartres Street in the Lower Ninth Ward. Leaders from the Lower 9th Ward will be joined by Mercy Corps and a delegation of over 100 community leaders from Oregon State including Mercy Corps CEO, Neal Keny-Guyer; The Honorable Tom Potter, Mayor of Portland; and The Honorable Ted Kulongoski; Governor of Oregon.

Pam Dashiell, President, Holy Cross Neighborhood Association (HCNA) said, "Mercy Corps and the Oregonians are wonderful, caring, insightful partners. Their support as the Lower Ninth Ward and Holy Cross neighborhoods advance to a sustainable future is very much appreciated."

The new Center for Sustainable Engagement and Development will provide many services including: Assisting returning residents with resources and training on rebuilding their homes and navigating the "paperwork process"; buying construction items in bulk to offer low-cost (or no-cost) rebuilding materials; educating residents on how to use sustainable – or "green" – technologies and materials; facilitating strong and positive civic engagement; and teaching stewardship of the lower ninth's unique natural systems.

For more information, call Pam Dashiell, President, Holy Cross Neighborhood Association: 504-430-9041 or Susan Laarman, Mercy Corps: 503-341-3799.

In This Issue:

Community Center Opens: Page 1

Sustainability:

Sustainability Planning & Implementation: Nearly One Year Later

By Charles Allen, Vice President, HCNA

Figure 9:3. *An example of one of the post-Katrina neighborhood associations*

14 Johnson, J.G., P. Cohen, B.P. Dohrenwend, B.G. Link, and J.S. Brook. 1999. A longitudinal investigation of social causation and social selection processes involved in the association between socioeconomic status and psychiatric disorders. *Journal of Abnormal Psychology 108 (3):490–499.*

15 Marmot, M.G., 2004b, *The Status Syndrome.* New York: Henry Holt & Company, LLC.

suggests that there exists a "social status syndrome" wherein it is not simply the lack of access to resources that SES predicts which negatively affects health, but in fact there is an independent effect of social status on health. This makes sense in light of the research reviewed above, i.e. low SES is associated with more life events stressors, greater exposure to daily hassles, and less positive social support. In addition, people tend to be aware of these differences between themselves and those who are seemingly better off, and the propensity to compare oneself with others, combined with the feeling of powerlessness that low self-efficacy and external loci of control bring, may contribute to serious distress[16].

Natural disasters create precisely the kind of extraordinary circumstances that induce generalized stress in the affected populations. A disruption as severe as a hurricane, flood, or powerful earthquake is likely to result in an acute stress response for most, if not all, of the people who experience it directly. This short-term response begins with the "fight or flight" reaction but may contribute to psychological symptoms, such as anxiety or distress in the short term. For many survivors of disaster the response and the ensuing symptoms abate within a few days or weeks following the event. However, some survivors may continue to experience distress, anxiety, depression or other psychological states long after the event, hampering their ability to recover and potentially leading to chronic illness. The people most likely to suffer from psychological and consequent or associated physiological illness after a disaster occurrence are generally those who had greater stress exposure, i.e. those of lower status, well before the event took place.

Q: How might the geographic scale of exposure to the disaster also affect what was said in the last paragraph?

Thus, stress is socially distributed. Ethno-racial minority populations and those of lower SES not only have fewer coping resources on average which they can call upon to deal with stressful incidents, but also face greater exposure to stress in the first place[17] This can lead to a stress related poverty cycle: lower SES predicts greater exposure to stress and stress leads to poor health outcomes, thereby hindering capacity to achieve greater social status. Mirroring what has already been described in the chapter on disaster related social vulnerability, ethno-racial minority and SES compound each other to create similar feedback loops of disadvantage and gender, age, and nativity are additional factors which may mediate both stress exposures and poor health outcomes related to stress.

---

16 It is because of these factors that it is vitally important to restore basic services (such as clinics and bus services) to the hardest hit, and lowest SES neighborhoods of New Orleans.

17 In a disasters context, these populations have fewer coping resources, and are more likely to suffer a disproportionate exposure.

Stressors, Mediators and Moderators

So what sets the stress process in motion to begin with? As we saw earlier, Selye blames "noxious stimuli," or what we would presently call stressors. A stressor triggers a physiological or psychological response setting the stress process in motion. But this simple definition belies the multitude of questions in the literature about what constitutes a stressor and how it impacts the stress response to contribute to negative health outcomes. These questions are quite complex as stressors can take on different forms in different circumstances and are often tied up in the very construction of theoretical frameworks for explaining stress. For example, in disaster research, the disaster itself is clearly the most prominent stressor; however a disaster not only induces stress for the experience of living through it, but also for the problems it causes in its aftermath. The concept of stress proliferation speaks to this issue—proliferation creates secondary stressors, which occur in the wake of the primary stress trigger and result in ongoing strain on the impacted individual that may lead to problematic health outcomes[18] . Additionally, a disaster may lead to a spike in the prevalence of PTSD symptoms for a short period of time before subsiding in most of the population, but for some people the PTSD festers and becomes chronic; teasing out the reasons why the same stressor has vastly different implications thus requires additional investigation into individual

Figure 9:4. *An example of a secondary stressor—the fear of losing what little Katrina didn't take.* Photograph by Curtis

18 Pearlin, L.I. 1999. The Stress Process Revisited. In *Handbook of the Sociology of Mental Health*, 395–415: Springer Science & Business Media B.V. / Books.

Figure 9:5. *A flood stained "Head Start" sign outside an abandoned community center in the Lower 9th Ward.* Photographs by Curtis

contextual factors. Unfortunately, as has been mentioned throughout this book, challenges to collecting data, or having access to data, make understanding these contextual factors difficult. For example, it is hard to assess to what degree the individual was suffering from a stress related illness prior to the disaster.

The level of trauma an individual or a population has experienced over time and previous responses to traumatic events may mediate or moderate the ensuing stress response. Mediators have an independent effect on the eventual health outcome while a moderator increases or decreases the likelihood that a certain health outcome will occur. To make matters more confusing, certain factors may act as both mediators and moderators. In the case of a disaster like Katrina, the relatively low SES of the inhabitants of the Lower 9th Ward may have mediated their experience of the disaster, i.e. these populations were more prone to disease and disability even before the disaster as a result of their social and minority status, thus increasing the potential for ongoing distress in the aftermath. Alternatively health insurance coverage may have acted as a moderator for those who had it—the knowledge that resources to address any consequent illnesses or injuries were available likely had the effect of lessening anxiety.

With respect to disasters, magnitude acts as an important mediator and moderator. Shrubsole[19] highlights three ways in which magnitude has been measured in past studies, according to: 1) the characteristics of the event; 2) magnitude scales; and 3) individual perception of the size and impact of the event. With regard to the first two, it is less the measurement of the magnitude itself and more what the measurement means for survivors of the disaster that affects health outcomes. In other words, the characteristics of an event can include both the type of disaster that occurred, e.g. earthquake, flood, hurricane, and the intervening factors that

19 Shurbsole, D. 1990. Natural Disasters and Public Health Issues: A Review of the Literature With a Focus on the Recovery Period. In *ICLR Research Papers*. London, Ontario: Institute of Catastrophic Loss Reduction.

Figure 9:6. *A banner hangs above the neighborhood around the 17th Street Canal Break on "Katrina day" 2006.* Photograph by Curtis)

partially determine the degree of ensuing damage. In the case of Katrina, for example, the hurricane itself rated a "3" on the Saffir Simpson Hurricane Wind Scale when it made landfall in New Orleans. This designation provides a general picture of the force of the hurricane, which in turn offers some idea of the potential damage that will result and the threat to human life—factors moderating the disaster's impact on health.

However, the categorization itself tells us little about health outcomes. Katrina was devastating in part because of intervening factors, such as the pre-existing vulnerability of populations along the Gulf Coast and the haphazard government response immediately before and after, both of which exacerbated an already catastrophic situation and ensured a prolonged recovery period.[20] It is therefore existing circumstances combined with individual perception that visit a direct impact on health. There are several aspects of event perception that may influence psychological health in particular. First, the extent to which a disaster is natural or manmade appears to influence psychological symptoms, with manmade disasters creating more distress. As both the Johnstown flood and Katrina suggest however, these are not discrete categories, the original "natural" disaster was made far worse by human engineering failure in terms of a dam and a levee break. Additionally, survivors of Katrina felt victimized by the government as they waited for rescue in the days following hurricane landfall. With many natural disasters people may have a sense that they can control their own fates, at least to a small degree, by engaging

20 Lee, E.-K.O., S. Ce, and T.V. Tran. 2009. Coping With Hurricane Katrina: Psychological Distress and Resilience Among African American Evacuees. *Journal of Black Psychology 35 (1):5–23.*

in preparedness activities, or evacuating when danger is imminent[21]. However, this sense of control is predicated on having access to appropriate resources and feelings of efficacy, which are not distributed equally across the population.

POST-TRAUMATIC STRESS DISORDER

Stress has the potential to severely disrupt individual well-being and contribute to chronic mental and physical illness. Disasters create conditions ripe for stress across the entire affected population, including not only those people who experienced the disaster directly, but their loved ones and associates, emergency responders and others charged with addressing resultant damage[22]. Furthermore, the capacity of disasters to wreak catastrophic damage on populations and their environments sets the stage for an unusually widespread and severe stress response among the affected population. The level of trauma experienced, therefore, and the size of the population exposed has led many researchers to focus on increases in the incidence of post-traumatic stress disorder (PTSD) following disasters. For example, 36% of the population living in the path of Hurricane Andrew developed PTSD in its wake[23], while reports have suggested a PTSD prevalence rate of 30% for those in New Orleans following Katrina and 12% for those in other damaged areas[24]. In a longitudinal study conducted after an earthquake in Northern China, the prevalence of PTSD was found to be between 20–30% within a period of nine months in the two villages sampled[25]. These rates are considerably higher than the PTSD prevalence rate of 3.5% found in the U.S. population as a whole, and given the propensity of disaster survivors to exhibit symptoms of PTSD even in the absence of any pre-existing psychological symptomatology, it is clearly an important outcome to track after a disaster.

However, PTSD is certainly not the only stress-related condition to emerge in post-disaster settings and while some researchers have also paid attention to the complex of mood and anxiety disorders outlined in the Diagnostic and Statistical

21 A person believing they had "control" in their own fate could also be tied to a societal belief that safety measures were in place to protect the citizens. The failure in the flood control system undermined an individual's perception of the "normal" safeguards he or she was basing decisions on.

22 The 911 calls generated during Katrina contained a large number of calls from family members reporting that their contact with a near and dear one had suddenly ceased—obviously this would be stress generating for the call irrespective of how close he/she was to the hurricane.

23 David, D., T.A. Mellman, L.M. Mendoza, R. Kulick-Bell, G. Ironson, and N. Schneiderman. 1996. Psychiatric Morbidity Following Hurricane Andrew. *Journal of Traumatic Stress 9 (3):607–612.*

24 Kessler, R.C., S. Galea, R.T. Jones, and H.A. Parker. 2006. Mental Illness and Suicidality after Hurricane Katrina. *Bulletin of the World Health Organization 84 (12)*

25 Wang, X., L. Gao, N. Shinfuku, H. Zhang, C. Zhao, and Y. Shen. 2000. Longitudinal Study of Earthquake-Related PTSD in a Randomly Selected Community Sample in North China. *The American Journal of Psychiatry 157 (8):1260–1266*

Manual of Disorders (DSM)[26], less attention has been paid to generalized distress measures or stress-related physiological outcomes. PTSD may provide a useful snapshot of the wide-ranging psychological implications of a disaster in the weeks after its occurrence, but it is also a very specific clinical diagnosis with symptoms that tend to subside in the majority of the population within weeks to months after a disaster. Furthermore, most studies focusing on the health effects of disasters have been conducted within a year after the disaster while it generally takes more than a year to determine the chronicity of PTSD[27].

Additionally, the psychological constructs presented in the DSM have run into a range of criticisms over the years, one of the most persistent relating to their tendency toward universalism despite being drawn from studies that have overwhelmingly sampled non-Hispanic white populations, and men in particular[28]. This may be especially problematic given the evidence suggesting that minority populations tend to have higher levels of psychological distress in general (National Center for Health Statistics 2007) and may be more vulnerable to stress post-disaster (Curtis, Mills, and Leitner 2007). For these reasons, and because PTSD is sufficiently covered in the disasters literature, the following section will look at post-disaster stress more broadly.

Domains of Stress and Disasters

Disasters create environments rife with potential stressors. From the fearful confusion immediately following to the eventual attempts to recover and rebuild, there are a variety of factors that may contribute to the onset or amplification of mental and physiological disorders. In order to explore some of these factors, it will be useful to separate them into three domains, those encountered in: 1) the immediate aftermath; 2) the intermediate recovery period; and 3) long-term reconstruction. Stressors and their impacts on health will almost certainly cross domains, but there are nevertheless differences in the situations encountered in each of these periods that may have distinct impacts on the health of different populations.

Immediate Aftermath: Death

Not surprisingly, death is the best predictor of post-disaster stress and mental disorder. This tends to be true with respect to witnessing a death, encountering

---

26 American Psychiatric Association. 2000. *Diagnostic and Statistical Manual of Mental Disorders* DSM-IV-TR. 4th Edition, Text Revision ed. Arlington, VA: American Psychiatric Publishing, Inc.

27 Shrubsole, D. 1999. Natural Disasters and Public Health Issues: A Review of the Literature with a Focus on the Recovery Period. In *ICLR Research Papers*. London, Ontario: Institute for Catastrophic Loss Reduction.

28 Kessler, R.C., S. Galea, R.T. Jones, and H.A. Parker. 2006. Mental Illness and Suicidality after Hurricane Katrina. *Bulletin of the World Health Organization* 84 (12).

dead bodies, or losing a loved one in the disaster[29]. The first two instances likely predispose people to PTSD or the onset of a mood or anxiety disorder, at least for the short term, but losing a loved one in any situation tends to be a traumatic experience, and in a disaster situation when the death of a loved one is the worst in a series of traumatic experiences, the impact is likely to be all the more substantial.

The bereavement process is culturally mediated and not always well understood, but it may take individuals one to four years to achieve a state of recovery after the late-life death of a loved one, and even then they may never fully recover[30]. Bereavement following the suicide death of a loved one differs from the grieving process in other situations due to an increased sense of survivor guilt and questions about how the death could have been prevented. This frame may be relevant to disasters for two reasons: first, there are plenty of opportunities for disaster survivors to question what they or others might have done differently that would have prevented a death (e.g. prepared more thoroughly, evacuated on time, responded more quickly)—a part of the grieving process not as likely to be experienced by someone losing a loved one to chronic disease for example; second, there is some evidence to suggest that suicides, and suicidality, tend to increase following disasters (Kessler et al. 2006).

Immediate Aftermath: Loss

Enduring the death of a loved one is the most severe of the types of loss survivors of disasters experience. Disasters are characterized by their capacity for destruction, and in their wake people often find themselves without homes or possessions. The conservation of resources model suggests that people strive to develop and accrue resources for themselves, including psychological, personality-related and socioeconomic resources, and when threatened with the possibility or actuality of resource loss, they experience increased psychological distress[31]. Several studies of disasters and health have since employed this model to describe the circumstances leading to psychological disorder among the populations affected by disaster, and findings have included that loss of home and possessions acted as prime indicators of post-disaster psychological illness[32]. It is important to note that the loss

29 Galea, S., C.R. Brewin, M. Gruber, R.T. Jones, D.W. King, L.A. King, R.J. McNally, R.J. Ursano, M. Petukhova, and R.C. Kessler. 2007. Exposure to Hurricane-Related Stressors and Mental Illness After Hurricane Katrina. *Archives of General Psychiatry* 64 (12):1427–1434.

30 Arbuckle, N.W., and B. de Vries. 1995. The Long-Term Effects of Later Life Spousal and Parental Bereavement on Personal Functioning. *The Gerontologist* 35 (5):637.

31 Hobfoll, S.E. 1989. Conservation of Resources: A New Attempt at Conceptualizing Stress. *American Psychologist* 44 (3):513-524.

32 Sattler, D., J. Sattler, C. Kaiser, B. Hamby, M. Adams, L. Love, J. Winkler, C. Abu-Ukkaz, B. Watts, and A. Beatty. 1995. Hurricane Andrew: Psychological distress among shelter victims. *International Journal of Stress Management* 2 (3):133–143. Regehr, C., A.R. Roberts, and T. Bober. 2009. On the

Figure 9:7. *Two homes where people died—still clearly marked a year later for neighbors to see—especially the bottom image which literally spells it out. In the top image a car sits outside the home where two died...* Photographs by Curtis

measured in these studies is not merely material but symbolic as well. Losing a home generally implies the loss of keepsakes and other emotionally valuable items. Furthermore, the U.S. population tends to prize homeownership as the realization of the American dream and therefore sink their finances as well as their dreams into their homes. Coping resources may diminish in the wake of a disaster with the dissolution of social networks and the severe disruption of routine. Thus there are many ways to experience loss after a disaster, and each of these working in combination with other stressors increases the chances of illness onset.

Immediate Aftermath: Displacement and Entrapment

Irrespective of the disaster type, people are often required to evacuate their homes either to escape its impending impact, or to clear a devastated region. This creates many stress generating opportunities. This book has already provided examples of why some people will prefer to "take their chances" rather than evacuate. People

Brink of Disaster. *Journal of Social Service Research* 34 (3):5–13; Kaiser, C.F., D.N. Sattler, D.R. Bellack, and J. Dersin. 1996. A Conservation of Resources Approach to a Natural Disaster: Sense of Coherence and Psychological Distress. *Journal of Social Behavior & Personality* 11 (3):459–476; Galea, S., C.R. Brewin, M. Gruber, R.T. Jones, D.W. King, L.A. King, R.J. McNally, R.J. Ursano, M. Petukhova, and R.C. Kessler. 2007. Exposure to Hurricane-Related Stressors and Mental Illness after Hurricane Katrina. *Archives of General Psychiatry* 64 (12):1427–1434

Figure 9:8. *This home had personal messages of the "loss" felt sprayed on both interior and exterior walls. This photograph, looking in from the front door, also shows the flood discarded property of the home, and the heavy mold stains by the ceiling. This was again one year after Katrina.* Photograph by Curtis

who do leave their homes and neighborhoods behind face the stress of waiting to find out what they will return to—a process exacerbated by their complete lack of control over the situation.

Before even reaching that stage however, there is the stressful process of evacuating itself, or perhaps worse, the panic associated with the incapacity to leave when the people around you are escaping to safety. As described in Chapter 6, those least likely to evacuate are the same populations that are also more likely to have health conditions prior to a disaster and more likely to develop psychological symptoms after a disaster. For those who do evacuate, there is a potential for long-term displacement or even permanent relocation. In studying the population of an Italian village several years after a series of earthquakes, it was found that those residents who had permanently relocated expressed increased depression, anxiety, hostility and sleep disturbance when compared with those who had returned to their homes[33]. People with extended social networks on which they can rely for a place to stay or material support after a disaster have an advantage over those without such networks; however the stress of relying or imposing on others for more than a short period of time may nevertheless contribute to psychological distress in evacuees. Additionally, the general experience of displacement may take a significant toll on the mental health of evacuees[34].

Immediate Aftermath: Adversity and Fear

Even those people who have evacuation destinations with friends or family may still be herded to temporary shelters during the initial evacuation period. The

33 Bland, S.H., E.S. O'Leary, E. Farinaro, F. Jossa, V. Krogh, J.M. Violanti, and M. Trevisan. 1997. Social Network Disturbances and Psychological Distress following Earthquake Evacuation. *Journal of Nervous & Mental Disease* 185 (3):188–195.

34 Carr, V.J., T.J. Lewin, R.A. Webster, P.L. Hazell, J.A. Kenardy, and G.L. Carter. 1995. Psychosocial Sequelae of the 1989 Newcastle Earthquake: I. Community Disaster Experiences and Psychological Morbidity 6 Months Post-Disaster. *Psychological Medicine 25 (3):539–556*

conditions of such temporary shelters may be enough to induce the onset of mental distress in evacuees. Conditions such as those described in the chapter on Health Vulnerability, along with the indignity and discomfort associated with having to use the bathroom or change clothes in the presence of strangers increased the likelihood of psychological symptoms among Katrina survivors. In addition, a sizeable portion of Katrina evacuees who spent time in either the New Orleans Convention Center or Superdome experienced threats of violence, as well as facing severe physical adversity following the hurricane in the form of hunger, thirst, lack of sleep and untreated injury, mostly while waiting to be rescued. Beyond the anxiety evacuees may feel for their homes and possessions, there is also the fear that accompanies waiting to hear the fates of friends and family members that have not yet been located.

Intermediate Recovery Period

Some of the stressors that first rear their heads in the days immediately following a disaster may persist and evolve as the evacuee moves into the next stage of regaining control and returning. The stresses of displacement and disruption, threats of violence and untreated injuries, for example, foreshadow larger social ills that emerge, or expand in the weeks and months after the initial onslaught. Moreover, the recovery period sees people returning home to confront the damage and loss visited upon their homes and neighborhoods. Those who choose not to return face the stresses associated with rebuilding their lives in a new location surrounded by new people. The economic hardship and health problems some experienced prior to the disaster only grow more troublesome afterward. Three broad issues encountered (or perceived) during this period that are most often addressed in the health and distress literature are: 1) damage to the built environment and critical infrastructure; 2) disrupted social networks; and 3) crime.

The opportunity to return home after a disaster may be a period of both relief and anxiety for disaster survivors. The first assessments of damage to one's home and neighborhood may prove quite traumatic depending on the level of destruction and amount of loss. Beyond the initial impact of seeing the disaster's effects for the first time, there are also ongoing issues of safety to consider[35]. Where homes remain standing they may also be unsafe to live in due to the amount of structural damage sustained. The federal government responded to such problems after Katrina by providing myriad trailers in which people could reside while their homes were detoxified and potentially rebuilt. However, the experience of living in a trailer when you previously lived in a home may be psychologically damaging in a similar

35 This was described for the after effects of the St. Louis Tornado where returning residents balanced the search for personal belongings with the risk for further structural collapse.

manner to that of living in a temporary shelter for a period of time. This may be especially true in cases where there is no identifiable time frame in which one's house will be rebuilt and suitable for inhabiting again. Furthermore, destruction to one's street or neighborhood may exacerbate stress, again especially when there are few visible signs of rebuilding or recovery in the area. As noted below, there is a dearth of research on these kinds of issues in the disasters and health literature, and geographers would do well to jump into the breach.

Beyond the immediate home environment, the critical infrastructure of the city may suffer substantial damage from a disaster as well. In New Orleans, the public hospital system was virtually destroyed as a result of Hurricane Katrina creating not only immediate barriers to health response for victims of the disaster but denying the lower income populations most affected by disaster their primary source of health care. Furthermore, there is the potential for commerce to suffer and businesses to close in the aftermath of a disaster, effectively rendering segments of the population unemployed at the very time they most need access to financial and material resources. In short, disasters have the capacity to punch giant holes in the city's resource base at a time when a substantial proportion of the population needs to access it and those most reliant on it under everyday circumstances find themselves completely without resources.

Informal social networks and formal social organizations, such as churches, political associations, and community groups also face disruption at best or dissolution at worst after a

Figure 9:9. *Trailers are often wedged into any available space—though in the bottom image, this was the only family who had returned—all other homes were abandoned with no sign of rebuilding.* Photographs by Curtis

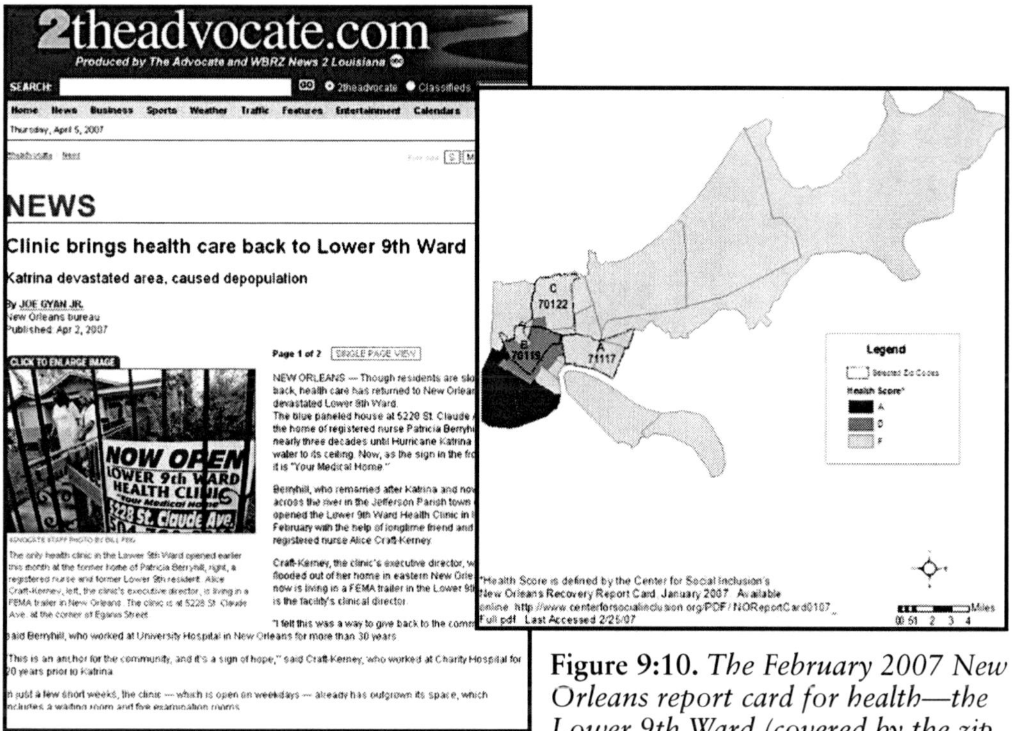

Figure 9:10. *The February 2007 New Orleans report card for health—the Lower 9th Ward (covered by the zip code 71117) as well as the majority of the rest of New Orleans still scored an "F" for health services provided to returnees. Next to this, a newspaper story reports the opening of a clinic in the neighborhood.*

disaster. We saw previously how short-term and permanent relocation can negatively impact psychological health; a primary reason for this health effect may be the loss of social networks. Participation in social groups and the presence of networks of support in peoples' lives tend to have positive effects on health, and in a post-disaster environment. The people affected are presumably in great need of social support. More generally, the re-establishment of a regular routine may help survivors begin to feel "recovered," and where one's routine involved encounters with neighbors and social groups, the absence of such regular meetings may prolong feelings of distress and anxiety about any real or perceived lack of recovery. Finally, social networks provide a feeling of safety and security, a need surely enhanced in the wake of a catastrophe, especially where a breakdown in the system of order lends itself to increased opportunities for crime.

Disasters create confusion and enhance feelings of desperation. The havoc they wreak also impacts the ability of public systems to perform their usual roles, including law enforcement. The evacuation of a large geographic area and the housing of evacuees in shelters where no private space is available, left remaining possessions vulnerable to theft. In other words, the post-disaster environment is one ripe for increases in criminal behavior. Researchers found a doubling in the reports of crime between their baseline survey of Hurricane Andrew survivors and the follow-up

Figure 9:11. *The Lower 9th Ward was full of churches before Katrina—most are still badly damaged and abandoned.* Photograph by Curtis

survey[36]. In terms of stress and mental illness, the perception of increased crime may be as damaging as any objective measure of crime after a disaster[37], and the combination of fear, doubt as to the potential for neighborhood recovery, the experience of living in temporary housing, and the inability to seek support from the social networks an individual has come to rely on, is enough to engender serious distress in any number of people.

Long-Term Reconstruction

Few studies of health in relation to disasters have measured psychological symptoms or other indicators of ill health beyond a year or so after the event occurred. For this reason it is difficult to know the extent to which psychological disorder or other negative health outcomes may persist among survivors, and which groups might be differentially affected well into the future. Most studies conducted more than a few months after a disaster occurrence indicate that psychological symptoms

36 McDonnell, S., R.P. Troiano, N. Barker, E. Noji, W.G. Hlady, and R. Hopkins. 1995. Evaluation of Long-term Community Recovery from Hurricane Andrew: Sources of Assistance Received by Population Sub-groups. *Disasters 19 (4):338–347.*

37 Working with different community groups in post-Katrina New Orleans often leads to discussions about crime. During the monthly Holy Cross neighborhood meeting a "crime report" is read giving details about what has happened and where. A community collaborator who accompanied us around Holy Cross "matter-of-factly" described the recent gunshots she had heard one block from her home, and the neighborhood itself now has "crime placards" placed where something has happened.

for most of the affected population decrease over time, though this may be less true for evacuees who permanently relocate after the disaster. In general, though, there has been too little research to make definite statements as to the length these post-disaster illnesses last. Returning to the discussion earlier in this chapter about allostatic load and the effects of unmitigated or chronic stress, it seems safe to speculate that rates of disability and chronic disease may be higher among those individuals and populations that continue to experience psychological distress or mental disorder as a result of living through a disaster, especially if that disaster is in effect ongoing, which many could argue is true until a pre-disaster state of normalcy has been reached. However, as indicated below, this is one of several areas of research that could stand to be expanded in order to better understand the impacts of disasters on health.

Research Gaps and Opportunities

The study of disasters comes rife with methodological problems, many having to do with the confusion and disorganization that accompany the disaster situation itself. Specifically, the potential study subjects have been scattered throughout the region or even the country following a disaster, and obtaining a sufficient, much less representative, sample may be very difficult. One solution may be to wait for a designated period of time, allowing people the opportunity to resettle before contacting them. However, there are a host of problems associated with this strategy as well, two of the most notable being bias, i.e. the people who return home may not be representative of all the disaster survivors, and the inability to track the evolution of symptoms from the time of the disaster. Another strategy involves surveying evacuees in temporary shelters immediately after an event; however this again lends itself to bias, and it does not offer any information about the persistence of negative health symptoms beyond the initial phase of shock. Additionally, conducting follow-up in a study such as this where the collection of contact information and names is necessary would require a research team to be prepared at a moment's notice for a disaster to strike and an expedient Institutional Review Board process[38].

Another problem with establishing a body of evidence on the health impacts following a disaster has to do with the varied types of events and their idiosyncrasies. Fires in the hills of Santa Barbara may have different health effects than a tsunami off the coast of Indonesia for many reasons, including the disparate resources allocated to the affected populations and the existing health services

38 For research projects involving people, an academic must submit a detailed research plan to a university committee that decides whether the subject is protected from harm and that enough safety protocols are included to protect the subject's confidentiality.

infrastructure in each place. One means of normalizing these differences to some degree would be to utilize control groups or compare the sample population to a non-affected comparison group. However, few studies of disasters and health have yet attempted this.

Beyond these issues of sampling, the bulk of the research on disasters and health has taken place in the near-term. Although it is certainly appropriate to investigate the immediate and short-term implications of disasters for health, what may be more crucial for understanding the health legacy of disasters are the types of outcomes that persist or even develop several years after an event. For example, Solomon and colleagues looked at the tendency toward immunosuppression in people who had lived through the Northridge earthquake in Los Angeles and found that the subjects suffering the most disruption as a result of the earthquake had greater immune function impairment[39]. In addition those people with fewer personal coping resources faced the greatest likelihood of impairment. While this study was also conducted within a month after the disaster, the findings offer a peek at what could become future disaster-related chronic health issues. Immune function is a potent measure of general stress, or allostatic load, and therefore a predictor of chronic disease outcomes[40].

Moreover, the disasters and stress literature does tend to agree on the idea that there is a substantial increase in the onset of mood (e.g. depression) and anxiety disorders (e.g. PTSD) following a disaster, the experience of which is enough to justify follow-up studies. But mood and anxiety disorders have also been well-connected to increased possibilities for morbidity and mortality. Ultimately, it is chronic illness—be it either psychological or physiological—that plays the greatest role in determining individual well-being. Gaining insight into the stress-related correlates of disaster over the long term will provide a window into prevention and treatment opportunities for the population as a whole as well as those most severely afflicted.

Two final shortcomings of the disasters and health literature are also two areas in which geographers may be able to contribute the most in the future. First, one of the difficulties with establishing a consistent evidence-base for the health impacts of disasters is the lack of available baseline data for the populations affected by a disaster. Without appropriate baseline data on the prevalence of disorder prior to an event, little can be said with any authority about the prevalence or severity of disorder after the disaster. Using a control or comparison group as mentioned

39 Solomon, G.F., S.C. Segerstrom, P. Grohr, M. Kemeny, and J. Fahey. 1997. Shaking Up Immunity: Psychological and Immunologic Changes after a Natural Disaster. *Psychosomatic Medicine 59 (2):114–127*

40 This still follows the normal "disaster" logic that the impacted area returns to normal within an acceptable time frame. This is certainly not true of post-Katrina neighborhoods.

above may be one way to correct for this situation, but this does not necessarily alleviate a broader problem having to do with health data in general.

The beauty of generalized measures of distress, like the construct of serious psychological distress[41], which is included in the National Health Interview Survey (NHIS) and the World Mental Health Survey, is that they can obtain information about the comparative stress levels of entire populations without being constrained to the clinical symptom lists contained in the DSM. However, region (i.e. Northeast, South, Midwest, West) is the greatest geographical specificity the NHIS offers. Surveys obtaining information on psychological distress and disorder at more local levels where significant geographical specificity may be included, on the other hand, have used so many different instruments and clinical scales in so many different types of populations and using so many different sampling methodologies that it becomes difficult to make any general claims about the psychological state of populations. A lack of spatial acumen or awareness pervades the literature here as well, and geographers therefore have a prime opportunity to rectify this shortcoming.

Secondly, geographers can put their skills in utilizing geotechnologies to good use in studying the health impacts of disasters. One of the shortcomings mentioned above is the inability to track health developments or symptomatology over time. In addition to this, the literature on disasters and health has tended to rely on self-report data rather than incorporating objective data on the state of the built environment, for example, into analyses. Geographers have a chance to expand the literature by addressing these drawbacks simultaneously with the application of GIS and GPS, especially when combined with local collaborators' "insight"[42]. Mixed methods approaches offer a great deal of promise for the advancement of this literature and geographers have the appropriate skills to bring to the table.

CONCLUSION

Stress is a multifaceted health issue, influenced by any number of contextual factors and poised to influence any number of negative health outcomes. When combined with the myriad stressors that arise in the wake of disaster, the potential for widespread disease becomes a formidable issue. Research on psychological disorder, in

41 Kessler, R.C., G. Andrews, L.J. Colpe, E. Hiripi, D.K. Mroczek, S.-L.T. Normand, E.E. Walters, and A.M. Zavlasky. 2002. Short screening scales to monitor population prevalences and trends in non-specific psychological distress. *Psychological Medicine* 32:950–976.

42 Curtis, A., J.W. Mills, B. Kennedy, S. Fotheringham, and T. McCarthy. 2007. Understanding the Geography of Post-Traumatic Stress: An Academic Justification for Using a Spatial Video Acquisition System in the Response to Hurricane Katrina. *Journal of Contingencies and Crisis Management 15 (4):208–219*

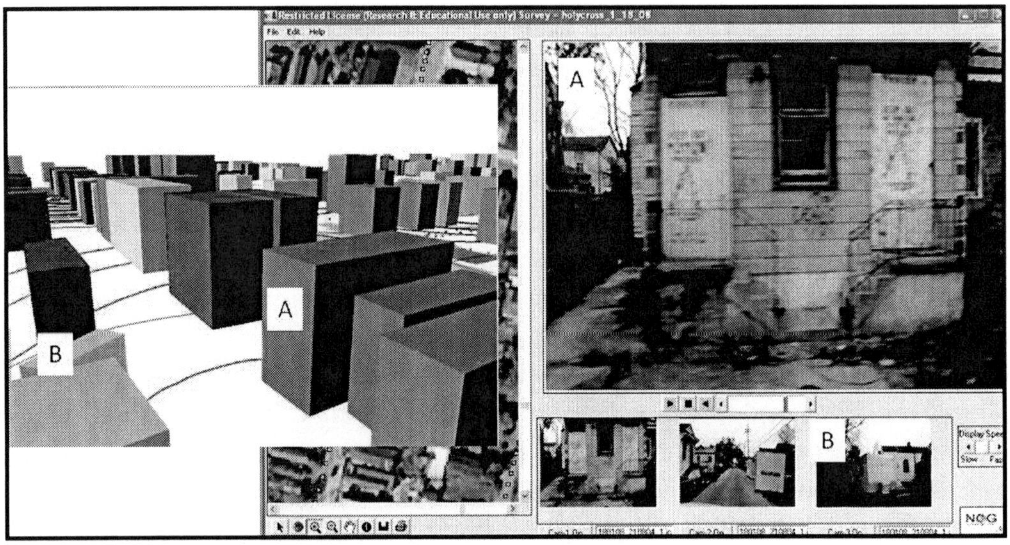

Figure 9:12. *The spatial video has been used to create a 3-D GIS allowing for the estimation of what the family living in trailer "B" sees from their window—for example, an abandoned and damaged home "A."*

particular after a disaster, is well developed in some ways, but the nature of disasters and the barriers to consistent measurement make it difficult to establish a firm evidence-base for any particular finding. Furthermore, the bulk of this literature focuses on clinical symptoms of psychological disorder in the period immediately following a disaster, and most of the research has neglected to account for spatial factors. This leaves many areas open to exploration and geographers, with their skills in spatial analysis and penchant for thinking spatially in general, will find many opportunities to leave their mark on this literature.

TEN

The Post-Disaster Landscape

PREVIOUS CHAPTERS HAVE described the way the built environment is changed by a disaster—buildings are flooded, sometimes moved or even completely destroyed. This changed landscape can also be viewed at multiple spatial scales—whole sections of a city may be destroyed and needing to be rebuilt as after the fire of 1906 in San Francisco, or maybe the spatial extent includes a one-street village as with the Tornado that struck Huntersville, Tennessee in 2008. Occasionally, the geographic extent is so large that whole towns or cities must be rethought and rebuilt, such as with Johnstown or New Orleans. However, irrespective of spatial scale, lives and landscape will have been changed to some degree.

Disasters are both visual and visceral—people often become involved in these landscapes as though they are either actors in a performance, or an audience to that performance. Think again to the cover of this book, this was a posed shot

Figure 10:1. *Sacramento Street, San Francisco April 18, 1906.* Arnold Genthe.[1]

1 Readers are pointed to Arnold Genthe's own biography "As I Remember" which details how he captured the events of 1906 on film.

Figure 10:2.

for anyone passing, but especially the disaster tourists. Starting almost as soon as vehicles were allowed entry into the Lower 9th Ward, buses would ferry people in who wanted to experience the disaster. Even today (2009) in the French Quarter, signs advertise disaster trips alongside plantation and swamp tours. Go to Expedia and search for a hotel in New Orleans, one of the entertainment options is also a disaster tour.

Posing for a disaster photograph certainly didn't originate with Katrina—nor did disaster tourism or the need to experience through physical observation. Older disaster photography often contain posed scenes of people proudly standing next to the ruins of their home—or bravely smiling as they "bond" together in a shelter. Possibly the most famous of these were taken by Arnold Genthe (1869–1942) who captured the 1906 destruction of San Francisco.

After the earthquake, with the fire beginning to take hold, people had actually sat down on chairs to watch the events as though at a performance. It could be argued that this was ignorance of the magnitude of the events, or even an indication of coping—taking in the destruction and all that it meant while also preserving a distance from it, viewing it as a performance and therefore not an actor on its stage.

Figure 10:2 shows other examples of disaster photography—the two families posed proudly in front of the ruins (presumably their own) left by a tornado. The inset map shows what in the United Kingdom has previously been called the "Battle of Britain" spirit—where citizens of London would make the best of a situation that would frequently send them into the Underground railway stations as the bombs dropped during World War II. In the photograph shown here, smiles abound as people eat and drink in the shelter after the Long Beach earthquake of 1933.

In the weeks and months after Katrina another form of disaster "performance" emerged on the landscape as the buildings became the focus of messages and artistic creations. For most of these messages the canvas was the building itself, the markings made directly onto walls, doors and any other visible surface. The audience included search and rescue workers, animal rescuers, friends, relatives and neighbors, the larger community, the city or Parish, official contacts (insurance agents), and just as with the man on our cover, all who pass by. Using the home as a message board makes sense if you deconstruct the components of communication; a source, a destination and message content. When traditional communication means are not available—maybe phone lines are down, maybe you are unsure of where a relative is or if they have a contact number, then one way to "anchor" the message, to make sure the intended person/people receive it, is to use the physical space which is common to both message source and destination. The source may be a search and rescue team, or a family member. The destination may be animal rescue, a body retrieval unit or another relative. The connector is the building where both teams are visiting or the address of the friend.

Figure 10:3 shows a selection of search and rescue spray painted markings. It was impossible not to notice these vibrant colors as you walked the streets of New Orleans, the messages arguably being as dominant on the landscape as the damage itself. Even today, four years after the storm, many homes still carry the fading markings. Vibrant colors sometimes covered entire walls ("G" & "H") or houses ("F"—these also extend around the side of the building). In the most common and one could argue, proper standardized version, the "X" is used to convey messages for the next search and rescue team to cover that neighborhood (for example in "B" 0 living found, 2 dead, the next team raised this to 3). No one really questions the right of these teams to deface property, even though sometimes the messages are excessively long ("A"), painted on brick, and sometimes even used by the search team for their own statements (the Fire Department of New York sprays "Never forget 9.11.01" in "C." Do they have the right to use someone's home in this way? What impact might this have on the returnee?) In "E" on the home of Fats Domino in the Lower 9th Ward, several messages were written because no one knew if he was alive or safe. The search and rescue team tasked to this location spelt out their name...possibly the only time we had seen the name rather than initials—look at

Figure 10:3. Photographs by Curtis

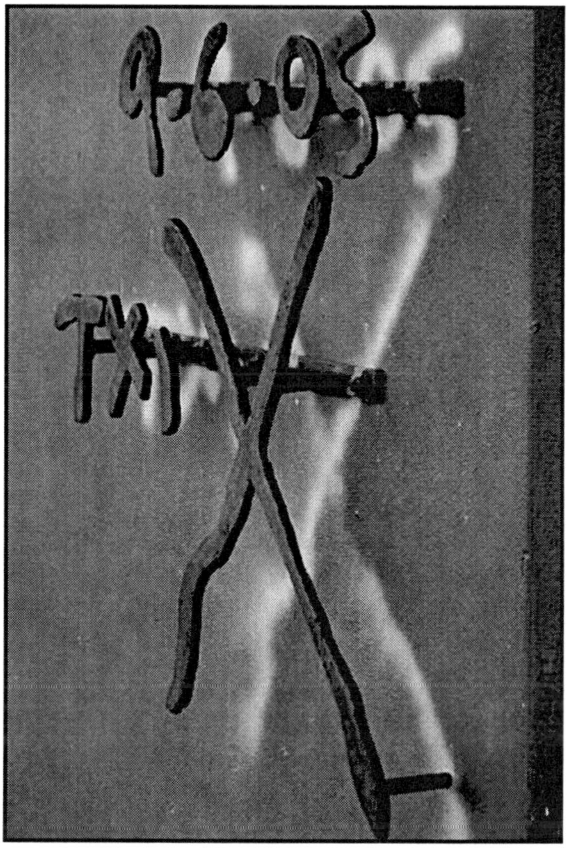

Figure 10:4.

the photograph and see if you can understand why.

As we conducted research in New Orleans in the years following Katrina we began to become suspicious that the "X" also became visual badges of honor. As homes were rebuilt, flower beds sown and "normalcy" returned, the "X" would sometimes remain. This suspicion was verified on one home, as seen in Figure 10:4 when the owner has made a metal "X" to sit above the painted version which has begun to fade.

As a geographer these messages are also a good spatial source of information and a temporal snapshot of what happened in each neighborhood. Just as we broke apart the component parts of a message—so we can use the same schema in a GIS. We know the source (given by the left hand side of the "X"). We also know the date of the visit, and attributes or actions (was food and water left, did a dog die, was a human corpse found). Most importantly, we have the spatial location of the action, either collected by a GPS or by the building's address. By adding all these into a database we can tell the geographic story of human and animal rescue.

Figure 10:5 shows six maps created from data extracted from photographs taken while walking the different neighborhoods of New Orleans and St Bernard Parish. A sample of the database created to generate these maps is also shown. Part of the skill in building a GIS is knowing how to split apart your data into attribute fields that can be manipulated and mapped easily. In Figure 10:6, each row represents a house. Each search and rescue team (as identified by a different "X" or similar marking in the photograph) to have visited that house has a section of the database, containing date, the search and rescue team, whether there was a death, any other special information for that address such as a gas leak, and the color of the "X." Animal rescue team information is also added. This database allows us to map the progression of the first search and rescue teams to visit homes in any

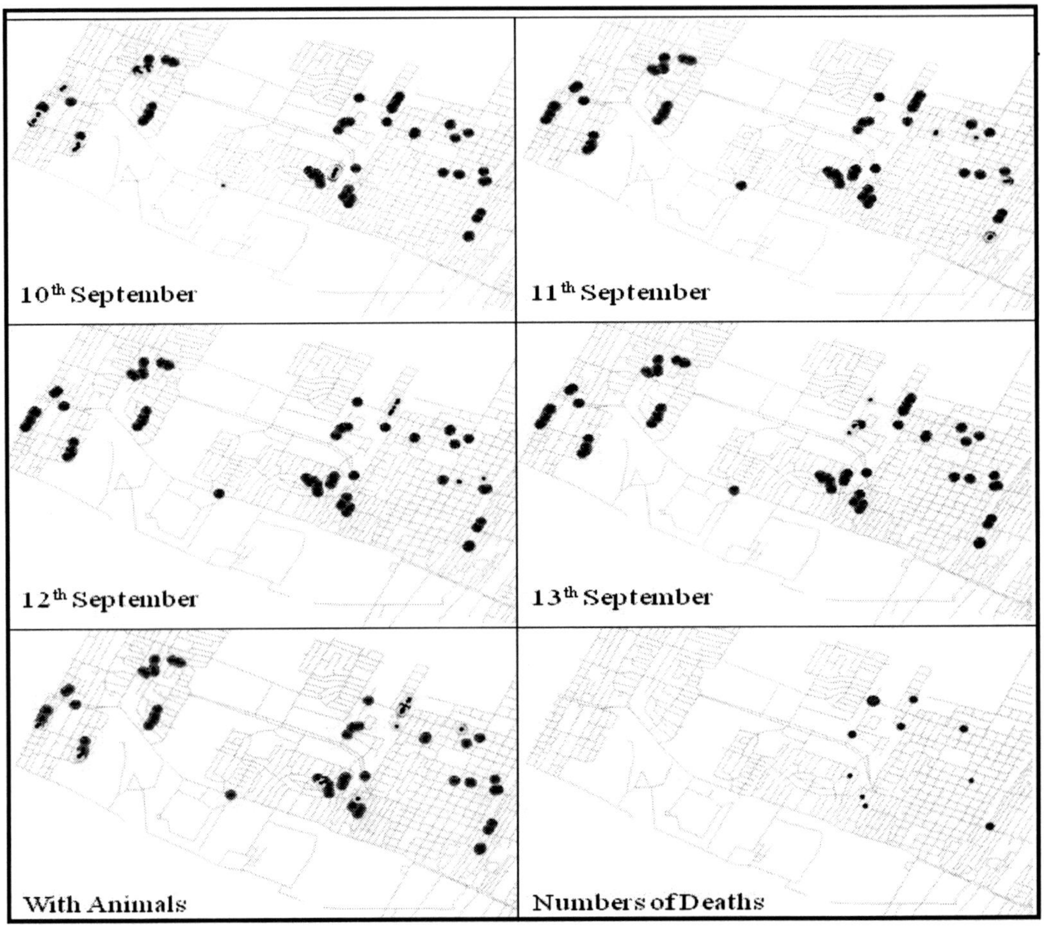

Figure 10:5.

neighborhood, whether any animals were found, and were there human mortalities (and how many).

Q: Given what you have seen in the search and rescue markings in this book—what other queries could you ask?

Q: Why might it be useful to create a database/GIS of this information?

Search and rescue teams were not the only ones to use the post-disaster landscape as a message board. Vendors, insurance adjustors, official notices from the city and federal agencies all resorted to posting information. And so did the returnees—sometimes leaving notes to communicate with friends and family, in other instances forming a dialogue with search teams, especially animal rescue, and finally as a forum to publically display their emotions. Some returnees reclaimed their individual space from both Katrina and the rescue teams who had marked up their homes with their own messages—sometimes positive, sometimes gut-wrenching in expressing loss—but having the last word appeared to be important. Why would

Death	Date	X1	Unit1	Date1	Death1	Detail1	Build 1	Color 1	Unit2	Date2
red box "1" on top	13-Sep		SF1	14-Sep	0		siding	black		
			CAN TF1	5-Sep	0		brick	orange		
			SF1	14-Sep	0		siding	black		
			CAN TF1	5-Sep	0		brick	orange	SF	14-Se
			CAN		0		brick	yellow	SF	14-Se
			SF1	14-Sep	0		window	orange		
			SF1	14-Sep	0		siding	black		
			SF1	14-Sep	0		siding	black		
			SF1	14-Sep	0		brick	black		
red box with "X" inside	11-Sep		SF1	14-Sep	0		brick	black		
			SF1	14-Sep	0		brick	black		
			CAN TF1	5-Sep			door	black		
			CAN TF1	5-Sep			door	black	ILSF1	11-Se
			SF1	11-Sep	0		window	blue		
			CAN TF1	5-Sep			board	black	SF1	11-Se
			CAN TF1	5-Sep			window	green	ILSF1	11-Se
			CAN TF1	5-Sep			board	green	ILSF1	11-Se
			CAN TF1	5-Sep			door	red	ILSF1	11-Se
			CAN TF1	5-Sep		water break	brick	black		
			GA2	30-Sep	0		door	black		
ken	25-Sep		SBFD	23-Sep	2		wall	orange		
			GA	30-Sep	0		siding	green		
			GA2	30-Sep	0		brick	black		
			GSAR	11-Sep	0	green square	brick	green		
			GSAR	11-Sep	0	green square	brick	green		
			GASR	12-Sep	0		door	blue		
			GSAR	12-Sep	0	white square	window	white		
			GSAR	12-Sep	0	orange square	window	orange		
			GSAR	12-Sep	0	orange square	window	orange		
black "1"	13-Sep		GSAR	12-Sep	0	black square	window	black		
			GSAR	12-Sep	0	black square	brick	black		
			GSAR	12-Sep	0	black square	brick	black		
			TXTF	6-Sep			wood wall	orange	DPD	14-Se
			TXTF	6-Sep	NE		wood wall	blue		
			TXTF	6-Sep	TFW	C	wood wall	blue		

Figure 10:6. *The database structure behind the search and rescue GIS*

someone do this? One possible explanation for this post-disaster landscape "art" is to revisit our discussion on the post-disaster state of an individual's mental health.

Although there are a number of contributing factors to the severity of mental illness following a traumatic event, the effects can be modified by using certain coping strategies. Although these can take many forms—the very public display of emotive messages painted across the home could be considered one example. These messages can be organized into classic categories of positive and palliative coping. As geographers it is interesting to see where these messages occur and who tends to use them. It is also interesting to dissect the content to see how frequently *place* emerges as a subject of "loss".

KATRINA RELATED PSYCHOPATHOLOGY

As we have already discussed, Katrina presents one of the most geographically complex mental health challenges the United States has ever faced. For example, the diaspora, especially those living in artificial FEMA villages are the most likely to carry the highest pre-event health burden, are the most likely to have suffered acute stresses during the storm, and continue to experience the greatest social distance from normalcy due to the artificial living environments of a FEMA trailer. Yet those who had the ability to return also face stresses associated with an uncertain future, a city with limited service provision, and neighborhoods where the devastation remains as daily reminders. When an individual, or family, is faced with this *return to nothing*[2], being forced to literally pick up pieces of their lives, either in the form of a destroyed home, fragments from inside, and lost loved ones and neighbors, it is likely that the importance of home and place will be reconsidered. For some it will be an opportunity for rebirth, while others will simply move away. These choices will vary from person to person, though the geography of origin, such as one neighborhood or town, may also place a role in the decision to return/ abandon depending on the strength of pre-event social, cultural and religious networks[3]. Geographic variation will also occur within any area as returnees rebuild next to apparently abandoned homes which may also result in a concurrent spatial pattern of psychopathology. This provides one example of how geographers could investigate patterns of a health outcome which is not traditionally looked at spatially.

Is it possible that post-disaster landscape "art" is a public expression of coping in the recovery process? If so, we can use a spatially modified conceptual theory taken from the Conservation of Resources (COR) model[4] which we discussed in the last chapter to help categorize the art. The premise of the original model is that an individual negatively reacts to a disaster because social and personal "resources" have been removed from his/her life. Another way of casting the COR model is to break it into four sections, each with an associated resource loss. These include the physical loss of the home (and possessions); the loss of social structures (including but not limited to the neighborhood); individual emotions (which can be focused on the self or on the neighborhood) and an individual's "buffer" (how finances, education or even previous decision making can help moderate the effects of the disaster). Losses to any of these categories singularly or in combination may result in psychopathology. However, if the individual can limit the extent of this loss, or even begin to replenish what has been lost by using positive coping strategies, then psychopathology may be avoided. An example of positive coping includes rebuilding, not only their own home but also those of neighbors.

Immediately we can see how the geographic scale of a disaster would impact the COR model—for if an individual loses a home through a wildfire, but the remaining neighborhood remains intact, although losses are still felt in terms of

2 We highly recommend Read, P. (1996). *Returning to nothing*. Melbourne, Cambridge University Press.

3 One of the few psychopathology studies to incorporate place found that smaller towns and rural areas fared more poorly than larger settlements after the 1993 Mississippi Floods Ginexi, E. M., K. Weihs, et al. (2000). "Natural disaster and depression: a prospective investigation of reactions to the 1993 midwest floods." *American Journal of Community Psychology* 28(4): 495–518.

4 Freedy, J. R., D. L. Shaw, et al. (1992). "Towards an understanding of the psychological impact of natural disasters: An application of the conservation resources stress model." *Journal of Traumatic Stress* 5(3): 441–454.

their home and emotions attached to it, it is probable that a supporting social structure remains, as do other neighborhood scale comforts (a favorite restaurant, gathering points with shared memories), and it is possible more of the safety "buffer" remains as employment may not be affected. When the extent of the disaster is geographically extensive, as with Katrina—the losses extend to every category, and so the risk for subsequent mental illness is far greater.

THE LINK BETWEEN GEOGRAPHY AND COPING

Coping strategies can be employed to modify the effects of an external shock. Although coping can be categorized in different ways, an admittedly oversimplification would split approaches into positive or negative actions. A typical positive coping approach would be to return and rebuild—either physically in terms of gutting, cleaning, reframing, or just making a commitment to these actions[5]. Being able to believe in something larger, for example a deity's plan or even a community spirit to rebuild, are also positive forms of coping. In general people who take such positive approaches after the disaster have been shown to fare better than those who feel lost and without control[6]. For example, in floods similar to those we have described through this book, people who had enough control in their lives to clean and rebuild, and especially if they could help others around them, have had better mental health outcomes[7]. This last point reemphasizes that disasters have both individual and community dimensions, and that these are linked. If we think back to the description of the community working as a team to sweep the flood waters out of each neighbors home in Ohio—this is a good example of both contributing and receiving these positive approaches. On the other hand, those who use negative or palliative coping strategies, for example involving avoidance, alcohol use or simply choosing to abandon their home where the disaster occurred, do not typically fare as well[8].

5 Asarnow, J., S. Glynn, et al. (1999). "When the earth stops shaking: earthquake sequelae among children diagnosed for pre-earthquake psychopathology." *J Am Acad Child Adolesc Psychiatry* 38(8): 1016–23; Baum, A. (1987). Toxins, technology, and natural disasters. Cataclysms, crises, and catastrophes. G. Van den Bos and B. Bryant. Washington (DC), *American Psychological Association* 5–54.

6 Solomon, S., E. Smith, et al. (1987). "Social involvement as a mediator of disaster-induced stress." *Journal of Applied Social Psychology* 17: 1092–1112.

7 Gleser, G., B. Green, et al. (1981); *Prolonged Psychosocial Effects of Disaster: A Study of Buffalo Creek*. New York, Academic Press. Erikson, K. T. (1976). "Disaster at Buffalo Creek. Loss of communality at Buffalo Creek." *American Journal of Psychiatry* 133(3): 302–5; Phifer, J. F. and F. H. Norris (1989). "Psychological symptoms in older adults following natural disaster: nature, timing, duration, and course." *Journal of Gerontology* 44(6): S207–17.

8 Chung, M. C., S. Farmer, et al. (2001). "Traumatic stress and ways of coping of community residents exposed to a train disaster." *Aust N Z J Psychiatry* 35(4): 528–34.

Figure 10:7. *My Home My Heartbreak.* Photograph by Curtis

Few studies describing disaster related coping have directly considered the geographic variation in how people cope, and tied to this, the importance of geography within coping. As the meaning of place varies with different people, it is not hard to believe that the impact of its loss will have different effects. We have already suggested in this book that many of the neighborhoods hardest hit by Katrina had strong cultural roots with the extended families often being geographically clustered together. For other "newer" cities, with a higher transitory population, the devastation of a neighborhood might only be an inconvenience, not a source of depression because of the loss of deep social networks. This raises an interesting geographic question—is the impact of a disaster not only tied to its spatial extent (how big it is), but also the strength of social and cultural heritage?

It is not that these messages are coping actions in themselves (though there may also be elements of this as well), but rather that they are indicative of the general frame of mind and as such could be seen as coping surrogates. Consider the following example of a building in St Bernard Parish (Figure 10:7), which has messages scattered on interior and exterior walls, on floors and across front steps, proclaiming how many memories were lost with the destruction of the home. Spray painted indicators of where heights had been measured as a child grew, on which steps the returnee had sat to recover from a first heartbreak, and the names and ages of those who had lived there were all painted on the walls.

What does this obvious outpouring of emotion communicate about the residents' state of mind—and equally, how are they dealing with these losses? From a geographer's standpoint, does this type of message vary in intensity or content by neighborhood? How is *place* used within the content of these messages? In terms of what was lost, is home a structure, a series of memories, a neighborhood, a culture, or even a nationality?

MARKING THE POST-KATRINA NEIGHBORHOODS OF LOUISIANA

As we have said, in the weeks following the storm, spray paint, signs and artistic constructions of debris were used to modify the majority of structures in Orleans and St Bernard Parishes. Human and animal search and rescue markings sometimes covered entire house fronts. Friends and relatives would also leave contact numbers and messages of hope written onto doorframes and pinned to doors, while impromptu local traders began to populate grass medians with an expanding array of services. Amongst these markings were also retorts, messages and soliloquies from the returnees themselves. For Orleans and St Bernard Parish, prior to Katrina, spray-painted defacement of property may have been endured, but it was not acceptable. Indeed, any such defacement of individual homes rather than communal spaces would have likely generated an immediate and angry response. This typical societal reaction changed after Katrina. Messages were written by home owners on paper taped to *their* walls, wooden boards covering *their* windows, and on all surfaces of the house itself; *their* walls, doors, windows, eaves, porch columns, even roofs. Messages were written in pen, marker and spray paint. Sometimes messages were even constructed from discarded items being arranged to convey meaning in the way they were placed.

One of the few papers to consider previous examples of post-disaster graffiti or "catastrofitti" as referred to by its authors, occurred after the Red River Valley flood in Minnesota in 1997[9]. Data collection for this disaster included field-based photography from which images were later extracted and categorized. This *catastrofitti* germinated from the officially sanctioned marking of flood damaged objects in order to protect against theft and illegal future resale (echoes of the Johnstown Flood from earlier chapters in this book). This initial use was expanded on by community members as a forum to voice frustration both as individuals and as communities. From this perspective there is a similarity to post-Katrina neighborhoods where an "official" rationale for spray painting on homes by search and rescue was co-opted by the residents as a means for self expression.

9 Hagen, C. A. and M. G. Ender (1999). "Graffiti on the Great Plains: A Social Reaction to the Red River Valley Flood of 1997." *Applied Behavioral Science Review* 7(2): 145–158.

COLLECTING DATA AFTER A DISASTER

An attempt was made to collect as much photographic evidence of the post-Katrina landscape as possible not only to record the messages written on homes but also to record visible damage, and the search and rescue information which was used to create the maps seen earlier in this chapter.

The spatial data collection strategy attempted a broad coverage of the devastated area, though it is accepted that bias is present based on familiarity with neighborhoods, encountering disaster-related impediments (flooded and debris filled roads), and finding new opportunities while in the field (in this regard it could be classed as "organic"—it grew or evolved spatially rather than followed a pre-set spatial strategy). Data collection was further complicated by the balancing with other disaster recovery-related duties, and respecting the needs and privacy of those who had suffered through the storm (though few returnees were seen). In addition, the field teams also suffered health problems during the data collection due to the emotional strain of walking through the devastation. As a result, several hours of data collection would result not only in respiratory problems and headaches but also what can only be described as melancholia. However, any biases present in these data are not considered serious especially as no statistical tests are used.

Data were collected on nine trips to two Katrina impacted Parishes, during a 21-day period in October 2005 (the first three weeks when the public were allowed re-entry into Orleans Parish). Although messages originate from any stage of the disaster time-frame, most are likely to come from the intermediate stage[10] when the person first returns home and is confronted with the aftereffects. There is also a temporal sensitivity to collecting these messages as rebuilding and weathering will change the landscape.

Data were collected on seven trips to Orleans Parish, and two to St Bernard Parish[11]. The first six trips to Orleans Parish involved the random selection of a neighborhood, and the subsequent walking of streets around a starting point. Each house with any Katrina related marking (usually graffiti), or other storm related event (excessive damage, belongings being removed to the curbside) was photographed. The seventh trip, accompanying a joint team from Louisiana State University CADGIS lab and FEMA concentrated on the levee breaks of the 17th Street, London Avenue and Industrial Canals, as well as other excessively damaged areas (Lakeshore). On this trip the area north of St. Claude in the Lower 9th Ward

10 Post-disaster stresses occur at different phases after the disaster in response to developing conditions, for example as the neighborhood recovers, or fails to recover, and these can result in different psychopathologies See the extensive review in Gibbs, M. S. (1989). "Factors in the victim that mediate between disaster and psychopathology: A review." *Journal of Traumatic Stress* 2(4): 489–514.

11 Nine data collection trips were taken in October 2005 (9th, 12th, 14th, 17th, 19th, 21st, 26th, 29th and 30th).

was entered, with FEMA authority, even though it was still officially restricted to public entry. The two trips to St Bernard Parish employed a different spatial strategy, with data collection being targeted towards the immediate residential areas surrounding known mortality addresses as published in the Times Picayune newspaper on October 23rd. On all these collection days, data were also captured on routes between neighborhoods.

In total over 3000 photographs were taken of post-Katrina homes, of which approximately two-thirds recorded graffiti or montages of purposefully arranged debris. The "messages" described in this chapter form a subset of 220 photographs which specifically captured resident expression (though many of these photographs contain more than one message). The total number of residences for each data collection site was calculated by overlaying the addresses of each path on post-Katrina flood imagery in a GIS– in this way properties that may have existed before the storm but had since been obliterated were not counted.

Although the majority of these messages were written, there were several examples of debris and belongings being "sculptured" into purposeful arrangements, something which has been seen before in post-traumatic landscapes[12]. Sometimes flood damaged discarded property and other debris create surreal images, for example a lone armchair on the space where a house stood, but these are not intentionally meant for an audience. The arrangements described here were obvious examples of a returnee wanting to convey a message.

CATEGORIZING THE MESSAGES

The few examples of Katrina-related graffiti originating from before the storm were restricted to boards nailed across windows and doors. Most were humorous, for example "Katrina please turn." Many houses in Orleans Parish, especially the older neighborhoods (for example Mid-City and Gentilly Terrace) are painted in colorful washes. The vibrant colors of the "official" spray-painted markings often contrasted dramatically with these pastel backgrounds. The largest number of separate messages was 10 (in 7 colors), and the largest extent covered the entire front and partial roof of a house. Although these color combinations may have appeared striking to an outside eye, it is hard to imagine the impact this violation would have had on the returnee. It is possible that this post-Katrina defacement may have contributed to the "resource loss" (from the COR concept) of the home beyond that caused by the storm itself. However, on all the data collection trips, there was only one example of a resident who took issue with the way his/her

12 Read, P. (1996). *Returning to Nothing*. Melbourne, Cambridge University Press.

Figure 10:8. *There are often "ironies" in the disaster landscape—such as the message of "Coming Back" (aimed at adults) being spray painted next to the children's "Bob the Builder"* Photograph by Curtis

property had been defaced (at least in terms of generating a public reproach); a board leaning against the house with "Please stop spray painting the bricks."

Unlike most homes marked by the different rescue teams, the majority of returnee spray painted messages occurred on temporary facades, especially wooden boards covering windows or doors, boards leant against the building, banners hung from windows (for example "No Vegas Mayor Nagin"), or on stand-alone signs/placards (for example "Bite me Katrina I'm home"). Others used discarded flood damaged property, especially refrigerators (for example "Looters be nice & leave"). Of those messages that were written by returnees directly on the home the majority were names and telephone numbers to use as contact information, or words of goodbye, almost as though the returnee had abandoned the home, the house now being only a vehicle for contact. A minority of the messages painted directly on the home by returnees could be classed as examples of positive coping, for example "Coming back" written in large pink letters across a window and wall, ironically covering various children's decals including Bob the Builder, and "St Bernard Parish Thanks Canadian Mounted Police" a fifteen foot high message on the side of a building.

Message categories

For some the use of the home was purely functional due to the absence of other viable means of communication. In this way the home became a geographically tied message board. These messages often had specific target audiences, the home being used to initiate or continue dialogue with officials associated with the disaster, to friends and family, and as warning to strangers. In the first category "officials" ranged from insurance adjusters (usually just contact telephone numbers), faceless officials who might bulldoze both home and neighborhood ("Lois came by...took what we could...do what you have to with house (followed by a contact number) in Baton Rouge"), to animal rescue. Targeted communication was at its most poignant when aimed at potential survivors. The majority of these were directed to the returnee, again the house (or surrounding property) being a geographic constant in all the confusion. For example, the message scratched into the floodwater dirt of a Taxi window or the note pinned to the home sprayed with "Ken" indicating a body had been removed from within.

Other returnees took control of their situation by issuing warnings to the outside world that pre-Katrina order had again been restored to their property. For store keepers, the perceived fear of looting led to windows being boarded and warnings written across, for example on the Super 10 just before crossing the Industrial Canal was: "WARNING TRESPASSERS ARE CONSIDERED LOOTERS AND ARE SHOT DEAD." The majority of residential returnee warnings against looting were found in the Arabi and Chalmette sections of St Bernard Parish, with people being willing to deface their own home to protect the contents within; remember the front of the caravan with "Do not touch, Will shoot!" More subtle messages of usurping power were scattered through the French Quarter; written on the side of a taped-up refrigerator; "If there are enough police to enforce curfew then isn't there enough to stay open keep us safe." And on another refrigerator, "LIFT THE

Figure 10:9. Photograph by Curtis

BLOODY CURFEW. The curfew is unnecessary, unAmerican and unhelpful...call Mayor Nagin (telephone number)."

The actual act of being able to write a message may be empowering, for example the reclaiming of property from those who had previously defaced it, or from authorities who potentially were deciding the fate of both home and neighborhood. The content of these messages can also be interpreted as indicative of coping strategies. For example there are different emotive categories which are traditionally considered as part of coping, especially humor and anger. Additional resources can be attained by having faith in something bigger, whether religious, or the belief of a neighborhood being rebuilt and its residents returning.

Emotive Categories: Humor

The ability to laugh has long been considered an indication of coping second only in effectiveness to talking about what happened[13]. Different types of humor also relate to the various post-disaster phases, with the period generating the most humor being the phase following the intermediate stage, before disillusionment has set in.

Various forms of humor were present in returnee's messages scattered across Orleans and St. Bernard Parishes, though most were concentrated in the Mid-City area. Forms of humor included comments about all stages of the disaster, from light hearted-comments before the storm made landfall (what could be considered pre-event anxiety coping), through humanizing the hurricanes, to laughing at the subsequent flooding and recovery process. Examples of humor after landfall also included humanizing the storm, for example as though being written by the storm itself "Welcome home! Love, Katrina. P.S. I redecorated." while elsewhere the storm was the target of a humorous attack; an almost life-size cardboard cutout of a Nun held a paper with 'Katrina is a bitch' on it. The subsequent flooding also provided a source of humor; "Lake Scott—No Fishing" (appearing in Mid-City on Scott Road), and on the front of one lakeshore cabin; "For sale needs "some" work. Closed due to weather Slight water damage" and next to these words a happy face with a

13 Indeed, its coping properties can extend beyond making light of a traumatic situation, using "gallows humor" or "survivor humor" to gain a form of control over the event (Moran 1990; Lipman 1991), to actually lowering stress elevated cortisol levels that can cause immunodeficiency (reported in (Pasquali 2003)). Different types of humor also relate to the different post-disaster phases, with the period generating the most humor being the phase following the event itself, but before disillusionment had set in (Pasquali 2003). For more information see Weinrich, S., S. B. Hardin, et al. (1990). "Nurses respond to Hurricane Hugo victims' disaster stress." *Arch Psychiatr Nurs* 4(3): 195–205; Pasquali, E. (2003). "Humor: An Antidote for Terrorism." *Journal of Holistic Nursing* 21(4): 398–414; Lipman, S. (1991). *Laughter in Hell: Use of Humor During the Holocaust.* Northvale, NJ, Jason Aronson; Moran, C. (1990). "Does the use of humour as a coping strategy affect stresses associated with emergency work?" *International Journal of Mass Emergencies and Disasters* 8(3): 361–377.

tongue licking upwards. Clever word interplays included "Make levees not war," "laissez le bon mold rouler," and on the side window of one minivan in Bywater "NOLA's ARK New Orleans...proud to call (crossed out) crawl home. GOING HOME!!!" On one home plastic bones hung from an open window, while next to these across the building's brick in large white spray painted letters and arrows pointing to them from all directions was "I died here waiting for an adjuster."

For two traditional sub categories of humor, sarcasm and satirical comments, it was hard to distinguish between humor and anger. For example, on the lighter side, written on a board nailed across a window in Mid-City "4 sale—seasonal water front property—boat slip in back—1800-FEMA," and on the front awning of a house in St Bernard Parish was "Kiss our "A" Katrina & levee board." An example of where anger and sarcasm merge was found on a city block in St Bernard Parish where homes had floated away leaving just the concrete foundation slabs. "Built by Tedesco" was written on the side of a half-house which had floated into the road before (probably) being cut in two by a bulldozer. On the side of a car was also written "All homes by Tedesco." The house it sits in front of had floated away into a distant ditch, only the concrete steps and foundation blocks remain. Even in this

Figure 10:10 Returnee humor. Photographs by Curtis

situation an element of logistical control has been established, as the address had been spray painted on the top step and on a sheet of corrugated steel where once the house had stood. The residents of this block attached at least part of the blame for their homes demise to the original building contractor.

Emotive Categories: Anger

Four common targets for anger emerged in the post-Katrina neighborhoods. The first were political entities, President Bush, Governor Blanco, Mayor Nagin, and other "elected officials." Most of these attacks were directed towards events concerning the storm, though one resident went further to comment on Iraq. Of politically elected committees, unsurprisingly the levee board came in for the most comments[14]. Although most attacks were non-specific, the previously mentioned message on a refrigerator in the French Quarter was critical of current curfew policies. The second category of attack involved private industry, for example Murphy Oil, likely due to the oil spill incident, the previously mentioned Tedesco builders, and the insurance industry in general. Interestingly, FEMA came under no real attack, the only comment mentioning the agency was the sarcastic *call 1.800.FEMA*. This provides an example of how the Katrina canvas is dynamic and would have changed over time, as the growing hatred for the agency began to make it a more frequent target of attack, as can be seen on French Quarter tourist t-shirts.

The third target of anger was the storm itself, which was humanized, and most frequently referred to as a "Bitch." On a house in St Bernard parish is "You didn't take my pride, bitch I'll be back." The final category of anger (or more accurately desperation) was directed towards the rescue teams, and especially animal rescue. These messages ranged from chastising the teams for defacing the brickwork, to desperate pleas to animal rescue that the owner was home and didn't need his/her pets being rescued.

FAITH IN SOMETHING BIGGER

Messages of Hope: Religion

Religious activity often follows disasters, both in terms of providing actual shelters and through sermons and scripture that offer solace and strength to returnees[15]. As

14 The levee board is comprised of elected officials who do not necessarily have engineering qualifications. They are supposed to oversee the general safety of the city's flood protection system.

15 Religious coping strategies are frequently adopted and for all the different phases of recovery. Just as with other coping strategies, these can include positive and palliative associations, using spirituality to help understand and provide support to recover, or aiming anger at an unjust god. Those who find strength from a spiritual message, and who can even spin the events as an opportunity to rebuild and remake the community into something better, have been found to have better psychopathological

Figure 10:11. *Returnee anger.* Photographs by Curtis

we have already mentioned in this book, the church plays a large role in the South in general, and especially to the African American communities of New Orleans. The majority of religious references in the post-Katrina neighborhoods fell into the first category of positive reinforcement often aimed at the surrounding community. The most frequent word usage was simply "God", usually in some form of thanks, for example "Owner is back God is good" and "God bless Texas"—presumably because the resident had been evacuated there. Several religious messages combined the faith in spirituality with a faith in social support and rebuilding, for example painted across the side of a building was "TAKE CARE 1 another" with a crucifix shape, and elsewhere on the back window of a car written in vivid yellow "Homeless but (crucifix shape) hopeful (crucifix shape)." This spiritual support

outcomes. See Smith, B., K. Pargament, et al. (2000). "Noah revisited: Religious coping by church members and the impact of the 1993 midwest flood." *Journal of Community Psychology* 28(2): 169–186; Pargament, K. (1997). *The Psychology of Religion and Coping: theory, research, practice.* New York, Guilford Press; Pargament, K. I., K. Ishler, et al. (1994). "Methods of Religious Coping with the Gulf War: Cross-Sectional and Longitudinal Analyses." *Journal for the Scientific Study of Religion* 33(4): 347–361; Ollendick, D. G. and M. Hoffmann (1982). "Assessment of psychological reactions in disaster victims." *J Community Psychol* 10(2): 157–67.

Figure 10:12. *Messages of faith—unfortunately even though this person was already rebuilding their home—it was abandoned again by August 2007.* Photograph by Curtis

was not solely Christian, for example in Gentilly Terrace one Rastafarian message written by a door frame read "Jah bless us all we need it," and in St Bernard Parish a Star of David was next to a "We shall rise again." Some messages actually invoked scripture to give weight (or credence) to the message of support. In another example a voodoo doll had been placed sitting on a car. One of the most overt messages was written on a board across a window: "Those who sow in tears shall reap with JOYFUL SHOUTING! He who goes to & fro weeping, carrying his suitcase (bag) of seed, shall indeed return with a SHOUT of joy bringing his sheaves with him—PS126 HOPE FOR OUR FUTURE." This last message was interesting as on the front of the house, a black sheet hung from a second story window with "No Vegas, Mayor Nagin! C'mon, Gentilly—ya'll rebuild!" This resident had used her home to display their Christian faith, make a political (or rather cultural) statement regarding the preservation of heritage, and issued a geographically specific rallying cry to her own neighborhood. When this picture was taken, the owner was already working on her house, her actions supporting her statements—yet on driving by the home in August 2007 it appeared deserted.

FAITH IN PLACE

Another source of positive messages came from those who issued a rallying cry to neighbors, and from places that were important to the individual (Figure 10:13). In this sense there was a feeling of "we can get through this together." Again, this positive outlook has been found to help the mental state of disaster survivors[16]. There were different geographic scales associated with "place"—whether the immediate neighborhood, a larger section of the city (such as "Gentilly"), the city itself or even the state. Two examples from St Bernard Parish included a board leant against a tree with "Our House Our home Our heritage We'll be back," and on another house a wooden board resting outside had "COONASS FOREVER." Several messages invoked the Southern Louisiana spirit of fighting through adversity which is a source of regional pride "We'll be back better and stronger" (in Gentilly Terrace); "It aint over...New Orleans will rise again," and "S.T.B. WILL RISE AGAIN!!!" One of the more elaborate messages in Mid-City was a teddy bear and an American eagle stuffed toy strung from a tree, beneath them a wooden sign had been nailed with "Hang in there N'awlins we are America's #1 city."

Figure 10:13. *Pride in place and promises of return.* Photographs by Curtis

COPING CONSTRUCTIONS

There were several examples of objects being used to convey meaning rather than just relying on words, and for most of these the common theme was a celebration of the life and events of Louisiana. On the street intersection of Chartres and the aptly named "Flood" streets in Holy Cross was an elaborate artistic creation, the centerpiece of which was a chair with various objects

16 Dougall, A. L., K. B. Hyman, et al. (2001). "Optimism and Traumatic Stress: The Importance of Social Support and Coping." *Journal of Applied Social Psychology* 31(2): 223–245.

Figure 10:14. *Sometimes the message board was not the home, but objects around it. In this case a van. Interestingly enough, it can still be seen, though slightly faded, in the Google Street View image for this location.* Photographs by Curtis

Figure 10:15. *An "altar" constructed in the Holy Cross neighborhood made up of a collection of items obviously personal to the residents, and indicative of New Orleans.* Photographs by Curtis

covering and surrounding it (Figure 10:14). A flood stained record album cover (*Blue Train* by John Coltrane) rested on the seat, and in front of this was a roll of film and some photographic slides. Next to the album was a white Mardi Gras mask and a single red rose. Beside the chair a large jagged piece of glass had a dolls head fixed on the point, a black piece of cloth over its head giving the impression of a nun. On the other side of the chair was a porcelain toilet with "KAT" written in orange and next to this a picture (in plastic protection) of a New Orleans marching bandleader. A muddied lampshade sat on a bus stop sign by the collection, while in the tree behind it an original multicolored painting on canvas was wedged between the braches. A torn American flag had been hung on the house behind the tree. On the other side of the road a house had been completely destroyed, and two boards had been laid at the base of the rubble with the message "Katrina the Bitch" and "and her sister." Beneath these signs was a magazine "The Blatant Image. A Magazine of Feminist Photography" Had the returnee wanted the passerby by to draw a connection between the storms and this feminist literature?

Q: Can you interpret any of these items?

Another arrangement was based around three stuffed animals in Mid-City; Tigger, Winnie the Pooh, and Eeyore (Figure 10:16). A bottle of Abita beer (a local Louisiana brewery) was placed next to Eeyore's mouth as though drinking, and

Figure 10:16. *It may just look like three cuddly toys posed on a log, but there were several Louisiana mementos incorporated into the "scene."*
Photographs by Curtis

another bottle was by Pooh's paw. Beneath the three was a sign warning owners against dogs defecating. A book had been placed on the other side (Tragedy is my Parish—working for GOD in the streets of New Orleans) while a Britney Spears DVD lay by Eeyore. Is Britney (and the beer) supposed to represent the sordidness that the hurricane was sent to cleanse (the singer being born in Louisiana)? Is the book a reflection of the sentiments of the returnee who created the image—or is this all a satirical comment with the meaning being the opposite?

The most striking of all the creations, and indeed all written messages, seen by the authors was found in St. Bernard Parish spanning four garage doors (Figure 10:17). Written in pen down one frame side of the first garage was; "Families lived and worked here" Next to the search and rescue "X" on the garage again in pen was "Recall Blanco" and "St. Bernard Pray 4 Us Save us from bureaucracy" on the other frame side was "Murphy Oil a plague on all our houses." On the second garage left frame side: "Murphy OIL + our elected did this." On the other right frame "Vote out all of them." On the garage frame header "At night people cried". On the garage door itself "Kathleen = Katrina" (repeated three times) and "Kathleen

Figure 10:17. *A voodoo installment in suburban St. Bernard Parish—notice the face mask on Elmo.* Photographs by Curtis

equals Katrina," "Voo Dieux XXX bureaucrats incumbents" "Appointed officials VOO DIEU XXX" (each "O" had a cross in it), "We had jobs—we worked—now we descend into bureaucracy for a living."

The third garage is also covered with messages. On the left frame side "How is this my USA?" On the other right border "RS430 victim Chalmette Angola an alternative to this." On the garage door itself, "My soul languishes St Anthony's lost cause," "The parish will not RIP," "Katrina = Kathleen." On the left frame side of the fourth garage door, "Our elected failed us XXXXXX sold to Murphy Oil," Above the door "Damned." On the door itself "St. Anthony's lost cause" "Voo Doo papa legba pray for us," "Recall Governor Blanco", and throughout the search and rescue "X" "Goodbye Chalmette 70043 LSU Arabi Diner, SAINTS, Rocky Carlos" (and various other local establishments) "we will not rest in peace." To the right of the garage door, written in pen was "Human lives were abandoned forgotten lost" On the door itself was "Impeach Bush" between these two words are three "X" with "1000" written at the base of the "X" (obviously meaning 1000 lives lost). On the other side of the garage door was "Kathleen's Katrina" with three similar "X" above and below. At the top right of the door was "(name) lived here (contact telephone number)."

In front of the garage door was a car. On the open driver window sat a boy doll on a Meals Ready to Eat brown pouch with "Voo" written on his head, and "Doo" on his chest. A Virgin Mary crockery plate hangs from the open window top. On top of the car was a box, a skull and a pumpkin head. Another skull rests by the car tail fin. In front of both wheels are two "toss and tickle me Elmo" dolls both wearing storm clean-up masks. On the side of the car was written "our elected failed us, Kathleen Katrina, Voodieux Blanco = joke" (variants of this are repeated on the door and front wing) and "Blanco failed". Beneath the car lies a dead bird.

From looking at these garage doors it is obvious all four were the work of one person (or one group). In many ways this one creation captures all the categories described in this chapter, except humor. Political figures (especially Governor Blanco), the failure of the rescue operation, and even a questioning of what is it to be an American are mentioned. Private industry and the quagmire of navigating the recovery process receive comments. Even religion, is mentioned, both Christian (St. Anthony patron saint of lost and stolen articles) and the widespread use of Voodo words and images. Some of the images are purely for effect—though again providing an insight into the desperate state of mind of the creator "At night people cried" and the face masks on the Elmo dolls. Finally, and for practical purposes, a contact telephone number is also left. Even a new lexicon is used in the creation, the search and rescue "X" representing death and loss.

Within this creation, several references are made to what this returnee considers his/her home and neighborhood, both from his/her residence "(returnees

Figure 10:18. *Saying goodbye.* Photograph by Curtis

name) lived here" to a larger geography "Goodbye Chalmette," and the naming of several local institutions, ranging from restaurants to the New Orleans Saints. The final geographic association is the comment "How is this my USA?" Was the use of four properties to show that this street, this neighborhood now abandoned had contained families? Why was St. Anthony mentioned, was it meant to be the search for lost memories of home and neighborhood and the hope to one day rediscover them?

COPING AS AVOIDANCE

Just as "optimists" employ positive "rebuilding" coping strategies, so pessimists adopt palliative coping, the resource loss being so great that all that the individual wants to do is move away. Examples of this type often mention the place of loss within the eulogy, followed by a message of good-bye. For example, in St. Bernard Parish; "God bless We'll miss everyone!" was spray painted in green across the front of a house. This was followed with "mom is okay" across a recessed front wall. On another wall was "CHS classes of 1966 1969 1978." The search and rescue "0" at the base of the "X" had also been turned into an unhappy face. For this returnee, three aspects of life are captured on the home: neighbors, family and high school. The sentiment here was of farewell and moving away.

Other examples included, "Barrow's 4-ever Bye Home"..."time for a new life"; "Been nice to know y'all God Bless Everybody"; and simply "We WERE HERE."

DISCUSSION

There is an implicit geography to psychopathology, for example many of the factors associated with symptoms can literally be mapped—the degree of devastation, previous social and health vulnerabilities, even the historical attachment to the community. If we accept that the public messages described here are indicative of the returnee's feelings and potential for coping, then we can also look at the geography to these responses.

Geographers could also investigate questions such as, for returnees *what exactly is home*? Is it a parish, a neighborhood, a street, a house, even a part of a house, or the belongings found inside? For some it was the building ("We survived Katrina House didn't," and "RIP House"). For others the demise of the home was linked to the plight of the returnee "For Sale by homeless owner." The content of other messages tied the destruction of the house to something deeper than property or belongings, a loss of culture or sense of belonging. Other messages named specifics; a sequence of graduations from a local high school, the demise of local physical gathering points (restaurants) and conversation topics (the New Orleans Saints). It would be interesting to interview these returnees, wherever they are now, and ask how their perception of home and place has changed in the years since Katrina

Is there a predictable geography to these messages? Why did St. Bernard Parish produce more (and more descriptive) comments than any of the other Orleans Parish neighborhoods? Of the Orleans neighborhoods Mid-City was the most active with messages, especially humorous comments. Again why? Sections of this neighborhood do tend to be perceived as "artsy"—to what degree do pre-event characteristics dictate post-storm messages, and therefore the post-storm landscape? Why was the majority of spray painted contact phone numbers also in St. Bernard Parish? Was there a diffusion process at work, once one returnee adopted this approach of contact, did it spread geographically? Is it possible to gauge the degree of individual or neighborhood recovery based on these messages? Indeed, to what degree are the messages representative samples from their neighborhoods?

EXERCISE: MARKING THE POST-DISASTER LANDSCAPE

Using Google Street View, pick ten locations in New Orleans. Use the overlay maps in Chapters Five and Six to help develop a spatial sample strategy for these locations. For each location look at each home on one city block (include the homes facing the block as well). Answer the following:

A. How did you devise your sample strategy?
B. What is the proportion of buildings per block that still display search and rescue markings?
C. How many homes on each block had multiple markings?
Which had the most?
D. Did you see any personal message? If so, give the address.

Answer questions B to D as a map.

ELEVEN

The Geography of Recovery

As a long time resident of Hollygrove, Patrice is realistic yet optimistic about the recovery of the neighborhood. She said "I see it coming back and it's good. It's not a magic wand or a secret weapon that can make it happen. It's just one family at a time rebuilding as best they can with what they can." (Interview with Hollygrove Resident, Patrice Boyd by Alan Gutierrez; available online http://thinknola.com/post/patrice-boyd/)

INTRODUCTION

Recovery is a process and a state. It occurs at a variety of scales, such as individual, street, neighborhood, city, state, and region. Recovery is both temporally and spatially dynamic. It is not necessarily a linear process and it often results in an uneven geography. Research that investigates these aspects is spread across several academic disciplines and takes many different forms. However, little is known about the effects these variations in recovery have on people and places. This chapter introduces the concept of recovery, impediments to this process and then presents another geospatial tool that can be used by researchers/community groups to capture fine scale data. To reinforce the ongoing struggle for recovery in the neighborhoods of New Orleans, all photographs selected for the first four pages were taken in June 2009.

WHAT IS RECOVERY?

Recovery can be defined in many different ways, which are in part dependent on who is studying recovery and at what scale. Recovery is the least studied of the four phases of emergency management (planning/preparedness, mitigation, response, recovery)[1]. However, when it is the subject of investigation, it is studied at either a coarse spatial scale and/or does not take into account the human geography involved. Perhaps the most often cited recovery study was performed by Haas, Kates, and Bowden in 1977[2]. Though some studies came before, this one is the most comprehensive as they use multiple study areas and cover both historical and currently (at that time) recovering cities. At the end of this chapter, we provide a timeline of books and scholarly articles that deal with recovery. You can tell from the titles alone that recovery is not necessarily the sole focus, though in the 1990s and 2000s it does become the primary issue of investigation for some of these works.

Q: Why do you think Katrina revived interest in recovery?

From what we are learning from the ongoing recovery in New Orleans, the scale at which this process occurs is much finer than a city or even a census tract. It occurs at the level of the neighborhood, the street, and from one house to its neighbors. Throughout the references listed at the end of the chapter, the definition of recovery varies. It can be conceived in terms of concepts such as reconstruction or return to normalcy, or it can be a personal definition that varies from individual to individual—when do they feel they have recovered from the disaster? However, in order to proceed, we do need an operational definition we can use for the remainder of this chapter. For the purpose of this discussion, we will focus on residential recovery and define it as the process of a resident returning to permanently occupy their pre-disaster home. In addition, from many conversations with New Orleans residents over the course of the past four years, this is an operational definition they use when describing the state of their neighborhoods: who is back, who is not back, how can we get them back in their home?

THE GEOGRAPHY OF RECOVERY

Recovery occurs over time and across space. The temporal aspect has most often been studied as researchers are interested in how long it should take for a place to recover. Some have conceived of recovery as a linear temporal process that has a positive trajectory. In effect, it has been proposed that after clean-up, a place moves forward toward recovery; that residents move back in and gradually resume their pre-disaster lives. However, others have provided an alternative

1 This is a "classic" text and all students of disasters are encouraged to read it. Mileti, D. S. 1999. *Disasters by Design: A reassessment of natural hazards in the United States*. Washington: Joseph Henry Press.

2 Haas, J. Eugene, Robert W. Kates, and Martyn J. Bowden. 1977. *Reconstruction Following Disaster*. Cambridge, MA: MIT Press.

framework that proposes recovery is not necessarily linear at all. Indeed, some places can decline rather than recover. Certainly, this non-linear conception visibly plays out in New Orleans and in many other places across the Gulf Coast where neighborhoods appear to recover, but then decline, or they never recover in the first place. Furthermore, recovery has been studied in the scale of one year to several years post-disaster. However, we know that recovery—particularly what is called long-term recovery—can last decades. This situation argues for more studies to undertake an extended longitudinal design in order to capture the entire recovery process.

When "patterns" of recovery are discussed in these studies, the focus is usually on temporal patterns and not the geography of this process. The seminal work of Haas and others in 1977 does acknowledge that there is geography to recovery. They discuss the recovery of economic sectors of a city, and at the time of their study these sectors were influenced by various city models proposed by scholars such as Park and Burgess[3]. So, the city was seen as either concentric circles, zones, or a hybrid of these ideas. As a result, recovery was believed to follow spatial patterns based on these models. However, these are coarse areas that deal with only aggregate units and with a focus on

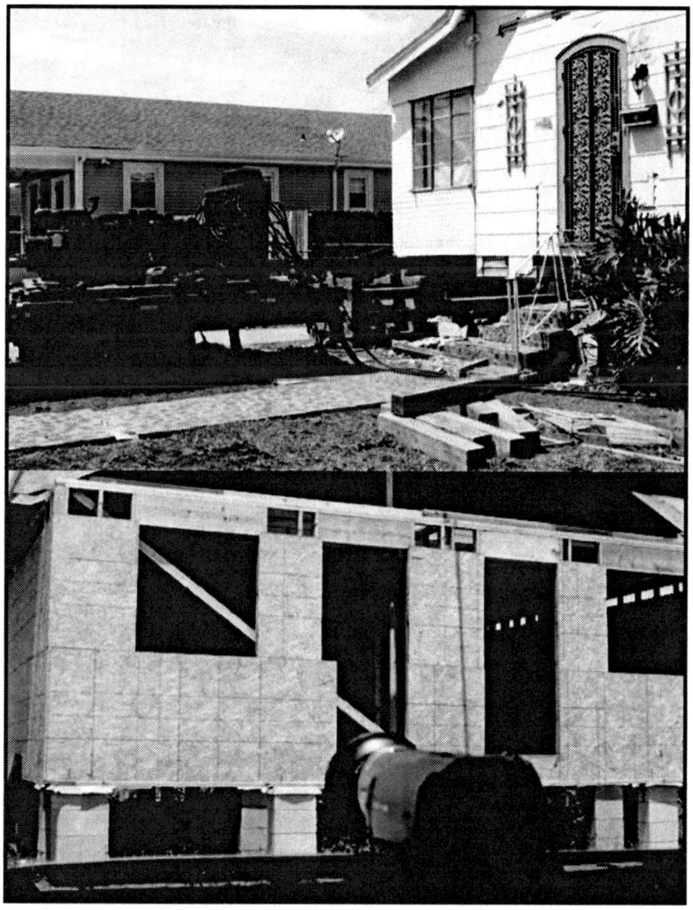

Figure 11:1. *Two examples of rebuilding/resiliency. In the uppermost photograph the home is being raised, and in the bottom image, a rebuild is captured using the spatial video system.* Photograph by Curtis, June 2009

3 Park and Burgess's work as presented by the Center for Spatially Integrated Social Science (CSISS) available online: http://www.csiss.org/classics/content/26.

Figure 11:2. *A U.S. mail van navigates between the potholes in the Upper 9th Ward.* Photograph by Curtis, June 2009

economic development/business recovery. Think back to our earlier discussions in this book where we showed how zip code boundaries do not effectively "map" onto what people consider to be their neighborhood. Therefore "recovery" data released at this scale has little meaning to the person who lives on a section of street with potholes, abandoned neighbors and no local clinic or store.

The spatial scale at which recovery occurs has not received a great deal of attention, that is, until Katrina. Once again, look at the timeline presented at the end of the chapter. There appears to be a post-Katrina boom in recovery studies (and these are only the scholarly activities). However, it is not just academics who are interested in advancing work on fine-scale recovery, government agencies and non-profits also play an invaluable role, especially in monitoring and mapping recovery in particular[4]. Partly because of issues of confidentiality, though, the scale of information release to the public is still coarse: planning districts, zip codes, census tracts.

What we have learned from four years of post-Katrina research is that recovery occurs based on processes at a variety of scales, from federal, state, and local policies, to the decisions of neighbors and family members. Traditionally, the finer scales are overlooked. Nevertheless, the results of the process are visible on the landscape. They are visible in neighborhoods and along street segments and in individual houses. Luckily this upswing in Katrina research has coincided with new geospatial technologies that also allow us to capture these spatial patterns. The spatial video has already been discussed, and it will be further developed in the next Chapter. Another form of field data collection will be introduced at the end of this Chapter.

4 For example, see the New Orleans Recovery Index developed by the Brookings Institution (http://www.brookings.edu/metro/katrina-reading-room.aspx).

Before we discuss recovery any further, it is useful to introduce some of the concepts and key works in the investigation of the built environment.

THE BUILT ENVIRONMENT

The built environment is the human-made structural environment in which we live. It is comprised of a wide variety of components such as buildings, transportation networks, greenspace, resources (grocery stores, social services), and advertisements to name a few. The built environment has potentially important relationships to the health and well-being of those who live their lives in its midst. One of the most studied aspects of the built environment is its relationship to health. A number of studies focus on identifying specific aspects of the built environment that either improve or damage our health, such as having walkable sidewalks and healthy food options as opposed to abandoned buildings and liquor stores.

Figure 11:3. *An example of one of the many neighborhood associations, in this case "rebuilding our community one unit at a time."*
Photograph by Curtis, June 2009

The Built Environment and Health

A variety of neighborhood physical and social characteristics have been investigated with regard to their relationship with health outcomes. These characteristics include abandoned or boarded buildings, healthy food outlets, traffic, noise, violence, alcohol or tobacco advertisements, recreation facilities, and quality of greenspace[5]. In addition, some of these studies focus specifically on the link between urban setting and racial disparities in health outcomes, just as we have discussed in earlier chapters. Also, as we have discussed in several sections, aspects of the built environment act as stressors, and that stress is linked to a variety of health concerns.

Evidence from preliminary fieldwork in the Lower Ninth Ward and Holy Cross neighborhoods of New Orleans corroborates the situation described by Saulny that the changing (degrading) environment generates stressors to those who are

5 Although there are many studies focusing on each of these built environment and health relationships, one excellent overview text is Kawachi, Ichiro and Lisa F. Berkman, eds. 2003. *Neighborhoods and Health*. New York: Oxford University Press.; see also Sooman, A. and S. Macintyre. 1995. Health and Perceptions of the Local Environment in Socially Contrasting Neighbourhoods in Glasgow. *Health and Place* 1:15–26.

attempting recovery[6]. In other words, aspects of the living environment place additional stressors onto the already stressful situation of trying to rebuild a life, home and community. This situation raises questions of how the recovery (or lack of recovery) is disproportionately impacting morbidity among those already facing health disparities. These factors suggest a negative health burden is being placed on those already disadvantaged and may therefore be inhibiting recovery.

Our own anecdotal evidence has linked health outcomes of returning residents to the visible characteristics of their neighborhoods[7]. Augustin et al. identify such signs as psychosocial hazards and define them as "visible characteristics of neighborhoods—such as violent crime, abandoned buildings, and signs of incivility—that give rise to a heightened state of vigilance, alarm, or threat." Furthermore, they note "daily exposure to psychosocial hazards in the neighborhood is known to activate a physiological stress response." [8]

Though neighborhood stressors were prevalent in poor African-American neighborhoods in New Orleans before Katrina, the hurricane added to the overall health burden by creating event-related stressors. Dead trees and boarded buildings have been identified as some of the environmental stressors that have emerged in the post-disaster built environment[9], but other stressors exist as well, such as illegal dumping of tires and trash and the environmental health concerns surrounding exposure to these debris piles[10]. In addition to a relationship with health, aspects of the built environment have also been linked to safety. These are two key issues that can encourage or inhibit recovery. If people feel that the environment is

6 Saulny, Susan. 2006. A Legacy of the Storm: Depression and Suicide. The New York Times. Online:http://www.nytimes.com/2006/06/21/us/21depress.html?_r=1&scp=1&sq=a%20legacy%20of%20the%20storm&st=cse&oref=slogin. Accessed October 17, 2008.

7 In one such story, a returnee told us how she and one other man were the only returnees on her street. Every morning they would cross each other's paths as they went to work. None of the other homes were being rebuilt, and their condition continued to decline. She didn't know the man to talk to, just acknowledge. Then one day he stopped his car and broke down in tears. The loneliness / stress / constant reminders had become too much for him and she was the only one who would understand because it was a shared experience.

8 Augustin, Toms, Thomas A. Glass, Bryan D. James, Brian S. Schwartz. 2008. "Neighborhood Psychosocial Hazards and Cardiovascular Disease: The Baltimore Memory Study." *American Journal of Public Health* 98(9): 1664–1670. p. 1664.

9 DeSalvo, Karen B., Amanda D. Hyre, Danielle C. Ompad, Andy Menke, L. Lee Tynes, and Paul Munter. 2007. Symptoms of Posttraumatic Stress Disorder in a New Orleans Workforce Following Hurricane Katrina. *Journal of Urban Health* 84(2): 142–152.; Galea, Sandro, Chris R. Brewin, Michael Gruber, Russell T. Jones, Daniel W. King, Lynds A. King, Richard J. McNally, Robert J. Ursano, Maria Petukhova, and Ronald C. Kessler. 2007. Exposure to Hurricane-Related Stressors and Mental Illness after Hurricane Katrina. *Archives of General Psychiatry* 64(12): 1427–1434.

10 Galea, Sandro, Chris R. Brewin, Michael Gruber, Russell T. Jones, Daniel W. King, Lynds A. King, Richard J. McNally, Robert J. Ursano, Maria Petukhova, and Ronald C. Kessler. 2007. Exposure to Hurricane-Related Stressors and Mental Illness After Hurricane Katrina. *Archives of General Psychiatry* 64(12): 1427–1434.

Figure 11:4. *The worst kind of abandonment, the building has many entry points, vegetation to partially hide it, and a dark interior. This would certainly be a concern for local residents because of its potential for crime.*
Photograph by Curtis, June 2009

unhealthy or unsafe, they may not return, or perhaps they will return, but only to leave again.

USING THE BUILT ENVIRONMENT TO MAP RECOVERY

The geospatial technology used to map recovery will be discussed in detail in the following chapter, so this section will focus on providing a justification for why such an approach is appropriate. In the previous section we introduced some of the reasons why the built environment is important for the well-being of people and places, and now we will turn to the specific aspects of the built environment in a post-disaster landscape that give clues about recovery. Table 11:1 provides an overview of characteristics that can be captured and mapped and which contribute to our ability to identify areas that are recovering and those that are in decline.

Additional research is needed to understand how recovery occurs in the built environment. For example, often we see abandoned homes with overgrown yards alongside cleared lots with neatly cut grass. What has created this pattern? We could make some educated guesses, but to know for sure the best source is the neighborhood residents themselves. We found out that if the curb was spray painted orange, this meant that the property is owned by the Louisiana Land Trust. The homeowner participated in the Road Home program and sold their property

Table 11:1. *Sample of Recovery/Decline Characteristics Observable in the Built Environment*

Recovery	**Decline**
Cars in driveway	Tall grass
Manicured yard	Broken windows
Construction crews	Trash dumps
Contractor sign	Demolition notice

to the state where it is now being held until it can be sold to a new resident or development group. In this way, state policy manifests in the built environment. These same data are available online. However, as you can see from Figure 11:5, the format in which these properties are provided is not user-friendly for a potential returnee gaining a sense of what is happening on his/her street or neighborhood. Although these can be mapped through the process of matching addresses in a GIS, these geographic skills are beyond most people.

Moving from state to local policy, the New Orleans Redevelopment Authority (NORA) has created The Lot Next Door program whereby a resident can purchase a blighted property if it is on either side of their home and the blighted property is owned by NORA (Figure 11:6). NORA has been purchasing homes from the Louisiana Land Trust and then selling them to participants in The Lot Next Door program. What impact might this program have on the geography of recovery? Theoretically, if returnees are financially able and are willing to participate in the

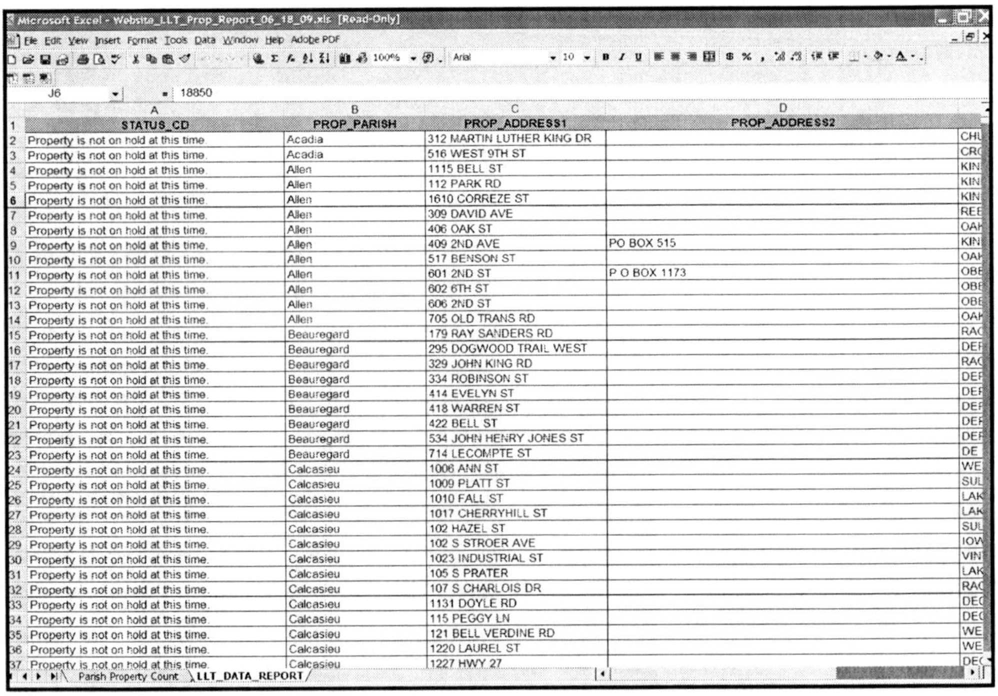

	A	B	C	D	
1	STATUS_CD	PROP_PARISH	PROP_ADDRESS1	PROP_ADDRESS2	
2	Property is not on hold at this time.	Acadia	312 MARTIN LUTHER KING DR		CHU
3	Property is not on hold at this time.	Acadia	516 WEST 9TH ST		CRO
4	Property is not on hold at this time.	Allen	1115 BELL ST		KIN
5	Property is not on hold at this time.	Allen	112 PARK RD		KIN
6	Property is not on hold at this time.	Allen	1610 CORREZE ST		KIN
7	Property is not on hold at this time.	Allen	309 DAVID AVE		REE
8	Property is not on hold at this time.	Allen	406 OAK ST		OAK
9	Property is not on hold at this time.	Allen	409 2ND AVE	PO BOX 515	KIN
10	Property is not on hold at this time.	Allen	517 BENSON ST		OAK
11	Property is not on hold at this time.	Allen	601 2ND ST	P O BOX 1173	OBE
12	Property is not on hold at this time.	Allen	602 6TH ST		OBE
13	Property is not on hold at this time.	Allen	606 2ND ST		OBE
14	Property is not on hold at this time.	Allen	705 OLD TRANS RD		OAK
15	Property is not on hold at this time.	Beauregard	179 RAY SANDERS RD		RAC
16	Property is not on hold at this time.	Beauregard	295 DOGWOOD TRAIL WEST		DER
17	Property is not on hold at this time.	Beauregard	329 JOHN KING RD		RAC
18	Property is not on hold at this time.	Beauregard	334 ROBINSON ST		DER
19	Property is not on hold at this time.	Beauregard	414 EVELYN ST		DER
20	Property is not on hold at this time.	Beauregard	418 WARREN ST		DER
21	Property is not on hold at this time.	Beauregard	422 BELL ST		DER
22	Property is not on hold at this time.	Beauregard	534 JOHN HENRY JONES ST		DER
23	Property is not on hold at this time.	Beauregard	714 LECOMPTE ST		DE
24	Property is not on hold at this time.	Calcasieu	1006 ANN ST		WE
25	Property is not on hold at this time.	Calcasieu	1009 PLATT ST		SUL
26	Property is not on hold at this time.	Calcasieu	1010 FALL ST		LAK
27	Property is not on hold at this time.	Calcasieu	1017 CHERRYHILL ST		LAK
28	Property is not on hold at this time.	Calcasieu	102 HAZEL ST		SUL
29	Property is not on hold at this time.	Calcasieu	102 S STROER AVE		IOW
30	Property is not on hold at this time.	Calcasieu	1023 INDUSTRIAL ST		VIN
31	Property is not on hold at this time.	Calcasieu	105 S PRATER		LAK
32	Property is not on hold at this time.	Calcasieu	107 S CHARLOIS DR		RAC
33	Property is not on hold at this time.	Calcasieu	1131 DOYLE RD		DEQ
34	Property is not on hold at this time.	Calcasieu	115 PEGGY LN		DEQ
35	Property is not on hold at this time.	Calcasieu	121 BELL VERDINE RD		WE
36	Property is not on hold at this time.	Calcasieu	1220 LAUREL ST		WE
37	Property is not on hold at this time.	Calcasieu	1227 HWY 27		DEQ

Figure 11:5. *Properties held by the Louisiana Land Trust downloadable in Excel*

program, then growth poles could emerge. The return and recovery of one household could then spur the same for one or two more households.

Q: What do you think is meant by a "growth pole?"

The policy of non-profits can also impact the geography of recovery. For example the Make It Right project which we have previously discussed in this book has identified a contiguous multi-block area of the Lower Ninth Ward for redevelopment. As a resident of this area commented and I paraphrase, it makes sense to rebuild in contiguous groups rather than allow it to proceed in a scattered pattern. During a recent data collection trip (June 2009) to the Lower Ninth Ward, we noticed that rebuilding is starting to occur in the areas peripheral to the Make It Right development. Once again, evidence from the built environment suggests that this non-profit's actions are leading to recovery both within and at the boundary of their project area. The idea of a growth pole may well be playing out on the landscape. Furthermore, this geography of recovery raises the issue of distance decay which has been discussed in a previous chapter. As we move away from clusters of recovery, do the signs of this process decline?

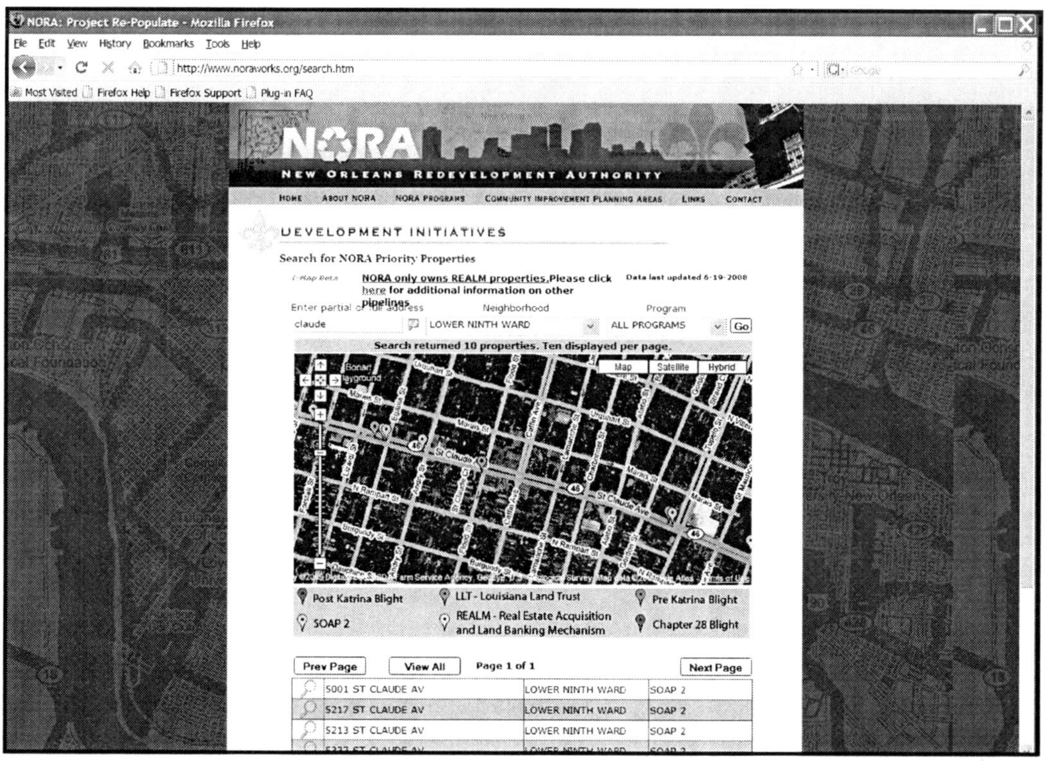

Figure 11:6. *NORA "The Lot Next Door" Program online mapping interface.*

IMPEDIMENTS TO RECOVERY: THE BUILT ENVIRONMENT AND SAFETY

As we have just discussed, the built environment is an integral component in the well-being of places. It can contribute a positive influence, such as encouraging physical activity through access to recreation spaces or it may enable less desirable behaviors such as criminal activity. Though the built environment may be a facilitator of behaviors, it can also signify the presence of less obvious activities. For example, the presence of litter or gang territory marked by graffiti may signify disorder in the neighborhood therefore making it more susceptible to further degradation, as is the premise of the broken windows theory[11]. In essence, the built environment may be a conduit or an inhibitor, or simply a signpost for activities occurring in the neighborhood.

Quantitative and spatial approaches have recently been advanced in terms of understanding the geographic relationship between crime and the built environment. Many variables have been examined as indicators of crime, such as broken windows, litter, graffiti, as have geographic locations such as rail stations. Recently, this research direction has expanded to include the current mortgage crisis. And yet even though many of these "incivility" indicators are natural byproducts of a post-disaster landscape, virtually no research exists linking these to upswings in crime, which in turn act as a further impediment to recovery.

Before we consider the post-Katrina crime situation more closely, it is worth mentioning Wilson and Kelling's (1982) theory of "broken windows" which is the seminal contribution to the relationship between crime and disorder in the built environment. Essentially, they introduce the hypothesis that physical disorder creates an environment that is conducive to criminal activity. However, it has been criticized for lacking empirical evidence to support this thesis. Many of the criticisms point to the inconsistency between various signs of disorder and their correlation with actual criminal incidents. However, the impact of these signs is not limited to actual crimes committed, but also to the residents' perception of safety. This perception can then influence their quality of life through increased stress and using coping mechanisms to deal with this stress. In 1989 the authors used a case from Newport News, Virginia to argue that reducing decay in the urban environment also reduces crime. They identified vacancy, abandonment, and "back alleys" as breeders of criminal activity, but addressing issues of trash, road and building deterioration (all of these variables are plentiful in many "recovering" neighborhoods of New Orleans) have received scant attention. They also note the power of graffiti as an indicator of gang activity and the use of street lighting to reduce criminal activity (we have already seen a resurgence of "tagging" in the

11 Wilson, James Q. and George L. Kelling. 1982. Broken windows. *The Atlantic Monthly* 249 (3): 29–38.

neighborhoods of New Orleans—and we also have to wonder how the remnants of the search and rescue "graffiti" fit into these previous theories). The challenge of a post-disaster landscape is that these indicators of incivility are far more temporally unstable as compared to typical inner city decline.

Unoccupied structures have long been noted as elements of the built environment that are conducive to criminal activity, ranging from prostitution, arson, drug trade and homicide. Such activities are shielded from public view in these structures. Though vacant non-residential buildings have commonly been identified as the conduits for these activities, residential dwellings are becoming an area of interest owing to the mortgage crisis, leaving many neighborhoods plagued with foreclosed properties. Foreclosure on single-family homes is linked to increases in violent crimes. Another way to think about the built environment and crime, Routine Activities Theory posits that the everyday movements of neighbors yields a natural observation or surveillance of people and activities in the neighborhood. Whereas vacant structures may serve as opportunities for criminal activity, routine activity acts as a defense to these behaviors. Again—in many neighborhoods of New Orleans, abandoned structures present the opportunity, while the low returnee rate also means a low degree of neighborly surveillance.

Crime has always been a problem in New Orleans; however, in late 2008 it officially received the honor of having the highest U.S. city crime rate:

> WASHINGTON (Reuters) - New Orleans had the highest crime rate of any U.S. city last year, with 209 murders and more than 19,000 reported criminal incidents, CQ Press reported on Monday in an annual ranking. New Orleans, where population declined dramatically after 2005's Hurricane Katrina, showed a jump in the murder rate, which rose from 162 in 2006 to 209 in 2007, the report found. The crime rate rankings were calculated using six crime categories: murder, rape, robbery, aggravated assault, burglary and motor vehicle theft.—From Reuters Mon Nov 24, 2008.

Although the city was relatively safe in the first few months after Katrina, the opportunities presented by large abandoned neighborhoods were not lost on the local criminal element. Bodies were dumped in these buildings knowing it would be a considerable time before they were discovered. This would eventually lead to the National Guard being recalled for neighborhood patrols.

Unfortunately, returning to the news story from Reuters, the situation has worsened, much to the despair of the returnees who are fighting to reestablish

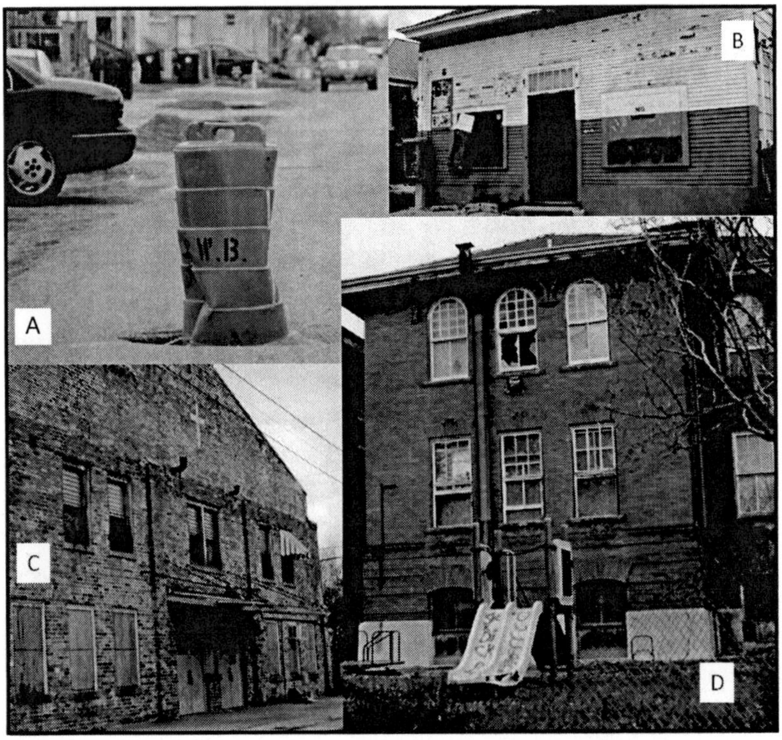

Figure 11:7. *Images from the Holy Cross neighborhood during early 2009 showing ongoing built environment disorders. "A" shows a large plastic barrel placed in one of the main potholes in the road. "B" (a corner store), "C" (a church) and "D" (a local school)—all important parts of the local social support system—now boarded up and abandoned. Even worse, these buildings also provide spaces for these crimes to occur. Notice in "D" the graffiti on the children's slide.*
Photographs by Curtis

their neighborhoods. In Holy Cross, placards are placed identifying where a crime occurred, and asking for help in identifying what happened (Figure 11:9).

Aspects of the post-Katrina landscape are often interwoven in crime reports when describing the location. For example, post-disaster landscape emphasis has been added to this recent newspaper story:

> Man found shot to death in *empty lot* on Louisa Street
>
> Police found this morning the body of a 23-year-old man who had been shot and abandoned near an *empty lot* in the St. Claude neighborhood. The man, whose identity has not been released, was discovered shortly before 10 a.m. in the middle of the 1600 block of Louisa Street, according to New Orleans Police Officer Shereese Harper. Detectives had cordoned off a large swath of the *blighted block*. The man's body had been discovered between an *unoccupied house and an empty lot of waist-high weeds*. Details on a possible motive or suspect were not immediately available. (Brendan McCarthy, The Times-Picayune, Monday, April 20, 2009)

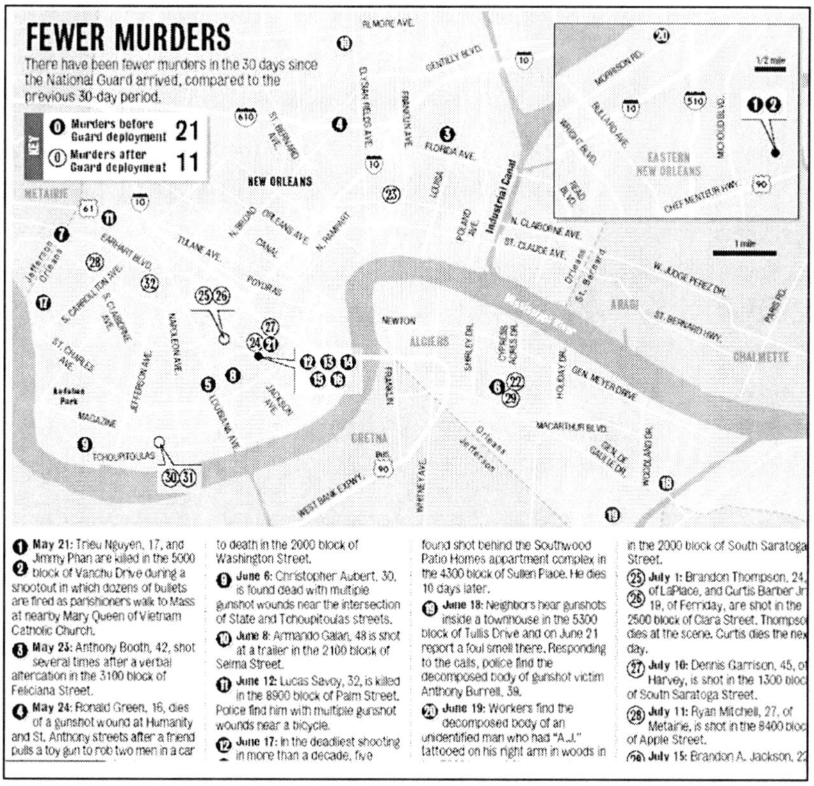

Figure 11:8. *A map from the New Orleans Times -Picayune newspaper showing the "success" of the National Guard in reducing murders was actually used by our field team to identify where crime had been to limit our risk during data collection.*

Figure 11:9. *An aggravated assault happened at this location in July of '08—this photograph was taken in March '09.*

By attending community group meetings it soon became evident that crime was one of the more pressing issues. Although local residents always believe there is a struggle to find accurate crime data, two online mapping engines are available as a resource for New Orleans.

Figure 11:10 shows the "New Orleans Citizen Crime Watch" mapping engine. This online geographic information site allows for searches by city, by neighborhood, by different crime types and as either points (often referred to as "pin maps" or "spot maps") or as hotspot densities.

Q: Which of our three exercise neighborhoods appear to have the worst crime problem?

One advantage of this web site is that each crime event can be "clicked on" to get detailed information and a source such as from the New Orleans *Times Picayune* newspaper.

Figure 11:11 demonstrates the online crime mapping service provided by the City of New Orleans. The three exercise neighborhoods from Chapter Twelve are mapped as an example. One advantage of this system is that data can be downloaded into Excel, which can then be imported to a GIS. What is interesting about the crime focused on in the Lower 9th Ward is its relative location. Again, Routine Activities Theory suggests that the everyday movement of neighbors yields a natural observation or surveillance of people and activities in the neighborhood. There would have been very little activity around this particular crime location.

Now that we have discussed the importance of the built environment for understanding recovery (and some inhibitors to recovery), let's turn to the mechanics of collecting spatial data from the built environment to monitor and map the fine scale human geography of this process.

COLLECTING INDIVIDUAL LEVEL RECOVERY DATA

As previously mentioned, spatial recovery has received little attention from the scientific community and as a result, there is still much to discover about this process. This situation is particularly true when considering the fine scale geographies believed to be operating and the longitudinal timeframe of these geographies. The best resource for understanding the geography of recovery is the people who are going through the process. As a result, recovery studies would do well to consider implementing the Community Based Participatory Research (CBPR) framework. Local knowledge is essential to the development of a comprehensive and meaning-

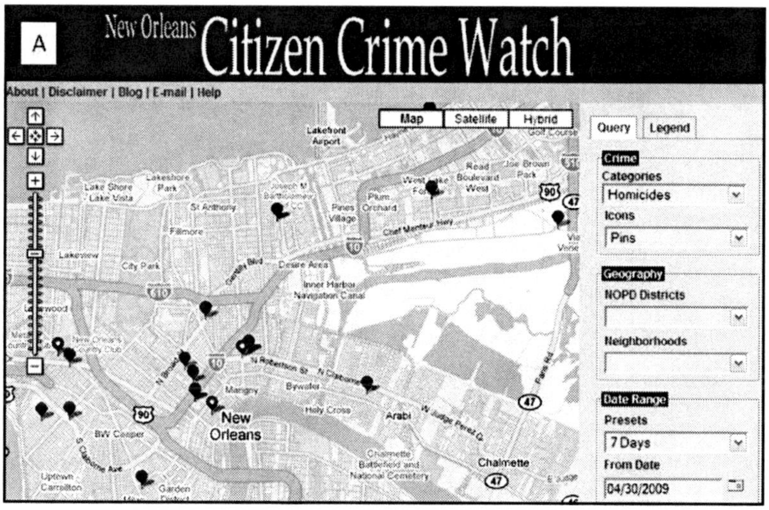

Figure 11:10. *The "New Orleans Citizen Crime Watch" mapping engine available at http://citizencrimewatch.org.*

ful understanding of what these people are exposed to in their environment; what they consider signs of recovery and signs of decline[12].

The following section introduces a collaborative project that focuses on capturing geographic information from the field and then making it usable in a GIS. As you read about this project, think about how it applies to understanding the geography of recovery. This approach could also be utilized in a final semester student project.

ENABLING COMMUNITY PARTICIPATION IN RECOVERY MAPPING: AN EXAMPLE

This book has introduced approaches such as mobile mapping for data collection, but there are other ways to create a spatial inventory of recovery. For example, we have been working with the Earthquake Engineering Research Institute (EERI) through FEMA to develop a tool for collecting post-disaster data from the field. Though it can be used for damage assessment or monitoring implementation of structural mitigation measures, it can also be used to collect data on recovery. In effect it is an all-hazards approach. Furthermore, it is easy to use so that neighborhood groups can have a "kit" to collect their own data. The GIS component, however, may require partnership with a university or necessitate a neighborhood member taking a class in GIS. In New Orleans, several non-profits have trained personnel on using GIS so that they can lead their recovery mapping projects.

12 Corburn, J. 2002. Combining Community-based Research and Local Knowledge to Confront Asthma and Subsistence-fishing Hazards in Greenpoint/Williamsburg, Brooklyn, New York. *Environmental Health Perspectives* 110 (Suppl 2): 241–248.; Israel, Barbara, Amy J. Schulz, Lorena Estrada-Martinez, Shannon N. Zenk, Edna Viruell-Fuentes, Antonia M. Villarruelm and Carmen Stokes. 2006. Engaging Urban Residents in Assessing Neighborhood. Environments and Their Implications for Health. *Journal of Urban Health* 83(3): 523–539.

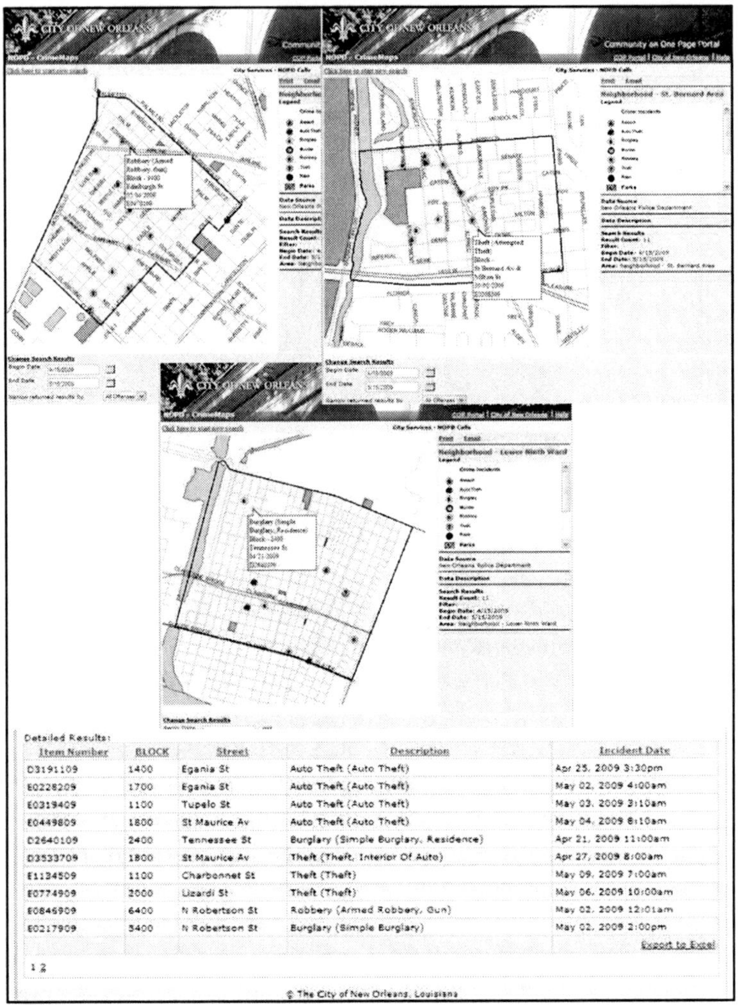

Figure 11:11. *An alternative mapping approach using more "official" data is the Community on One Page Portal run by the City of New Orleans and populated by New Orleans Police Department (NOPD) data (http://cno-gisweb02.cityofno.com/crimemaps).*

The Earthquake Engineering Research Institute's (EERI) objective is to "reduce earthquake risk by (1) advancing the science and practice of earthquake engineering, (2) improving understanding of the impact of earthquakes on the physical, social, economic, political, and cultural environment, and (3) advocating comprehensive and realistic measures for reducing the harmful effects of earthquakes." (http://www.eeri.org/site/about-us/our-mission) It achieves this objective through working with diverse communities, promoting research, facilitating information dissemination, and through education and advocacy activities (http://www.eeri.org/site/about-us/our-mission).

A collaborative project between USC and EERI has led to the development of the EERI Field Tool which directly contributes to the EERI project "Learning from Earthquakes" which facilitates information dissemination and education about past earthquake events via the use of post-event field photos displayed in a Google Maps or Google Earth interface (http://www.eeri.org/site/ projects/learning-from-earthquakes/earthquake-images). This project builds on this framework to improve quality in data collection and expand utility of these data through their development into a GIS-ready format.

Field data capture is composed of a visual image, coordinates of the image, and attribute data associated with the image. The techniques used to achieve field data collection can vary, however. For example, this project employed the use of the ATP PhotoFinder (http://photofinder.atpinc.com/), the Trackstick (http://www.trackstick.com/), and the Phototrackr (http://www.gisteq.com/PhotoTrackrLite.php). This geotagging technology is constantly improving in quality and usability. For users who are not comfortable with the new geotagging technology, some manual methods can also be used to achieve a geotagged image. These include using a handheld GPS receiver to record the location of each photo, then writing this in a field notebook to correspond with the order in which photos are taken. Also, the user can take a photo of the subject of interest and then take a picture of the coordinate reading on the handheld GPS. In addition, even when GPS fails (e.g. battery problems in the field, multi-path error, etc.), through Picasa (http://picasa.google.com) and Google Earth (http://earth.google.com), for example, photos can be geotagged retroactively. The result of this new technology is that it is easy for anyone to take a camera and a geotagger (or other GPS) into the field to collect geographic information. The EERI Field Tool then enables these data to be organized and integrated in a GIS.

Case Study: Hurricane Katrina (2005). Holy Cross provides an ideal case study in which to test the EERI Field Tool in a post-disaster environment. This case study demonstrates the way data can be repurposed for multiple uses, from validation of recovery models to use in mitigation studies. Though these data were collected to measure recovery status of the neighborhood (e.g. returned, rebuilding, vacant), once a photograph is collected, geotagged, and uploaded into a database, it can be integrated into a GIS for many uses. In this case we are interested if returnees are elevating their houses to mitigate against future floods. Therefore, we created an inventory regarding which structures are on slabs and which are elevated (Figure 11:12). In other words, this approach can be used to capture the dynamically changing landscape as shown earlier in Figure 11:1.

Q: How can information about where houses are built on slabs (not raised) and where they are elevated affect planning for the post-disaster needs of people in this neighborhood? What other data could be overlaid to improve planning? How would these maps be useful?

Such an inventory can be used to assess implementation of mitigation measures or to predict loss in a variety of scenarios (e.g. hurricane, flood, levee failure). These same images, however, can also be used to catalogue cultural resources such as historical properties, or to add detail to housing stock inventories (e.g. stucco, wood, brick structures). In this case, the houses are mapped as having either a slab foundation or being an elevated structure. A density analysis was then performed on the distribution of slab houses to identify hot spots of this risk (Figure 11:13a). The contours of this surface were extracted and converted into a KML. Figure 11:13b demonstrates the spatial context that is added by using Google Earth (e.g. visualization of slab houses in relation to the levee and the Mississippi River).

The entire process (from data entry to KML) was performed quickly and without complication because the EERI Field Tool processes field data into an organized GIS-ready output. In this case data were used to examine structural mitigation near a New Orleans levee but they can also be utilized to predict losses in the next hurricane or flood event.

Q: How would you structure a community/academic collaboration to assess recovery incorporating this tool?

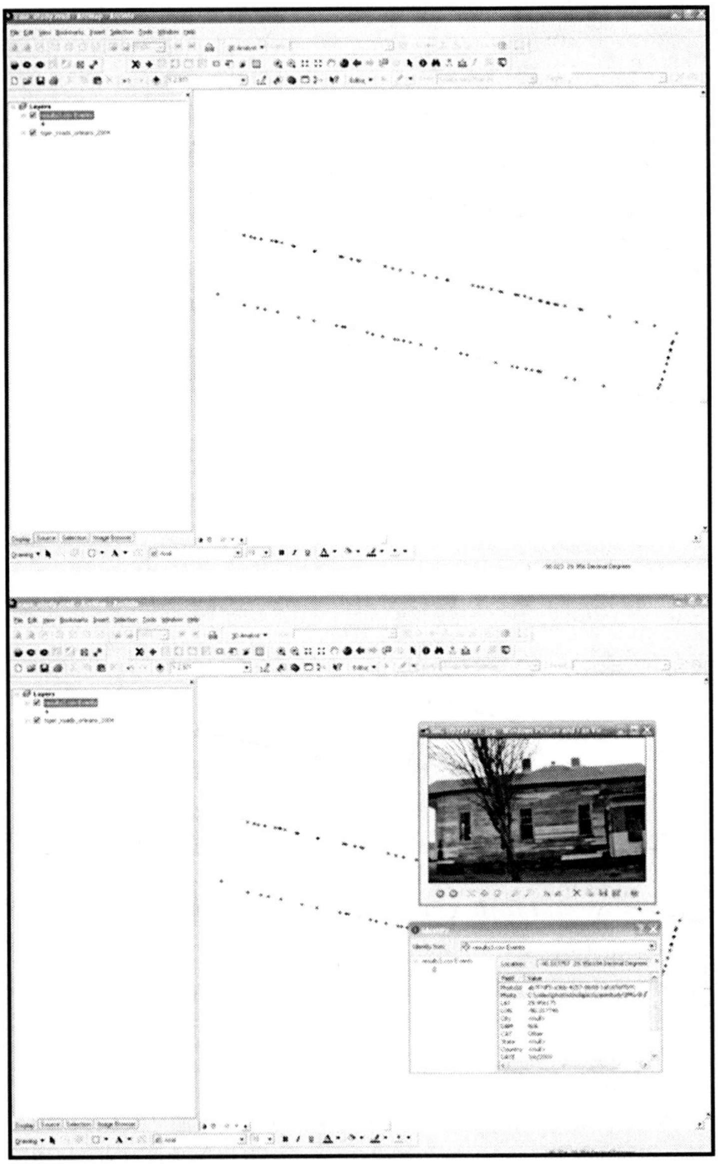

Figure 11:12. *Data collected by using the EERI Field tool for built environment survey.*

CONCLUSION

In this chapter we have focused on introducing you to the human geography of recovery. As previously mentioned, recovery is a broad topic that can be defined in different ways based on the specific interests of those conducting the studies. However, for this chapter, we use a definition of recovery that is consistent with our community partners in New Orleans, the permanent return of households to their pre-Katrina homes. There is still much work to be done to improve our knowledge about recovery. For those of you interested in pursuing this topic as a final project, the following list of refences should prove useful.

Figure 11:13. *(a) Hot spots of houses built on slabs in a section of the Holy Cross neighborhood of New Orleans. (b) Contours of the hot spots from (a) are presented in Google Earth in order to add geographic context.*

TIMELINE OF BOOKS AND PEER-REVIEWED ARTICLES ON RECOVERY

1. Baker, George W. and D. W. Chapman, eds. **1962**. *Man and Society in Disaster*. New York: Basic Books.
2. Dacy, Douglas C. and Howard Kunreuther. **1969**. *The Economics of Natural Disasters: Implications for Federal Policy*. New York: Free Press.
3. Barton, Allen H. **1970**. *Communities in Disaster: A Sociological Analysis of Collective Stress Situations*. New York: Anchor.
4. Harding, D. M., and Parker, D. J. **1974** . "Flood hazard at Shrewsbury, United Kingdom." *Natural hazards: Local, National and Global*, G. F. White, ed., Oxford University Press, New York, pp. 43–52.
5. White, Gilbert F. and J. Eugene Haas. **1975**. Assessment of Research on Natural Hazards. Cambridge, MA: MIT Press.
6. Bolin, R. **1976**. Family Recovery from National Disaster: A Preliminary Model. *Mass Emergencies*. 1(1976): 267–277.
7. Haas, J. Eugene, Robert W. Kates, and Martyn J. Bowden. **1977**. Reconstruction Following Disaster. Cambridge, MA: MIT Press.
8. Friesema, H. Paul, J. Caporaso, G. Goldstein, R. Linberry, and R. McClean. **1979**. Aftermath: Communities After Natural Disasters. Beverly Hills: Sage.
9. Wright, James D., Peter H. Rossi, Sonia R. Wright, and Eleanor Weber-Burdin. **1979**. After the Clean-Up: Long Range Effects of Natural Disasters. Beverly Hills: Sage.
10. Rubin, C.B. and D.G. Barbee. **1985**. Disaster Recovery and Hazard Mitigation: Bridging the Intergovernmental Gap. *Public Administration Review*. 45(Special Issue): 57–63.
11. Bolin, R. **1986**. Disaster Impact and Recovery: A Comparison of Black and White Victims. *International Journal of Mass Emergencies and Disasters*. 4(1): 35–50.
12. Bates, F.L. and W. G. Peacock. **1989**. Long Term Recovery. *International Journal of Mass Emergencies and Disasters*. 7(3): 349–365.
13. Bryant, E. A. **1991**. *Natural hazards*, Cambridge University Press, Cambridge, England
14. Berke, P. R., Kartez, J., and Wenger, D. **1993** . "Recovery after disaster: Achieving sustainable development, mitigation and equity." *Disasters*, 17, 93–109.
15. Berke, P. R. **1995** . "Natural-hazard reduction and sustainable development: A global assessment." *J. Plan. Lit.*, 9, 370–382.
16. W. G. Peacock, B. H. Morrow, and H. Gladwin, eds., **1997**. *Hurricane Andrew: Ethnicity, gender and the sociology of disasters*, Routledge, London.

17. Berke, P.R. and T. Beatley. 1997. After the hurricane: Linking recovery to sustainable development in the Caribbean. Baltimore: Johns Hopkins University Press.
18. Birkland, T. 1998 . *After disaster: Agenda setting, public policy, and focusing events*, Georgetown University Press, Washington, D.C.
19. Comerio, M. 1998. Disaster hits home: New policy for urban housing recovery. Berkeley: University of California Press.
20. Mileti, D. S. 1999. Disasters by Design: A reassessment of natural hazards in the United States. Washington: Joseph Henry Press.
21. Webb, G.R., K.J. Tierney, J.M. Dahlhamer. 2002. Predicting long-term business recovery from disaster: a comparison of the Loma Pieta earthquake and Hurricane Andrew. *Environment Hazards*. 4(2002): 45–58.
22. Adams, B.J. ; Huyck, C.K.; Eguchi R.T.; Yamazaki, F.; Estrada, M., C., Herring. 2003.
23. QuickBird Imagery of Algerian Earthquake used to Study Benefits of High-Resolution Satellite Imagery for Earthquake Damage Detection and Recovery Efforts. Earth Imaging Journal: 2003-00-00.
24. Cutter, C.T., J.T. Emrich, B.J. Mitchell, M. Boruff, M. Gall, M.C. Schmidtlein, C. G. Burton, and G. Melton. 2006. The long road home: race, class, and recovery from Hurricane Katrina. *Environment: Science and Policy for Sustainable Development*. 48(2): 8–20.
25. Kates, R.W., C. E. Colten, S. Laska, and S. P. Leatherman. 2006. Reconstruction of New Orleans after Hurricane Katrina: A research perspective. *Proceedings of the National Academy of Sciences*. 103(40): 14653–14660.
26. Olshansky, R. B. 2006. Planning After Hurricane Katrina. *Journal of the American Planning Association*. 72(2): 147–153.
27. Runyan, R. C. 2006. Small Business in the Face of Crisis: Identifying Barriers to Recovery from a Natural Disaster. *Journal of Contingencies and Crisis Management*. 14(1): 12–26.
28. Alesch, D.J., L.A. Arendt, and J.N. Holly. 2008. *Managing for Long-term Community Recovery in the Aftermath of Disaster*. Fairfax, Virginia: The Public Entity Risk Institute.
29. Zottarelli, L.K. 2008. Post-Hurricane Katrina Employment Recovery: The Interaction of Race and Place. *Social Science Quarterly*. 89(3): 592–607.

TWELVE

Using the Spatial Video for Neighborhood Recovery

THROUGHOUT THIS BOOK different examples of the spatial video have been used to illustrate aspects of the post-disaster landscape. This chapter concentrates on the tool as a means to assess street level recovery and then presents an exercise that mimics the spatial video using Google Street View.

In the summer of 2007, PolicyLink[1] collaborated with the CADGIS lab at Louisiana State University to use the spatial video to map out four Churches Supporting Churches (CSC) neighborhoods identified for economic development[2]. There were several reasons for this first survey—not least of which was a "damage assessment" of the neighborhoods almost two years after Katrina. Among the information to be extracted and mapped were locations of recovery, and which street sections continued to be abandoned and in serious disrepair. In addition, changes in parcel ownership were mapped where possible, such as when the orange curbside marking in front of a home was visible (meaning "Road Home"[3] had purchased the property).

In the summer of 2008 we returned as part of a USC team to again help CSC—this time to revisit the same four neighborhoods to create "change maps"—to show

1 PolicyLink is a national research and advocacy group dedicated to reducing economic and social disparity. They have a substantial presence in New Orleans.

2 From their website: Churches Supporting Churches (CSC) is a comprehensive strategy to assist at least 36 African American congregations in twelve areas of New Orleans where Katrina destroyed or seriously damaged their facilities to "Restart, Reopen, Repair or Rebuild the Churches" in order for them to become "Agents for Community Development" in a comprehensive way. The strategy is designed to go beyond food and shelter issues and focus on expanding the capacity of these congregations to rebuild their communities. (http://www.cscneworleans.org/intro.html).

3 From their website: *The Road Home* program is designed to provide compensation to Louisiana homeowners affected by Hurricanes Katrina or Rita for the damage to their homes. *The Road Home* program is the largest single housing recovery program in U.S. history. The program's objective is to provide compensation to Louisiana homeowners affected by Hurricanes Katrina or Rita for the damage to their homes (http://www.road2la.org/about-us).

how much improvement had occurred during the previous year. One of the sample neighborhoods included at the end of this chapter as an exercise corresponds to one of these CSC neighborhoods in the Lower 9th Ward. Our friend and collaborator for this particular neighborhood is the Reverend Charles Duplessis who has recently been included in the book "Overcoming Katrina: African American Voices from the Crescent City and beyond" by D'Ann R Penner and Keith C Ferdinand. This book also includes information about CSC.

While attending the CSC monthly meeting we were also introduced to Julius Lee, the Deacon/Trustee of the Greater New Saint Luke Baptist Church in the Upper 9th Ward. Another collaboration was organized whereby the Greater New Saint Luke Baptist Church would provide a vehicle and a driver, and we would video the route and create a GIS. A third video would also capture an inboard commentary along the route—this would prove to be an important advance as it both helped add context and history to the streets, and also helped clarify whether buildings were really being rebuilt, something which is hard to assess from a single data collection trip. On this first trip two other ministers accompanied us, and we would record data for three further neighborhoods, the largest of which was around the St. Bernard Housing Development in the 7th Ward.

We would make further trips to both the Upper 9th Ward and 7th Ward during 2008 and 2009 in order to assess the progress in rebuilding activities for different time periods.

This chapter will include the results of some of these surveys to illustrate the geography of a recovering neighborhood. However, although a focus will be placed on how we used a GIS, this is an example of a mixed methods approach to research where "hard" data and quantitative approaches are acquired and applied in concert with qualitative data and analysis[4].

In order to provide a little background on why these neighborhoods are important, at this geographic scale, Julius has included the following description:

OUR HISTORY

> Greater New St. Luke Baptist Church was founded in the early 1950s by Rev. Coleman and Rev. Nathan Thomas in the Upper 9th Ward of New Orleans. In 1954 we built our first sanctuary, at 2201 Montegut Street, on the corner of Montegut and N. Miro Streets. In 1970 our current pastor,

4 An interesting mixed methods approach for an urban environment can be found in Kwan, Mei-Po & Ding, Guoxiang (2008). Geo-Narrative: Extending Geographic Information Systems for Narrative Analysis in Qualitative and Mixed-Method Research. *The Professional Geographer*, 60 (4), 443–465.

> Rev. M. H. Littleton came and we experienced growth from the early 70s through the mid 80s. We are a very warm, friendly, welcoming and giving church to our members and those who lived in our community. We conducted multiple services to accommodate everyone. However, in the late 80s through 90s we experienced tremendous growth and in 1994 we constructed our new sanctuary next door on Montegut Street, seating approximately 500 members. As part of our mission through our outreach ministries we've provided food and clothing to those in need in our community and we've supported national organizations in Mission and Christian Humanitarian works.
>
> On August 29, 2005, Hurricanes Katrina came through the Gulf Coast and reeked major destruction to our neighborhoods, communities and churches. Our sanctuary and fellowship hall sustained well over $500,000 in building damages and contents destroyed. Hurricane Katrina left us with a big challenge to rebuild, but by the grace of God, we will overcome.

On each of our three data collection trips to the neighborhood, it became evident that a rebuilding "growth pole" was extending out from the church. Not only was the Church and its associated buildings in the best condition compared to many of those surrounding, but also other properties along the block face were beginning to be rebuilt—mostly with church money and help. There is little if any spatial theory regarding how rebuilding occurs after a disaster. In many ways it is an organic process, with different pockets stimulating growth, while other streets remain in disrepair. There are different impediments to rebuilding, most are financial; not being able to navigate the various recovery assistance programs, not receiving enough money to be able to properly rebuild, or not having flood insurance. There are other non-financial factors, such as falling ill, not wanting to live in a devastated area with inadequate service provision. Imagine what it is like to return to a street where you are the only family, the street has huge potholes that keep getting deeper, and the grass on your neighbors' devastated homes is ten feet high.

Julius has detailed some of these challenges in a neighborhood information handout:

> *The Current Situation*
>
> We had flood insurance but we did not have enough to cover the cost of the flood damage to repair our buildings and replace our contents. To date, we have spent close to $400,000 on repairs and restorations to both

> buildings. We are in need of at least another $150,000 to complete the restoration of our buildings, parking ground resurfacing and sanctuary contents replacement. We restored our fellowship hall and held our first worship service in it on January 28, 2007. We currently have 85 to 120 members worshipping on Sundays. We anticipate significant growth this next year as we are in constant communication with our members who are displaced around the country. Many are working on restoring their homes and some are looking for an affordable place to rent as they are preparing to return to New Orleans. So as our community grows, it is vitally important for Saint Luke to complete its restoration on our main sanctuary building within the next six months. Our Fellowship Hall will serve as a community resource and learning center after the rebuilding of our main Sanctuary building.

As previously described in this book one challenge for collecting data in a post disaster environment is the need to be able to collect fine scale information with temporal repetition in order to capture change.

This coincides with a recent theme for research, how to effectively include local context in urban environment studies. Whether the research involves identifying geographic patterns of crime, environmental health and the built environment, urban recovery in the post-disaster phase, social and environmental justice in the city, or most recently the impact of foreclosures in the housing market, there is a need to move away from larger units of aggregation (such as census tracts, and especially zip codes), to concentrate instead on neighborhood context and effect. "Neighborhoods" are usually not drawn along the same boundaries of data commonly available for typical GIS use, and as a result for the investigation of within-neighborhood interactions is virtually impossible. Part of this movement to a finer scale analysis is the understanding that many urban processes may be unique to each locality, and that although approaches of inquiry may be transferable, results may not be. Therefore emphasis must be placed on being able to collect the right type of data to perform these local analyses. Simply put, Julius wouldn't learn that much about the changing state of his neighborhood from using either zip code or census tract level data (even if it were available for the post-disaster period).

Figure 12:1 shows six of the neighborhoods that have been covered using the spatial video. These have been placed within a zip code map to illustrate how this commonly used spatial aggregation does not match the scale needed to assess neighborhood recovery. The three exercise maps included at the end of this chapter fall within the 7th ward, Lower 9th Ward and Holy Cross neighborhoods.

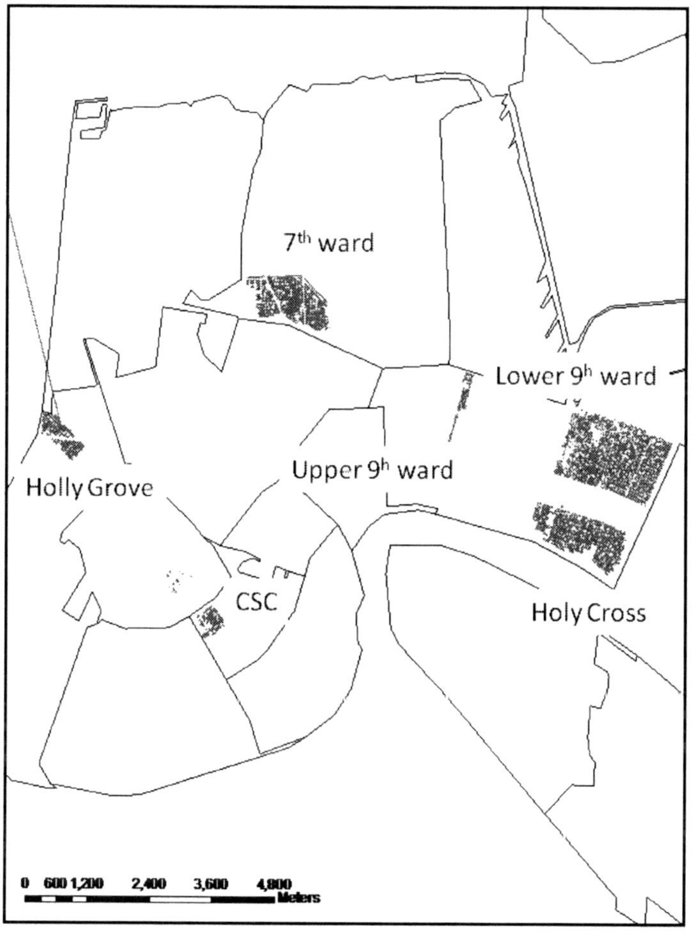

Figure 12:1. *Each outline polygon represents a different zip code while each point in the shaded areas actually indicates a house captured from the spatial video and included in our neighborhood recovery GIS.*

1: Assuring Data Quality

A great deal of geographic research is reliant on secondary sources of data, the most obvious of which is the census. However, when working with secondary data, the researcher needs to have confidence in where and how data were collected. For example, are there known errors or biases in collection, or were inappropriate sample schemes used that again could introduce geographic bias? In addition, as most non-census secondary data release occurs in an aggregated form, manipulation error may also exist, the most obvious of which is the inaccuracy of the initial geocode. Collecting primary data allays most of these fears, or at least keeps the control of potential error in the hands of the researcher. Unfortunately, the costs involved—both pecuniary and the immense amount of labor needed—tend to limit such data collection, especially when working with large geographic areas of a city. Compromises usually involve spatial sample schemes or single case study areas.

2: Inappropriate Spatial Aggregation

Secondly, the researcher, again if reliant on secondary data, may be forced to work with inappropriate spatial units. If the purpose of the research is to understand local spatial interactions, for example how the foreclosure of one house impacts the lifestyle, health or crime vulnerability of neighbors, then data made available at only the census block group level are of little use. For example, relatively little can be gained by considering the number of returnees to New Orleans by Census Tract if the purpose of the research is to see how home rebuilding on a blockface affects decisions made by immediate neighbors. A street may contain examples of return, rebuilding, and abandonment, and the pattern of these in relation to one another, and even the temporal pattern of how one state changes into another, needs data collection at the level of an individual home.

One consideration when working with data at finer aggregations is the violation of spatial confidentiality. This can occur either through the collection of sensitive data, or by revealing confidential information in maps.

3: Inappropriate Temporal Resolution

A third data problem is temporal resolution. Many urban processes are temporally dynamic—they change annually; with the seasons, by the day of the week, or even during the hour of the day (transport flows being the most obvious example). Sometimes multiple data sets important to the research display such temporal sensitivity, for example home foreclosures and mosquito presence[5]. This is even more problematic in a post disaster urban environment where change occurs at a faster rate than normal. Data collected in September may not reflect conditions in December.

Unfortunately the reliance on secondary data means that at best the analysis will capture a single snapshot of a process at work, and at worse will find no or spurious patterns if these data are out of phase. Census data are decadal; many other forms of data are only available by month, quarter, or year. In order to perform urban analyses where such temporal intersection is important, the ability to collect primary data is paramount.

If we combine data problems two and three, a researcher may be forced to run analyses on data that have inappropriate spatial aggregations and time frames. Using zip code health data taken from pre-Katrina New Orleans to assess post disaster health vulnerability at the neighborhood level is an obvious example[6].

4: Research Applicability

A research concern linked to issues one through three is the timetable of data acquisition, analysis and interpretation if any intervention is required. A well-executed

5 A recent article in the LA Times details how the rise in foreclosures in San Bernardino County, CA has caused concern in terms of mosquito increases, which in turn might result in increased West Nile Virus cases. Any investigation would require temporally specific information on foreclosures, as well as the resulting disease cases.

6 Infant mortality data were provided to us as a health marker we could use to predict where neighborhood stresses were greatest in the past, and which in turn might have been useful to predict where future health problems might emerge. Unfortunately the geographic scale of the data, the zip code level, meant that any meaningful prediction was limited. Neighborhood specific data would have made the analysis far more meaningful, as would continually updated health event data.

data rich analysis, extending over several years will always be important in terms of revealing processes that can help future populations. However, there must also be room for analyses that still maintain academic rigor, but that can provide insight for intervention. This is important on two levels. Firstly, there are many urban concerns where such an intervention is possible and therefore an ethical responsibility exists to provide societal feedback. Three easy to understand examples include crime surfaces (the most extreme case being a serial offender), the impact of house foreclosures on a neighborhood, and environmental health findings where changes in lifestyle might potentially prevent years of illness.

The second reason for expedited research leading to intervention is when a collaboration exists with a community—as our example here with the Upper 9th Ward, where the research is important to both the academic and the neighborhood itself. A common concern with community-based research is that the findings do not benefit the community, rather they just advance the researcher's own career agenda. The perception is of researchers "swooping" in for their own benefit and then leaving without any follow-up once the funding has dried up or their attention shifts to another topic. Public Participatory GIS (PPGIS) helps to partially solve this problem by placing more of the research power within community hands[7]. From this participatory framework an alternative research process emerges that is dynamic, where both data collection and findings occur continually with the participation of community representatives. In order for such a dynamic research environment to exist—a community has to "buy in" meaning the community needs to identify a problem they consider to be of importance. The work with CSC provides one such situation where findings are being used by community representatives to advocate for policy change. For example, a map was produced by PolicyLink in coordination with CSC for an advocacy meeting; data on the map came from multiple sources, including the output from the spatial video.

Of course, it is imperative for academic integrity to be preserved in such quick turn around research. In order to achieve this, standardized data collection protocols and research methodologies are important so that the amount of "on the fly" decisions are minimized. This is particularly important in post-disaster research where the data collection environment is often stressful, resulting in compromises

7 There are several papers describing how PPGIS has been used in community collaborations. Readers are pointed towards Elwood, Sarah (2006). Negotiating Knowledge Production: The Everyday Inclusions, Exclusions, and Contradictions of Participatory GIS Research. *The Professional Geographer*, 58 (2), 197–208; Elwood, Sarah (2006). Beyond Cooptation or Resistance: Urban Spatial Politics, Community Organizations, and GIS-Based Spatial Narratives. *Annals of the Association of American Geographers*, 96 (2), 323-341; Sieber, Renee (2006). Public Participation Geographic Information Systems: A Literature Review and Framework. *Annals of the Association of American Geographers*, 96 (3), 491–507.

rather than well designed research schemes. Of course there also have to be advances in suitable field technologies that allow for these types of approaches.

The spatial video approach initially developed to improve damage assessment data was described earlier in this book. It has also proven to be a useful tool to monitor neighborhood recovery after a disaster. It is simple to operate and portable allowing for systems to be used in multiple environments and by relatively inexperienced research aides. Spatial video can be used to capture a large portion of the visible built environment which means that data can later be extracted at fine spatial resolutions, from a single house to a part of that property, such as a trash pile. The flexibility of the system also allows for the data to be collected for different time periods, and for multiple runs; monthly, weekly, or even different hours of the same day.

Spatial video data can be processed and analyzed, potentially, within hours of the run. Although initial data extraction may take longer in terms of creating building footprints, or other built environment information, such as the location of trees or bushes, these can be more quickly updated on subsequent runs. In addition, unlike other built environment research using videography[8], data can be transferred to other researchers for additional investigation where different aspects of the built environment are extracted. For example, initial data extraction may involve identifying continuing aspects of abandonment, such as building disrepair, the presence of search and rescue spray painted markings, or evidence of rebuilding. A subsequent investigation, possibly conducted months later, might be more interested in the activity space surrounding neighborhood stores or churches, which of them are open and whether they are acting as recovery catalysts to their streets. At an even later stage, the researcher may want to "go back in time" to see which search and rescue groups covered which territories, or the extent of damage in the neighborhood[9]. These data can also be used as "benchmarks" against which future spatial video runs can be compared.

What makes these data so transferable is that the spatial video contains appropriate metadata[10]. The images are tagged to coordinates, and are also shown on an accompanying map. The date of data capture is also recorded on the browser. All of this information is useful in terms of providing useful archival data. At the same time quality control in the extraction process can be assured by revisiting the video allowing for the possibility of recreating the findings. These points make spatial video a better prospect than more typical windshield surveys, or even

8 The classic paper on using video to assess neighborhood characteristics is Sampson R, J and Raudenbush S W 1999 Systematic Social Observation of Public Spaces: A New Look at Disorder in Urban Neighborhoods. *American Journal of Sociology* 105: 3 603–51.

9 A good example are the maps of the Lower 9th Ward discussed in Chapter Two.

10 Remember that metadata generally are "data about data"—these tell us when, how and what nuances were involved in the data collection process.

house-to-house surveys where there is limited ability to maintain quality control across multiple teams, and where the initial selection of what data were recorded limits the flexibility of future research questions.

SPATIAL VIDEO—THE TECHNICAL DETAILS

A typical system consists of multiple cameras mounted on a vehicle's windows, with each being connected to a central global positioning system (GPS) receiver. This receiver encodes a GPS coordinate onto the audio track of each camera. The most common camera configuration for collecting urban data consists of one to the left and right of the vehicle, while an optional third camera mounted to the front window captures the road ahead. Each camera is fastened to a half-rolled down window by a suction mount. Figure 12:2 displays a camera being set on the left side of a vehicle in New Orleans during June 2008. The car is parked in front of a house being raised. Although cameras used for recovery analysis in New Orleans have recorded data directly to tape, it is also possible to record to the hard

Figure 12:2. *The portability of the spatial video system is obvious here, with a single small camera on one side window (a second on the other side) both connected to a GPS receiver no bigger than each camera. The entire system can easily be carried in a small back-pack.* Photograph by Curtis

Figure 12:3 A comparison of the same road segment captured one year apart. *In the top image the road is seen within the NCG spatial video browser—the bottom image is a straight digital capture from the video.*

drive of HD cameras, reducing recurring costs of media. Multiple recovery data collection trips have shown that a high (and rugged) vehicle works best, ideally an SUV, while speeds of no greater than 15 mph allow for good data extraction from the video image. The ideal team consists of three members, a driver, navigator and a third continually checking all cable connections and tape life. Again, experience has shown that a 1.5 mile square neighborhood can be exhaustively covered in 3 hours when running cameras to both left and right of the vehicle.

The processed output varies according to the software used. The National Centre for Geocomputation (NCG) has created a browser display which we have used extensively throughout this book. Although the original video feeding into this browser has been split into multiple captures, these are seamlessly displayed as a smooth video display. If the user pauses the video, the particular frame on display can be read by both the date of capture and a photograph ID. One possibility is for the user to return to the original data files and extract this image. Similarly, an accompanying text file can be accessed which gives the exact location for each picture ID. Figure 12:3 displays the same road section from the Lower 9th Ward for June 2007 and June 2008. The upper most image shows the browser, while the image capture below this displays the same road segment extracted from the image library for approximately one year later. Notice how the pothole has worsened, it is now marked with a road warning sign.

An alternative software application, the GeoVideo toolbar, developed by Red Hen Systems Inc, allows the video to be displayed directly within a GIS[11]. The path is also shown allowing the researcher to again move to any point along the route. The benefit of this system is that all work is performed within the GIS. The benefit of the NCG approach is being able to see all video simultaneously.

Although the video itself is of immediate research value, not least because it provides an archived data source with full metadata allowing for easy distribution to third party researchers, it can also be used as a research tool in itself. By projecting the browser onto a screen, local experts, such as community leaders, can watch the video and provide additional context to the images; how one street corner is now known to be a crime hotspot and therefore should be avoided, or why one home has remained vacant for the entire year even though rebuilding had progressed on either side, or the impact a church has on the fabric of the local community.

Additional analytical capability can be gained by extracting attributes from the imagery. A simple editing process within the GIS allows for information to be added—either as simple locations (such as a point)[12], or as more complex objects that match the actual shapes of (for example) buildings. A further benefit of extracting information from the spatial video into a building footprint is that height can be estimated. Although this can be interpolated on the video image itself, a simpler but still effective approach consists of counting the number of floors, or if more accuracy is needed, using standard building "benchmark" heights, such as door frames. These heights are used to create a three dimensional display of the structure. In this way, in addition to mapping the neighborhood, it can also be rendered as a three dimensional model in a GIS. This is particularly powerful for research involving the psychological impact of what a returnee to a Katrina neighborhood sees from

11 For more information see http://www.redhensystems.com.

12 The first academic paper to describe the spatial video work in the lower 9th ward used a "point" attribute display. See Curtis, A., Mills J.W., Kennedy B., Fotheringham S. and T. McCarthy. (2007) Understanding the geography of post-traumatic stress: an academic justification for using a spatial video acquisition system in the response to Hurricane Katrina. *Journal of Contingencies & Crisis Management* 15(4): 208–209.

his/her window (for example neighborhood destruction). Figure 12:4 displays one such three-dimensional scene for a CSC neighborhood in the Lower 9th Ward. The spatial video has been used to extract the images of each house on the same block segment as a FEMA trailer. Each of these has an open door, search and rescue markings, and two of them display considerable vegetation overgrowth. The image to the right shows the trailer, and a makeshift road sign "Egania."

The type of information that can be extracted from the video is only limited by what can be seen from the road. Therefore buildings at a considerable distance may be hard to decipher, while excessive shadows or even the presence of parked cars may also be problematic. In addition, typical GPS problems can occur, such as data collection during rainstorms, or under particularly heavy cloud or tree cover. In general though, a considerable amount of detail can be extracted from the video; damage assessment of the buildings, how much vegetation over growth had occurred, the number and type of search and rescue markings that remained, whether there was any evidence of rebuilding, and whether the home had been "returned to". In addition, other features can be created in the GIS that previously

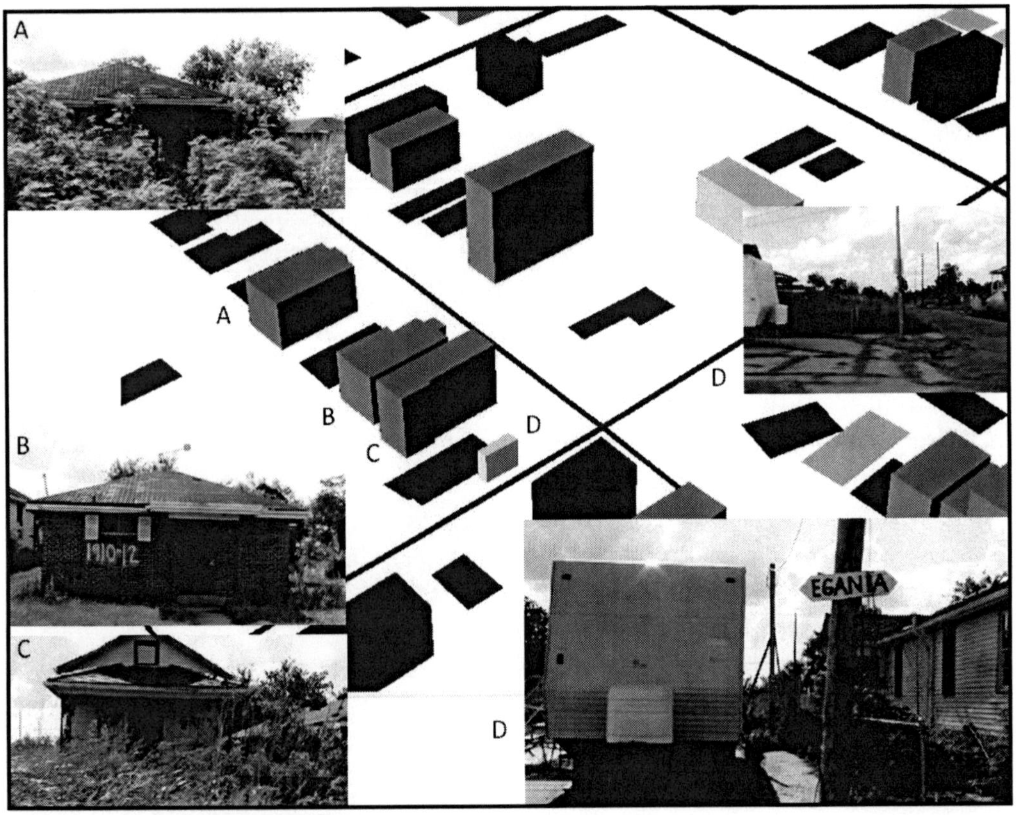

Figure 12:4. *The amount of damage surrounding a returnee's FEMA trailer. Source, 2008 spatial video data collection.*

did not exist on any map, for example the presence of FEMA trailers. These can also be given a height and actually placed as new buildings within the urban landscape. This is an important advantage as many of these trailers now have skirting and wooden frames, as well as sloped roofs—they are now part of the new urban landscape but one for which no data exists.

One of the trickiest aspects of this research is maintaining consistency in the way data are extracted from the video into a GIS. Table 1 displays the coding used for a 1 to 10 ranking of building damage and abandonment. Even within this scale further thresholds were established, for example some of the recovery maps only included buildings exceeding 5 on the scale. Of course it is also possible to return to the video and not only recode buildings to assure quality, but also revisit these structures for different interpretations. For example, for leaders from the CSC group, a property that had its doors and windows boarded shut was seen as being positive in that at least an attempt had been made to secure the property.

Table 1:

0 = no visible damage
1 = 1 element of damage (such as broken gutter)
2 = 2 signs of damage (e.g. hole in the wall plus 1 other)
3 = 3 signs of damage (e.g. no porch stairs plus 2 others)
4 = 4 signs of damage (e.g. collapsed car port roof plus 3 others)
5 = obvious signs of abandonment and neglect (boarded/open windows and doors)
6 = same as 5 with signs of damage
7 = multiple signs of damage but building shell appears sound
8 = major damage with at least 1 wall/roof ¾ to totally destroyed
9 = unrepairable damage—multiple walls/roof destroyed
10 = house has been cleared or just a pile of rubble remains.

Other data that have been extracted in the past from the video include:

The amount of vegetation overgrowth on a score of 1 to 5, where:
1 = some overgrowth, uncut grass, dead grass
2 = sizeable overgrowth, path covered, knee-high grass
3 = overwhelming cover—trees/grass—obscuring views and passageways
4 = vegetation has "overtaken" the property
5 = extremely high vegetation[13]

13 It should be noted that on similar work for the Holy Cross neighborhood of New Orleans that a "4" was the highest category. However the subtropical climate of New Orleans, as well as the Lower

The number of search and rescue markings visible on the building

The amount of rebuilding on a scale of 1 to 10 whereby:
1 = gutting/clearing left as a pile in front of the house
2 = building material (for example wooden planks) left outside the house
3 = 3 aspects of rebuilding (also include portable WC, tools, ladder)
4 = 4 aspects of rebuilding
5 = signs of rebuilding on the house (new deck/wall/fencing)
6 = rebuilding vehicles around the home
7 = multiple building materials and obvious work to the home (no people)
8 = people in and around the home
9 = single person working on the home
10 = team of people working on the home

Characteristics of individual buildings were also recorded such as whether the house was reoccupied, whether it was a church (and if it was open or abandoned), or a store/clinic (and if it was open or abandoned).

A separate GIS layer was created of all temporary trailers in the CSC area.

Finally, all buildings had a height estimate recorded, based on "1" per floor, with raised floors and attics scoring incrementally according to their extrusion.

Figure 12:5 shows various photographs from around New Orleans taken in October 2008. The following "scores" would be attributed to each image:

A = "8" damage score (major damage with at least 1 wall/roof ¾ to totally destroyed) "2" vegataion score, path covered, knee-high grass
B = "7"damage score (multiple signs of damage but building shell appears sound)—this property also displays the "Ken" marking usually meaning a body had been found inside.
C = "7" (possibly "8") damage score (multiple signs of damage but building shell appears sound). "3" vegetation cover (overwhelming cover—trees/grass—obscuring views and passageways)
D = "5" vegetation cover (Extremely high vegetation)
E = "5" rebuilding (signs of rebuilding on the house (new deck/wall/fencing))
F = "10" rebuilding (team of people working on the home)
G = "1" damage score (1 element of damage such as broken gutter)

9th Ward being a wetland in its natural state, have resulted in this area reverting to a grassland that in places is many feet high, sometimes completely engulfing the remaining structures.

Figure 12:5.

Even with these images there is room for doubt in terms of a correct scoring, for example, image "G" demonstrates a home that still has search and rescue markings, but no other obvious sign of damage. Interestingly, on a single video pass image "E" would score a "5" on rebuilding because the house had been reframed. However, through the video interview recorded during data collection we established that the owner had been taken sick six months earlier, and he no longer had the money to continue the rebuild and had abandoned the property. In subsequent video runs the inactivity of the rebuild was established.

Figure 12:6 shows the finished three-dimensional map created for the area around the Greater New St. Luke Baptist Church in the Upper 9th ward using the previously described scoring scheme. The following detailed descriptions match to the areas A to D in the figure. The small images are taken as screen shots from the video. These are included to illustrate how the transfer from video to mapping occurs, and they may also help you with the methodology for the exercises at the end of this chapter.

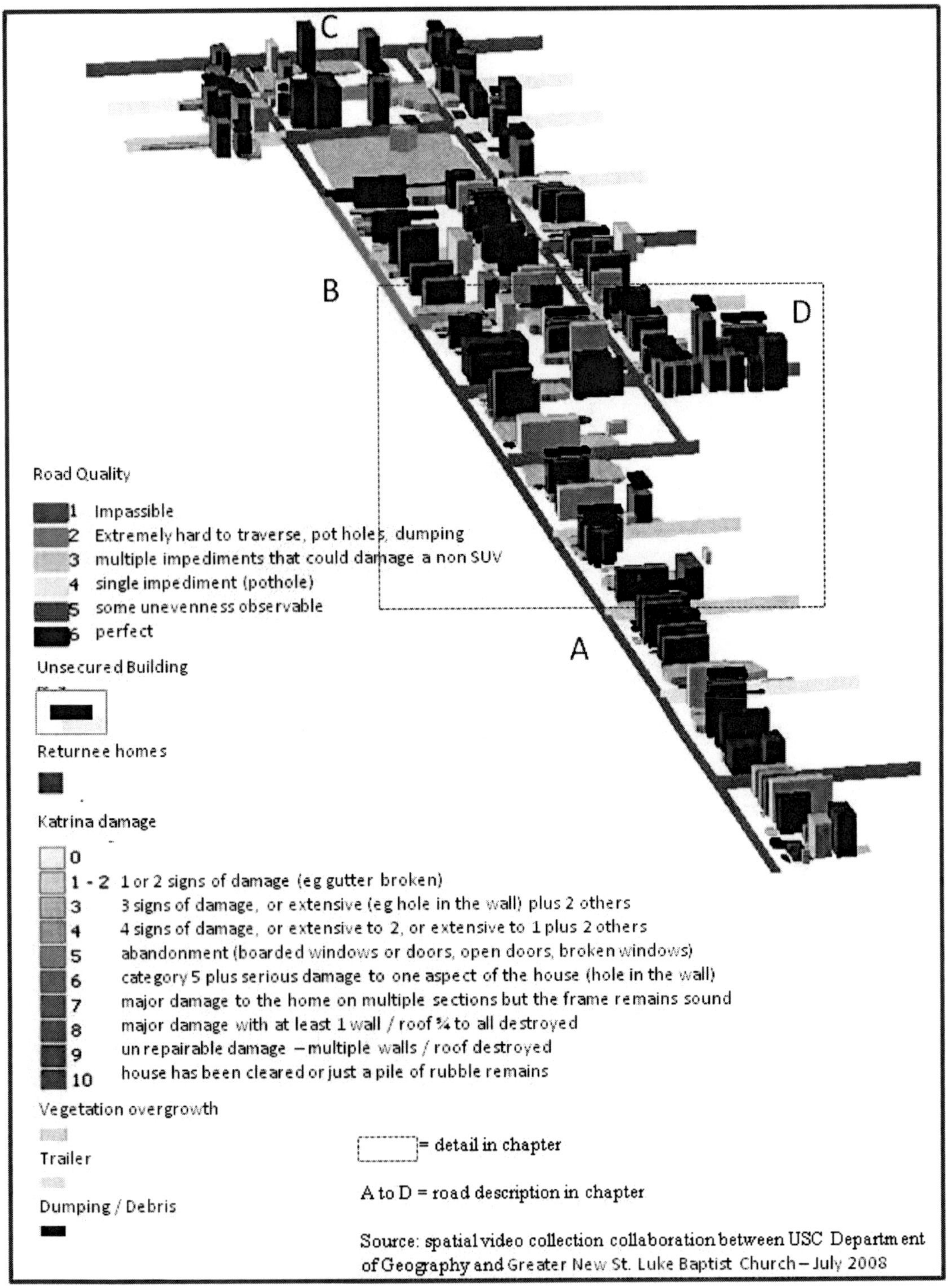

Figure 12:6. *Different examples of rating damage.*

Area A

The roads in this area display considerable damage to their surfaces. Several properties are especially damaged, with four also being open and unsecured. Hidden beneath the vegetation overgrowth are building foundations where a home once stood. There are five bulldozed properties. In amongst this damage, there is also some evidence of return. The sequence of images below shows the four properties occurring at one corner in area A (Figure 12:7). Considerable damage is evident. The two buildings to the left are unsecured while the two to the right, even though both are damaged, have been boarded up. Vegetation overgrowth can be seen in three images. It is also possible to still discern the search and rescue markings on the buildings.

Figure 12:7

Area B

Nine properties have been bulldozed, while a further five remain standing but damaged, three of these are also unsecured. The quality of the main north-south road along this stretch is also exceptionally poor. This area displays considerable spatial heterogeneity in terms of damage and return. The parallel stretch of road is of good quality, and there is evidence of returned and rebuilding homes (including FEMA trailers). However, across the block face from these are another series of abandoned and damaged buildings and the only section of non-traversable road.

The following series of images display a sequence of buildings from this area. This section contains one boarded and apparently abandoned home (4) two others show rebuilding work (6 and 3) with two more possibly being returnees or advanced renovation (1 and 2) while (7) has been returned to and next to this is a three-trailer cluster (8).

Compare these with the buildings on the other side of the same block, the locations being approximately the same as Figure 12:9. What is obvious is the amount of open land where the buildings have been bulldozed, with various gradations of vegetation overgrowth, from none (1) to considerable (3). The series of slides also shows a property which is badly damaged and unsecured (2), and a home where apparently the residents have returned (8).

Figure 12:8

Figure 12:9

Area C

The streets in this area all display large amounts of damage and abandonment, with most of the properties not being secured. Although not seen in Figure 12:10, the house to the left of (1) is badly damaged and burned out, while the two houses just before (2) are abandoned and unsecured. The road scene in (1) and (2) show further abandoned properties, the poor state of the road and debris/dumping on the side of the road. There is some evidence of return activity, a portable toilet can be seen in (1), and a vehicle sits outside a house in (2), though the perception while driving past these roads is one of desertion. Image (3) is an indication of how high the vegetation is here.

Figure 12:10

Area D

This again shows how clusters of damaged buildings often appear together. In Figure 12:11, both (1) and (2) show considerable damage, vegetation overgrowth, unsecured premises, and search and rescue markings, especially (1) where "gas off" is still clearly visible. Image (3) is also interesting as the property, which is again damaged, has a contact telephone number sprayed across its side. The tilt of these images is because the video camera was jolted because of the quality of the road.

Figure 12:11

A similar level of damage can be seen on the other side of the road to Figure 12:11. The first four homes of Figure 12:12 all show signs of clear damage, vegetation overgrowth, with (2) also being unsecured. Of interest on this section of road is that (5) has been returned to—the residents having to look out on Figures 12:11 and 12:12.

Figure 12:12

Figure 12:13. *Although the figure includes a color description, it is obviously presented in black and white*

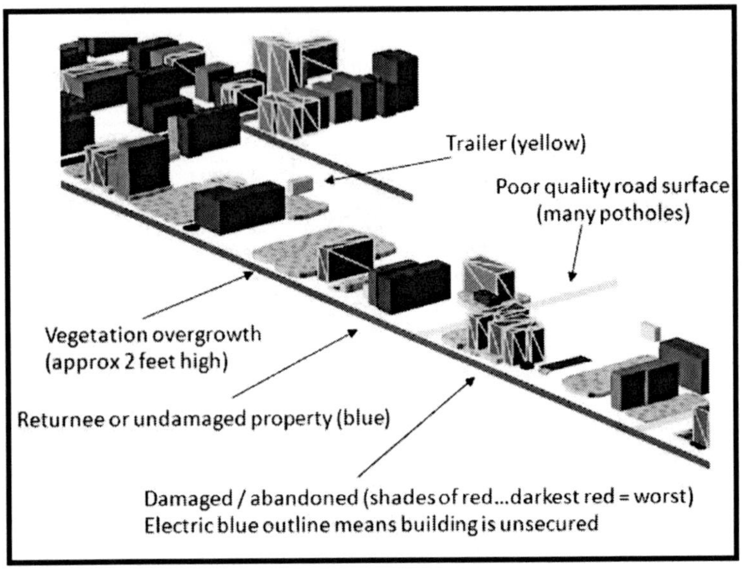

It is possible to mimic this field work by using Google Street View, which as New Orleans was driven after Katrina provides (almost) city-wide base level data. At the end of this chapter three maps have been provided for different sections of New Orleans showing building footprints corresponding to different CSC collaborative projects in New Orleans. The reader can use these to recreate recovery maps for different neighborhoods in New Orleans. The exercise maps also contain a single address that can be used to zoom into Google Street View. The 7th Ward and Hollygrove neighborhood maps also contain footprints for trailers—the Lower 9th Ward map contains no trailer footprints. The reader could attempt to add these into the map from the video. Within a GIS, layers could also be added for vegetation overgrowth, evidence of dumping, or this information could just be attached to each property.

Figure 12:13 shows an exploded view which corresponds with the area between A and D. Both Figure 12:6 and 12:13 were prepared for our community collaborators, and the only way to present all the attribute data is by using the full color palette. Unfortunately this leads to problems when printed in black and white, which is a common problem when multiple Xeroxes are needed for meetings.

In addition, although the breakdown of damage or vegetation suggested here is useful for creating detailed maps of a neighborhood, three further limitations exist; the possibility of misclassification by different researchers[14], the time taken

14 This is an important concern when working with large teams of data encoders. It is prudent to have a supervisor check all work on a random basis, and have encoders all classify the same street to ensure consistency.

to perform this classification, and the difficulty it poses in making comparisons between different time periods. In response to these concerns a new classification scheme was developed which improves consistency, is quicker to encode, and can be used to create simple add and subtraction maps between time periods[15]. Under the new classification scheme:

1 = abandoned (any property not obviously lived in)
2 = the property has been demolished
3 = active signs of rebuilding
4 = the home has been returned to (using at least three visible measures—garden furniture, cars, human activity, curtains etc).

For the category "1" three further sub-classes are used:

2 = property looks to be fine (no obvious damage/has been rebuilt) but empty
5 = there is obvious abandonment (the most tell-tale being boarded windows and doors)
10 = the building shows multiple signs of blight

Using this new scoring scheme, Figure 12:5 would be re-coded as follows:

A = 1 & 10
B = 1 & 10
C = 1 & 10
D = 1 & 10
E = 3
F = 3
G = 1 & 2

These last three sub-categories of "1" can help the map maker distinguish between areas of serious blight—and that would have a negative impact on any resident who had returned. The score of "5" also matches to the previously described abandonment threshold value.

By using this scoring system, it is possible to start to ask questions about recovering landscape. For example, imagine a family decides to return home after a disaster, there are several considerations that extend beyond whether they have the financial resources to rebuild or not. What about the rest of the street? Has anyone else returned? In the initial recovery process this may not be as important to someone who displays a "pioneering spirit" and are happy to be the first to

15 This simply means the difference between an encoded home in two different time periods actually makes a meaningful score.

rebuild on their block. Can we used the spatial video (or in this case Google Street View) to identify these pioneers?

Secondly, is there a minimum number of families that have to return on a block face before they act as a catalyst to the whole block face? There is some truth in the old saying "misery loves company"—there is solace to be found in a neighbor who is going through the same hardships but who is generally pushing towards the same goal of reestablishing the community. In Chapter Ten there were examples of people who used their buildings as message boards—some as a rallying cry to the rest of the neighborhood—others as their intention to give up. Can we see the evidence of these sentiments on the landscape? Does another entity, such as a church, act as growth stimulus, and can this be seen on the map? It has become evident from our work in the neighborhoods that rebuilding, returnees, and continued abandonment tend to cluster together—not so much as large patterns (making it hard to be revealed by spatial analysis) but rather along the same street segment. One block face will have a high proportion of returnees and then the next will have none. Why? Can we map these processes?

When we compare neighborhood maps, for example data collected using a spatial video and then using Google Street View—we can also see if the return has been successful for some of the recovery pioneers. Have homes that have been rebuilt again been abandoned? Does this re-abandonment coincide with a lack of other rebuilders on their street?

EXERCISE 1

At the end of this chapter, three maps have been provided for different sections of New Orleans; the 7th Ward, Lower 9th Ward and Hollygrove. Make four paper copies for each map. Give each building a unique ID. For each of the following building scores add the value into an Excel sheet linked to the building ID of the first map. Also, add the value directly onto a copy of the map.

For each building use the abridged (4 point scale) system to assign a score. For those buildings receiving a score of "1" add a subcategory of 2, 5, or 10 (for a description of these scores, re-read the chapter).

In addition, if a building is a store, give it a value of "S" and a "1" if it is open, and "0" if it is not. If a building is a church, give it a value of "C" and a "1" if it is open, and "0" if it is not.

Record the number of visible search and rescue markings on each building. Also record the number of floors (give a value of "1" for each floor, "1.25" for a home with an attic, and use "0.75" for a trailer).

When you have scored all the buildings, make a graduated color map for abandonment & recovery. Now answer the following questions:

Q: Draw areas on your map where you see severe abandonment and rebuilding/recovery clustering. Describe the geographic scale of this clustering (neighborhood, street segment, block, etc).

Q: Select five different block faces and describe what you see happening in terms of severe abandonment and rebuilding/recovery?

Q: Do you see any recovery pioneers? Where? How many?

EXERCISE 2

Use Google Street View to mark the locations of the trailers for the three neighborhoods. Again using Google Street View, and your damage scores, estimate the sight line for each trailer and tally its stressor score (using 1, 5, and 10 as the damage scores).

Q: How many trailers appear in each neighborhood?

Q: What is the highest average "sight-line" score for trailers in any neighborhood?

Q: What is the highest individual score for a trailer in a neighborhood?

Q: Do you notice anything interesting about the placement or style of the trailers?

Q: What final assessments on stress can you make (also referring back to any other map for these neighborhoods contained in this book)?

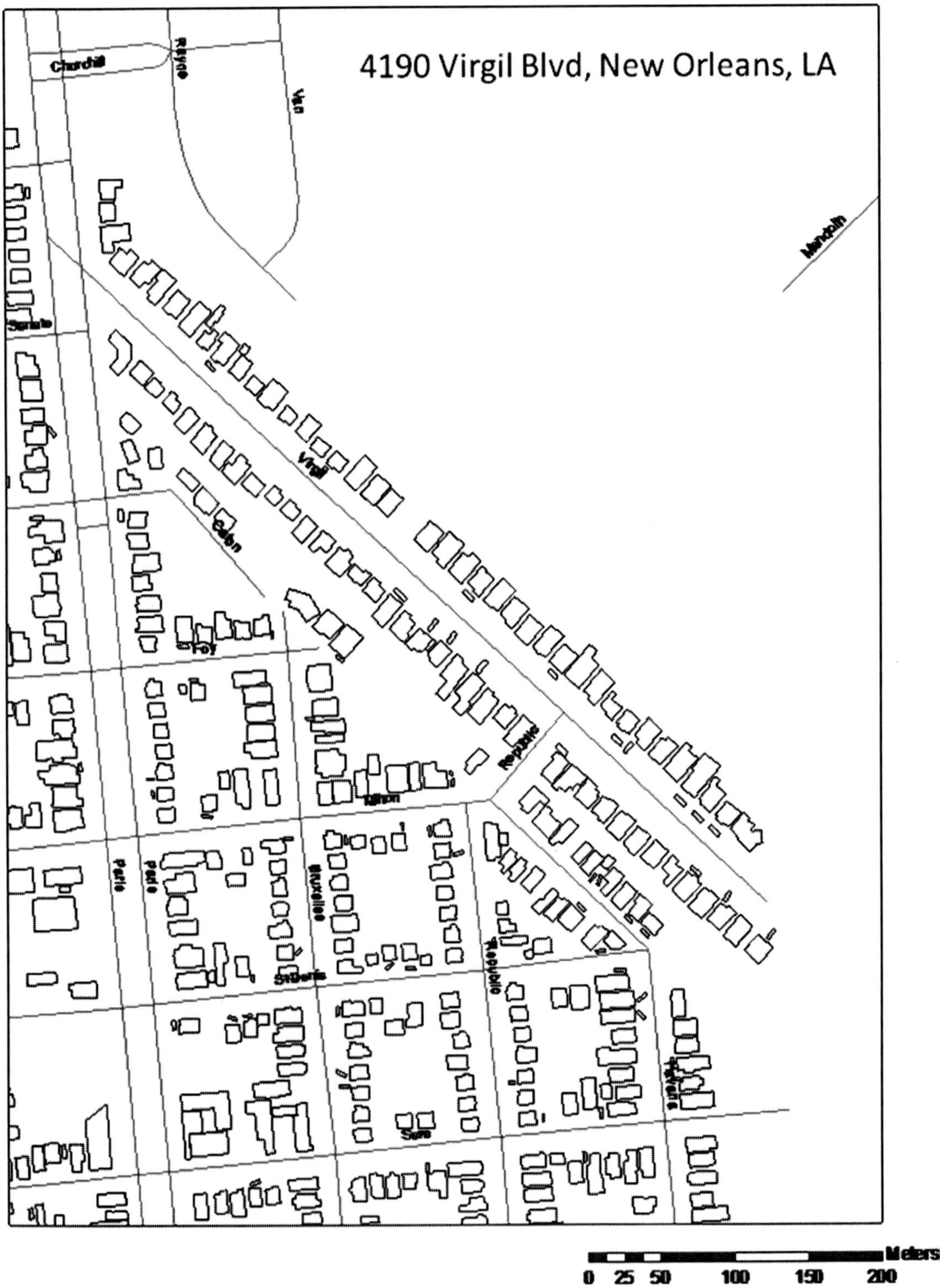
4190 Virgil Blvd, New Orleans, LA
Virgil
Republic
Foy
Paris
St Denis
Meters
0 25 50 100 150 200

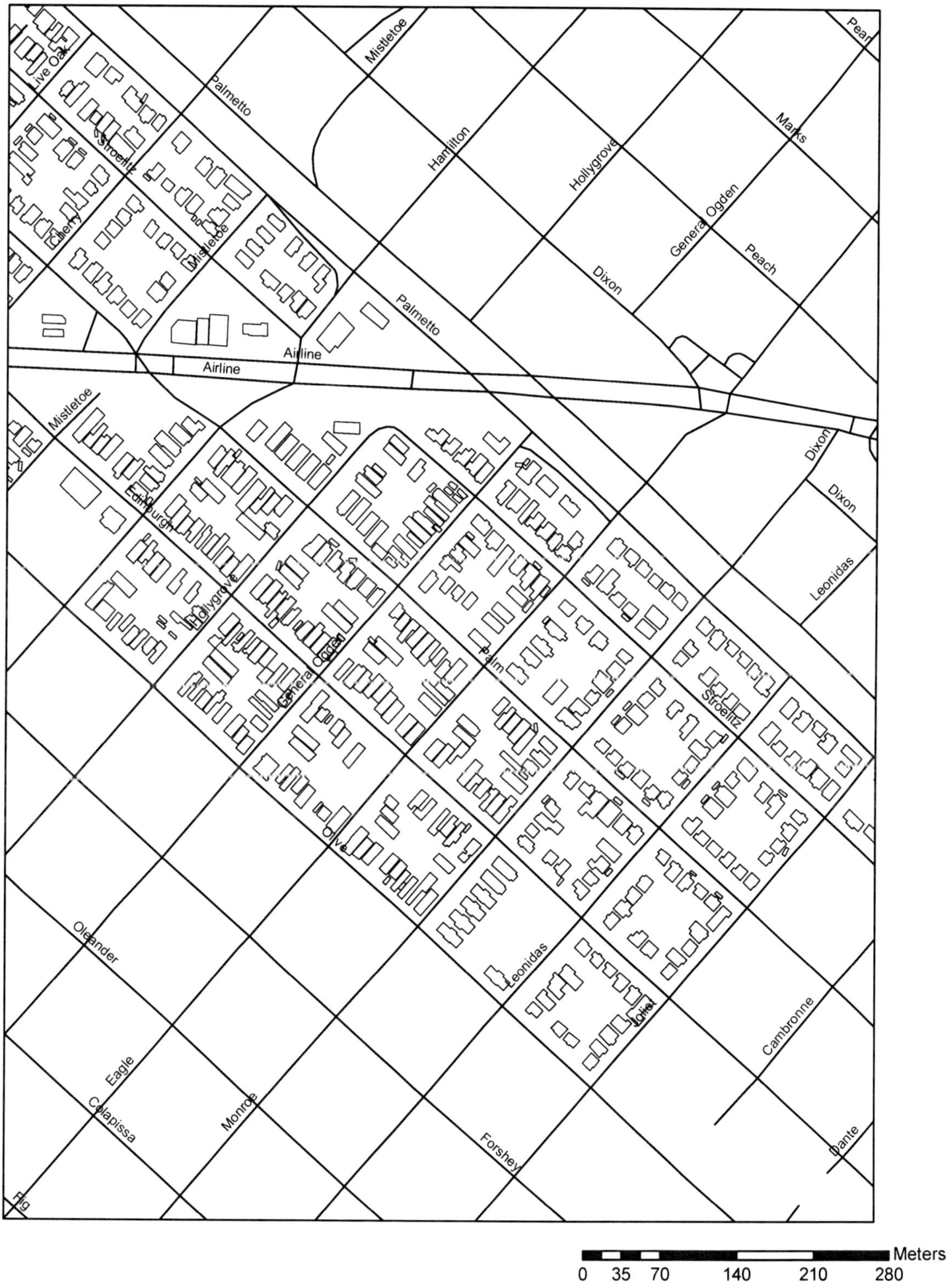
Live Oak
Mistletoe
Pear
Palmetto
Marks
Hamilton
Hollygrove
Stroelitz
Cherry
General Ogden
Peach
Dixon
Airline
Edinburgh
Leonidas
Palm
Olive
Oleander
Joliet
Cambronne
Eagle
Colapissa
Monroe
Forshey
Dante
Fig
Meters
0 35 70 140 210 280

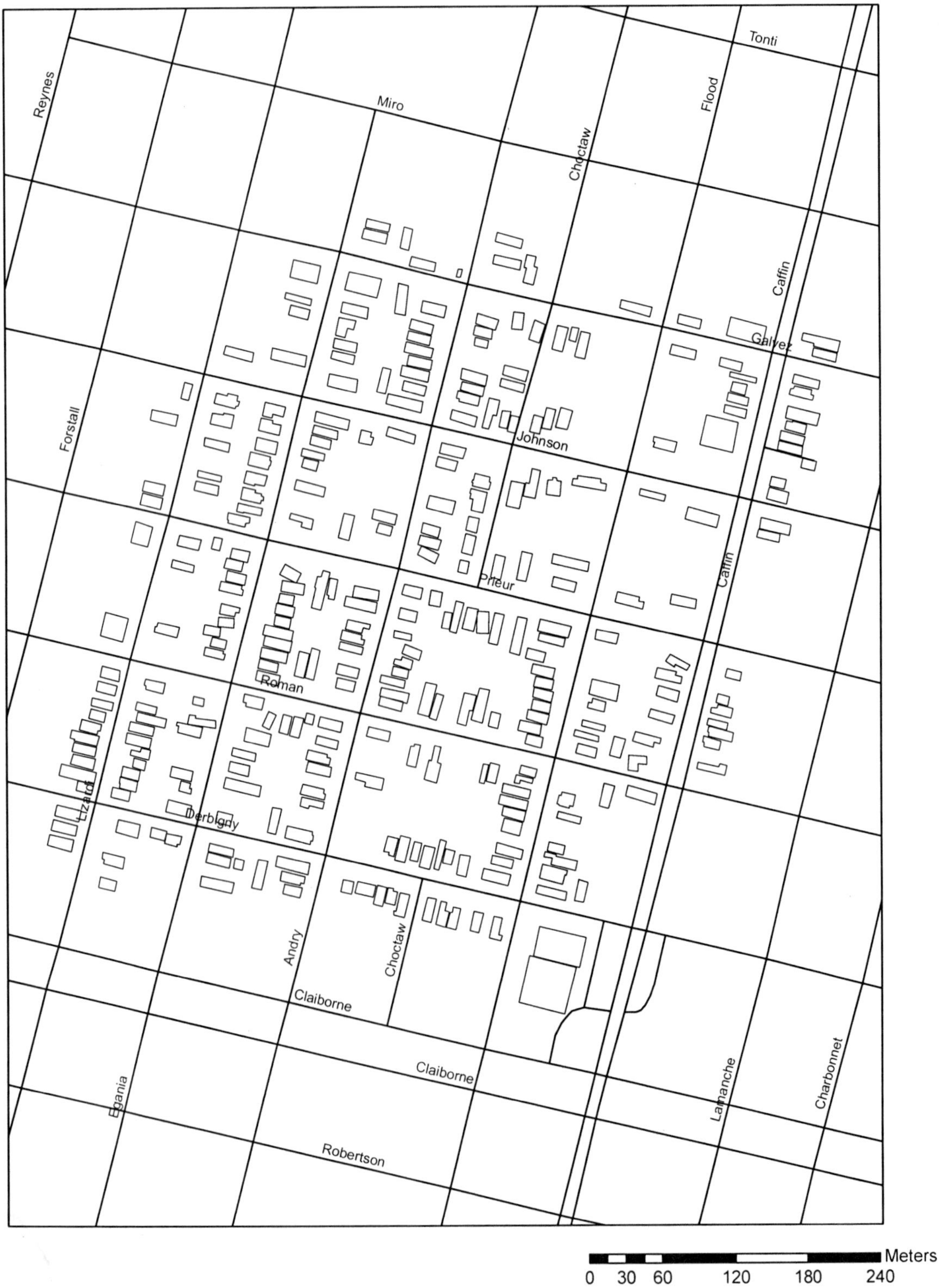

Tonti
Miro
Reynes
Flood
Choctaw
Caffin
Galvez
Forstall
Johnson
Prieur
Caffin
Roman
Lizardi
Derbigny
Andry
Choctaw
Claiborne
Claiborne
Robertson
Egania
Lamanche
Charbonnet
Meters
0 30 60 120 180 240

Conclusions

NOW THAT WE are at the end of the book, let's return again to the image on the front cover. The man and his message should now be clearly emblematic of many issues regarding the human geography involved in disasters. A geographer observes the environment and asks questions about what he/she sees: Why is this man here? Why is he in this location as opposed to another? Why has he written this message? Why does he use these particular words? This image is filled with the human geography of disasters.

As we discussed early in the book, natural disasters are most commonly presented in terms of their physical processes. However, they are best understood through a frame of human-environment interaction where these two worlds intertwine. Unfortunately, the human component of this interaction is underrepresented in natural hazards textbooks. This book has attempted to fill this gap by presenting concepts and cases that highlight society and space in the study of disasters. We have covered issues of health, environment, community, social vulnerability, policy, planning, mitigation, response, and recovery. Each one of these aspects is deserving of its own book and we encourage students to seek out more specialized texts in any topic where they have a particular interest.

As stated at the beginning of the book, disasters have a long-term impact on people and places. As geographers, we are interested in both. We are interested in the spatial aspects of disasters and therefore this has been our focus. Surrounding this focus, the book has developed themes which were interwoven throughout the chapters: understanding why geography is important, comparing the past to the present, issues of vulnerability, the psychological impact, and the road to recovery. We have used both historical and ongoing disasters to present these themes. In particular, studying the past is important for understanding the potential situations we may face in our future. However, in the United States, we have not been very good about systematically capturing, archiving, and sharing data from disasters. This is especially true for spatial data suitable for GIS analysis. We have also emphasized the importance of fine scale human geography in disasters.

Chapters One and Two dealt with spatial concepts, GIS, and other geospatial technologies. Chapters Three and Four used historical records from the Johnstown Flood to compare issues from Katrina. Chapter Five focused on the 2007 firestorm in Southern California. Chapter Six dealt with the complex issue of vulnerability, while Chapter Seven delved into the understudied topic of health vulnerability. Chapter Eight introduced you to Renaissance Village and the use of art therapy for the youngest victims of Katrina, and Chapter Nine dealt with stress and related health issues in disasters. Chapter Ten vividly described the human geography in the post-disaster landscape. Chapter Eleven discussed recovery and the spatial patterns of this process. Finally, Chapter Twelve focused on the use of spatial video technology to map recovery. Overall, this is an eclectic collection, but it is consistently linked through the common theme of human geography, GIS, and disasters.

You may be thinking that all of this is interesting (hopefully) for this particular class, but that it has little to do with you once the semester ends. Well, let me give you a final lecture that I have been giving to my natural hazards class for years. Regardless of your major, minor, or future career interests, understanding the complexity of hazards and disasters will be useful to you. For example, if you go into business, then understanding risk and business continuity in the face of disaster will enable your success in turbulent times. There is great need for understanding hazards and disasters for those who go into law, medicine, engineering, education, sociology, political science, history, communications, etc. Indeed, most of my students are not geography majors, but come from these and other disciplines. This situation is ideal as disasters are interdisciplinary. One perspective alone is not enough and contributions and insights are needed from all of these disciplines, as well as many others.

Furthermore, regardless of your disciplinary choice or future career, you need to be aware of the many aspects to disasters as an educated person, as a homeowner, as a parent, as a voter, and as a taxpayer. Disasters impact many aspects of your life, whether you actually experience one or not. Events that occur throughout the country impact where your tax dollars go, or how your representatives vote. As global warming introduces uncertainties into our near-future environment, the cause of sustainable development has gained ground. To this end, you should be aware of how your children and grandchildren will be affected. What hazardscape will they inherit? These are monumental issues and are certainly beyond the scope of this one book, but they are related to what we have presented here.

Therefore, we hope that this is a useful classroom tool to provoke discussion, but also to educate about the many social aspects of Katrina and disasters of its kind. Perhaps it will even interest a few to carry on with research in this area. There is more than enough work to go around!

Finally, we would be remiss not to again mention the many neighborhoods in Louisiana which continue to struggle and rebuild their communities. If you think you have the energy and commitment, seek out ways in which you can help. They would be very grateful.

Breinigsville, PA USA
17 August 2009
222347BV00003B/1/P